Weld Pool Dynamics in Deep Penetration Laser Welding

Shuili Gong · Shengyong Pang · Hong Wang ·
Linjie Zhang

Weld Pool Dynamics in Deep Penetration Laser Welding

航空工业出版社 Springer

Shuili Gong
AVIC Manufacturing Technology Institute
of China
Beijing, China

Shengyong Pang
Huazhong University of Science
and Technology
Wuhan, Hubei, China

Hong Wang
Taiyuan University of Science
and Technology
Taiyuan, Shanxi, China

Linjie Zhang
Xi'an Jiaotong University
Xi'an, Shaanxi, China

ISBN 978-981-16-0787-5 ISBN 978-981-16-0788-2 (eBook)
https://doi.org/10.1007/978-981-16-0788-2

Jointly published with China Aviation Publishing & Media Co., Ltd., Beijing, China Ms. Jingmei Li.
The print edition is not for sale in China Mainland. Customers from China Mainland please order the print book from: China Aviation Publishing & Media Co., Ltd.

This Springer imprint is published by the registered company Springer Nature Singapore Pte Ltd.
The registered company address is: 152 Beach Road, #21-01/04 Gateway East, Singapore 189721, Singapore

Foreword

With the continuous improvement in, the output power, beam quality, and energy efficiency, high power lasers as light and heat sources of high-energy density beams have been widely applied to material processing, forming a significant part of special or non-traditional material processing technologies. The laser materials processing technology has various advantages, including non-contact, precise energy control, wide material suitability, greater flexibility, high quality, and environmental friendliness. For this reason, this technology can be used not only for efficient automatic mass production, but also for a large variety of small batch processing, and further, for customized production. Thus, laser processing has become an essential technology for the manufacturing industry to transform and upgrade. After years of R&D, laser processing technology has become one of the most important technical means of high-energy beam processing in five major manufacturing fields: welding, cutting/drilling/marking, surface engineering, micro/nano fabrication and additive manufacturing (3D printing).

Laser welding is a process to join similar and dissimilar materials by fusion. It has many advantages over traditional arc welding, including higher speed, higher energy density and depth-to-width ratio of weld joints, smaller heat-affected zones and deformation, better quality and performance of the joints, more efficient production, and more flexible control and operation, which shows enormous potential as an advanced welding technology. Nevertheless, given the high energy density of laser beam, the dynamic state of melt pool at very high welding speeds is more complex. Welding defects elimination, welding process stability and deformation control have become key technological challenges that need to be understood. To establish a fundamental theory of laser welding, particularly the keyhole welding, understanding of laser weld pool dynamics is essential.

Professor Shuili Gong and his research team have long been engaging in laser welding basics and engineering application research. They have systematically carried out in-depth research on interaction mechanisms between lasers and materials, weld pool behavior and its influence on welding processes, having established a series of theories to aid the development of laser welding technologies and applications. This book is an extract and summary of achievements made by the author's team through many years of research, and is a collection of theoretical findings on the topic of laser keyhole weld pool behavior. Professor Gong's research team has proposed and established various theories on the dynamic behavior of laser welding pool, which provide a theoretical basis for revealing welding defect mechanisms, welding process stability mechanism and its equilibrium conditions, guiding the practical engineering applications. At the same time, the theories also underpin the research and application of other laser process technologies.

This book is a monograph on the basic study of laser welding, particularly on laser keyhole welding, laser welding with wire filler, laser welding of thick section materials and laser vacuum welding. It would be a valuable reference for researchers and engineers. I am very happy to recommend this book to the readers.

July 2020

Prof. Lin Li
Fellow of the Royal Academy of Engineering
The University of Manchester
Manchester, UK

Preface

In 1917, Albert Einstein, put forward the concept of stimulated emission and discovered the photoelectric effect, laying theoretical foundation for the invention of laser. In May 1960, Dr. Maiman, an American physicist, created the world's first ruby laser based on the development of quantum electronics, marking the beginning of research on laser and its application. In the early 1960s, reports began to appear about the technical applications related to laser drilling and welding. With continuous improvement on laser technology and laser beam quality and increasing output power, lasers are quickly applied to material processing as the light and heat sources of high-energy density beams, forming a special material processing technology cluster of great significance—laser processing technology. This technology has a wide range of advantages such as non-contact, precise energy control, suitability to a variety of materials, strong flexibility, high quality, resource saving, and environmental friendliness. It can be used not only for efficient and automatic mass production, but also for small-scale processing of multiple varieties of materials, and even for product customization. This makes the laser processing technology essential for transformation and upgrading of the traditional manufacturing industry. After years of development, the laser processing technology has developed into a high-energy beam processing technology, becoming one of the three important technical means in three major manufacturing technologies—welding and removal, surface engineering and additive manufacturing (3D printing). The laser processing technology can be divided into many categories, such as welding, cutting, drilling, rapid prototyping, etching, micro-nano processing, surface modification, spraying and vapor deposition, and play an important role in many fields of national economy and defense development.

Compared with traditional arc welding technology, the laser welding technology, typical of the laser processing technology, has an array of advantages such as high-energy density (up to 10^{15} W/cm^2), high depth-to-width ratio of weld joints, small heat-affected zone, small deformation, high quality and performance of the joints, high production efficiency, and flexible control and operation. The laser welding technology has enormous potential for welding.

The laser welding technology develops along with the advancement of laser technology. The laser welding method moves from heat conduction welding towards deep penetration welding as the laser output power increases, beam quality improves, and theoretical research and technology continues. The welding technology constantly develops. Laser butt welding, laser superposition welding, hybrid laser-arc welding, laser welding with filler wires, dual-beam laser welding, multi-beam laser welding, dual-beam laser welding with filler wires, ultra-narrow gap laser welding, all-position laser welding, long-focal-length dynamic laser welding, and other technologies emerge and develop one after another. Accordingly, welding equipment, welding systems, and process detection and control technologies are improved, laying a solid foundation for the application of laser welding technology in the national economy and defense development.

Since the laser technology has been applied to welding, the theoretical research of laser welding moves along. Systematic research has been carried out from different modes of the interaction mechanism of lasers and materials, the weld pool behavior and its influence on the welding process, which provides solid theoretical support for the development and application of the laser welding technology. The scientific and technical workers have studied the interaction mechanism of lasers and materials and the physical state change process of materials, obtaining two typical welding modes of laser heat conduction welding and deep penetration welding. They realized that the laser energy density is the key parameter of the mutual transformation of the two welding modes and found that the weld pool generated in deep penetration laser welding is characterized by the dynamic keyhole behavior, and the thermodynamic equilibrium in weld pool keyholes of deep penetration welding is a necessary condition for the stable formation of keyholes. Through application research and engineering practice of laser welding, welding scientific workers discovered that the weld pool behavior in deep penetration laser welding plays an important role in the stability of laser welding. Studying and mastering the dynamic behavior of weld pools in laser welding becomes the key foundation for the optimization of technical parameters and stability control in welding.

Over the years, with the support of government agencies and organizations, Beijing Aviation Manufacturing and Engineering Institute, Huazhong University of Science and Technology, Xi'an Jiaotong University, Harbin Institute of Technology, Dalian University of Technology, etc., have conducted in-depth research on the technology basis of laser welding, and have carried out continuous study on the weld pool behavior in laser welding, systematically revealing the weld pool behavior in laser welding and its influence, providing theoretical support for the practical application and promotion of the laser welding technology, and laying a solid foundation for drafting this book.

This book systematically describes the main research achievements on the weld pool behavior of domestic laser welding in recent years. It refines and summarizes the relevant research work from the institutions such as National Defense Science and Technology Key Laboratory of High Energy Beam Processing Technology in Beijing Aviation Manufacturing and Engineering Institute, Huazhong University of Science and Technology, Xi'an Jiaotong University and Taiyuan University of Science and Technology. The research work of the authors' research team especially provides great support to this book.

This book consists of 9 chapters, which systematically describe the weld pool behavior during laser welding and its influencing factors, including the experimental research, theoretical calculation and process simulation technology, physical state transformation behavior of weld pools, and the impact of technical conditions on the weld pool behavior. The research achievements of this field in China are intensively reflected in the research results, some of which present the latest frontier research work of the authors' research team, such as the weld pool behavior of dual-beam laser welding, the weld pool behavior of laser welding with filler wires, and the weld pool behavior of heavy-thickness full-penetration laser welding and laser welding in vacuum and low vacuum conditions, etc. Those who participated in compilation of this book are Shuili Gong, Shengyong Pang, Hong Wang, and Linjie Zhang, among which Shuili Gong is responsible for the content of preface and Chapter 1, Shengyong Pang for the content of Chaps. 3–5 and Chaps. 7–9, Hong Wang for the content of Chap. 2, and Linjie Zhang for the content of Chap. 6. The whole book was reviewed and finalized by Shuili Gong. This book not only can be used for reference for engineering technicians, scientific researchers in the field of laser material processing, scientific researchers majored in material science, and teachers and students in colleges and universities, but also have high reference value for research and development personnel in the field of laser equipment, optical scientific researchers, scientific and technical personnel in the related fields of laser application physics, and teachers and students in colleges and universities.

During the compilation of this book, we received great support and help from our research team. We would like to thank National Defense Science and Technology Key Laboratory of High Energy Beam Processing Technology in Beijing Aviation Manufacturing and Engineering Institute, Huazhong University of Science and Technology, Xi'an Jiaotong University and Taiyuan University of Science and Technology, and especially thank Professor Jianxun Zhang of Xi'an Jiaotong University for his guidance for this book, Professor Jianzhong Xiao and Professor Jianxin Zhou of Huazhong University of Science and Technology for their strong support for this book, and our colleagues' support and help. We would like to thank the authors listed in the bibliography of this book for their works and papers which have contributed greatly to the compilation of this book.

We'd like to comfort Professor Lunji Hu and Professor Liliang Chen of Huazhong University of Science and Technology with this book. Without their support and guidance over the years, the authors may not have completed this book.

Due to the authors' limited knowledge, it is inevitable that there are inadequacies and shortcomings in the book. You are welcome to criticize and correct us.

Beijing, China Shuili Gong
July 2017

Introduction

This book systematically describes the weld pool behavior in laser welding and its influencing factors from the perspectives of testing technology, theoretical calculation and process simulation technology, physical state transformation behavior of weld pools, and the impact of technical conditions on the weld pool behavior. The book covers extensive research achievements made by China in this field, some of which represent the latest cutting-edging research conducted by the authors' research team. These latest research efforts mainly relate to the weld pool behavior of dual-beam laser welding, laser welding with filler wires, full-penetration laser welding of very-thick parts, and laser welding in vacuum and low vacuum conditions.

This book can be used by teachers and students in universities and colleges and engineering technicians for reference.

Contents

Chapter 1
Laser Welding Basics

Abstract This chapter introduces the basics of laser welding process. The principles of laser material interactions, the characteristics and theories of laser welding, and the research outlines of weld pool dynamics of laser welding process are presented.

1.1 Interaction of Lasers and Materials

The interaction process between laser and material is complicated, involving multiple disciplines such as laser physics, heat transfer, plasma physics, non-linear optics, thermodynamics, gas dynamics, fluid mechanics, mechanics of materials, and solid-state physics. Laser welding, which is one of the major applications of the laser processing technology, also utilizes the physical phenomenon generated by the interaction between laser and material, and these physical phenomena may come down to the thermal and mechanical effect of the material being processed. The main research content and direction include the reflection, absorption, and energy conversion of the laser by the material, and the heating, melting, thermocompression shock wave, vapor ejection, plasma expansion, and shock wave of the material by the laser. Nevertheless, during the laser welding research, the focus is on the absorption of the laser by the welded material and the thermophysical and mechanical effects of the material itself (such as heating, melting, vaporization, and plasma effect). The main physical phenomena are shown in Fig. 1.1.

1.1.1 Laser Absorption by Materials and Material Heating

The regular motion of damped oscillation of the charged particles in the atom is described by the classical mechanics; that is to say, the charged particles are regarded as the oscillators following the classical mechanics. The oscillators vibrate as bound close to a certain equilibrium position by the elastic restoring force directly proportional to the displacement. In case of deviating from the balanced position, the oscillators will be acted by a restoring force. In terms of metal, oscillators in the metallic

S. Gong et al., *Weld Pool Dynamics in Deep Penetration Laser Welding*,
https://doi.org/10.1007/978-981-16-0788-2_1

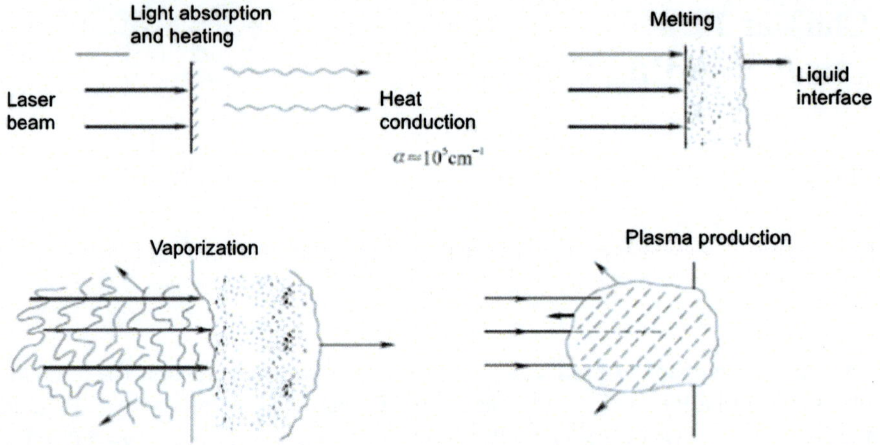

Fig. 1.1 Physical effect on the material surface under the laser action

material are mainly free electrons. Absorption in the metal is mainly completed by the vibration of free electrons which can move freely under the electric field action without the effect of the restoring force. The natural vibration frequency of the oscillators is $\omega_0 = 0$, and the relationship between refractive index *n*, extinction coefficient κ, and damping factor γ, plasma frequency ω_p is as follows:

$$n = \left\{ \frac{\sqrt{\left(1 - \frac{\omega_p^2}{\omega^2+\gamma^2}\right)^2 + +\left(\frac{\omega_p^2\gamma}{\omega^3+\gamma^2\omega}\right)^2} + \left(1 - \frac{\omega_p^2}{\omega^2+\gamma^2}\right)}{2} \right\}^{1/2} \tag{1.1}$$

$$\kappa = \left\{ \frac{\sqrt{\left(1 - \frac{\omega_p^2}{\omega^2+\gamma^2}\right)^2 + +\left(\frac{\omega_p^2\gamma}{\omega^3+\gamma^2\omega}\right)^2} - \left(1 - \frac{\omega_p^2}{\omega^2+\gamma^2}\right)}{2} \right\}^{1/2} \tag{1.2}$$

where

$$\omega_p = \sqrt{\frac{Ne^2}{m\varepsilon_0}} \gamma = \frac{e^2\omega_0^2}{6\pi\varepsilon_0 c^3 m}$$

The plasma frequency ω_p is the strength parameter, which determines the relative strength of the oscillators; the damping factor γ is the parameter of the band width. As we can see from Eqs. (1.1) and (1.2), when the laser frequency is far less than the plasma frequency, *n* and *k* increase rapidly, most of laser is reflected, and very little is absorbed; when the laser frequency is close to the plasma frequency, *n* shows the local minimum value, and k value monotonically decreases. Therefore, the laser can be absorbed well when the laser frequency is near the plasma frequency; when the laser frequency continues to go up and far more than the plasma frequency, n is quickly approaching 1, and k quickly becomes 0, in which case, the metal is transparent. The plasma frequency of the metallic material is between ultraviolet and near-infrared band, so the laser from near-infrared, visible, to ultraviolet band is relatively favorable to metal processing; the far-infrared laser is almost reflected by the metal. For the infrared light with low photon energy, the light frequency electromagnetic wave only works on the free electrons in the metal, while for the visible light or UV-light with high photon energy, the optical frequency electromagnetic wave can also work on the bound electrons in the metal as the inherent frequency of bound electrons in the metal is within the visible or UV-light frequency band. Under the action of the bound electrons, the reflecting capacity of the metal reduces, the transmitting capacity increases, and the laser absorption capacity increases, indicating certain non-metal optical property.

Due to interaction between the laser and the electrons, ions, lattice vibration, impurities, and defects in the material, the laser can be absorbed by the material. Therefore, the optical properties of the material are closely correlated to laser absorption.

Transmission of the laser action in the material can be described by Maxwell's equations. When electric field intensity of the laser in the material is substituted into the Maxwell's equations, the relation between the complex refractive index $\hat{n}(\hat{n} = n - ik)$ and the material's physical constant, which can reflect the electromagnetic wave propagation features, can be obtained.

$$n^2 = \frac{\mu}{2}\left[\sqrt{\varepsilon^2 + \left(\frac{4\pi\sigma}{\omega}\right)^2} + \varepsilon\right] \tag{1.3}$$

$$\kappa^2 = \frac{\mu}{2}\left[\sqrt{\varepsilon^2 + \left(\frac{4\pi\sigma}{\omega}\right)^2} - \varepsilon\right] \tag{1.4}$$

where: ω—Frequency;

ε—Dielectric constant;

μ—Magnetic conductivity;

σ—Electrical conductivity;

n—Refractive index;

κ—Extinction coefficient, which reflects the attenuation characteristic of the electromagnetic wave amplitude.

Equations (1.3) and (1.4) show that the refractive index and extinction coefficient of the material are closely correlated to the permeability, dielectric constant, electrical conductivity, and laser frequency.

For the isotropous homogeneous substances, according to Lambert–Beer-Bouguer law, the laser intensity I decreases exponentially with the transmission distance z

in the material, $I_z = I_0 exp\ (-\alpha_0 z)$, where α_0 is the absorption coefficient. The relationship between the absorption coefficient and the laser frequency and extinction coefficient is as follows:

$$\alpha_0 = \frac{2\omega\kappa}{c} = \frac{4\pi\kappa}{n\lambda} \tag{1.5}$$

where, c—Laser speed in the vacuum;

λ—Laser wavelength in the vacuum;

$1/\alpha_0$—Light absorption length, defined to be the light beam propagation distance when the light beam intensity reduces to 1/e (37%) of the original due to the photon energy being absorbed.

The feature of the absorption coefficient related to the laser wavelength is called the selective absorption, while the absorption where the absorption coefficient does not change with the laser wavelength is called the general absorption. Normally, the absorption coefficient has nothing to do with the laser intensity.

Absorption of the laser in the material depends on the permeability, dielectric constant, electrical conductivity, and laser frequency. During laser processing, once the laser frequency is determined, the absorption coefficient is only correlated to the extinction coefficient. Therefore, analysis of the absorption of the laser in the material can be converted to analysis of the extinction coefficient and physical characteristics of material. The higher the dielectric constant of the material is, the lower the electrical conductivity will be, the higher the refractive index will be, and the less the extinction coefficient will be, so the less the absorption coefficient will be; the higher the material permeability is, the higher the electrical conductivity will be, and the lower the dielectric constant will be, so the higher the absorption coefficient of material will be. In addition, laser absorption is also correlated to the microstructure and surface status of the material.

1.1.2 Laser Heat Source Model and Laser Heating Effect

When the substance absorbs the laser, excess energy appears first in some mass points (kinetic energy of free electron, excitation energy of bound electron, or excess phonon), other than the heat. These orderly original excited energies can be converted into the heat energy through two steps: Step I is the stochastic motion of the excited particles in both space and time. This process is completed during particle collision time (momentum relaxation time), which is a very short moment; Step II is the uniform distribution of the energy between each mass points. This process contains a mass of collision and intermediate state, especially non-metallic materials. Finally, a certain form of energy distribution will take shape in the material. In conclusion, the heating process of the material under the laser action is extremely short.

To study the thermal action due to interaction between laser and material, it's generally assumed that a heat source with its laser energy distribution the same as the absorbed laser energy distribution applies to the material surface, and under this precondition, the temperature field model is built to analyze the heating and cooling process during the laser processing. The heat transfer phase from the surface material into the matrix mainly follows the Fourier law of heat conduction. The heat source model of the laser varies with different materials. For metallic material, the laser absorption length is very limited. Laser absorption occurs within 1–5 μm of the material surface. The heat source model can be expressed as follows:

$$Q_{V(x,y,z,t)} = AI_0(x, y, t)\delta(z) \tag{1.6}$$

where: A—Absorptivity of the laser by the material;

I_0 (x, y, t)—Distribution of the laser intensity on the material surface;

δ (z)—Dirac function.

The laser intensity I_0 $(x, y,$ and $t)$ is usually represented as the production of spatial distribution I_0 (x, y) and dimensionless time waveform B (t). Typical wave forms B (t) include step wave, rectangular wave, triangular wave, trapezoidal wave, and Gaussian waveform, etc.

During the laser heating process, the thermal physical parameters of material (absorption coefficient, specific heat, thermal diffusivity, and thermal conductivity coefficient) vary with the temperature, however, for most of materials, the change of heat physical parameters is relatively little as the temperature changes, nearly equal to a constant, or the temperature in the process may be regarded as average value. In the following discussion, it's assumed that the thermal physical parameters of material are independent of the temperature.

When the laser beam of Gaussian distribution stands still relative to the material surface, the maximum value distribution of the temperature field on the semi-infinite material surface is as follows:

$$\Delta T = \frac{AP}{k\pi^{3/2}r}\arctan\sqrt{\frac{4\alpha\tau}{r^2}} \tag{1.7}$$

where: A—Laser absorptivity;

P—Laser power;

r—Equivalent radius of the laser beam;

k—Heat conductivity coefficient of material;

α—Thermal diffusion coefficient of material.

Heating of the material by the laser is correlated to various factors, such as laser power density, equivalent radius of the laser beam, and material heating time by the laser. The equivalent radius is generally defined as the distance from the center of the laser beam to the position when the light intensity drops to 1/e of the central light strength. For the Gaussian beam, the equivalent radius is $\omega/\sqrt{2}$ (ω is the spot size of the Gaussian beam). If the time characteristic constant $\tau_0 = r^2/4\alpha$ is introduced, Eq. (1.7) can be rewritten as:

$$\Delta T = \frac{AP}{k\pi^{3/2}r}\arctan\sqrt{\frac{\tau}{\tau_0}} \tag{1.8}$$

If the laser-material action time τ is far less than the time characteristic constant τ_0; that is to say, the laser irradiation time is very short, then

$$\arctan\sqrt{\frac{\tau}{\tau_0}} \approx \sqrt{\frac{\tau}{\tau_0}}$$

Equation (1.8) can be simplified to be:

$$\Delta T = \frac{AP}{k\pi^{3/2}r}\sqrt{\frac{\tau}{\tau_0}} \tag{1.9}$$

If the laser-material action time τ is far more than the time characteristic constant τ_0, the laser irradiation time is relatively long and the temperature field approaches steady state, thus,

$$\arctan\sqrt{\frac{\tau}{\tau_0}} \approx \frac{\pi}{2}$$

Equation (1.8) can be simplified to be:

$$\Delta T = \frac{AP}{2k\pi^{1/2}r} \tag{1.10}$$

If the laser-material action time τ is close to the time characteristic constant τ_0 (generally $0.1\ \tau_0 \leq \tau \leq 3\tau_0$), then

$$\arctan\sqrt{\frac{\tau}{\tau_0}} \approx \frac{\tau}{4}\left(\frac{\tau}{\tau_0}\right)^{3/8}$$

Equation (1.8) can be simplified to be:

$$\Delta T = \frac{AP}{4k\pi^{1/2}r}\left(\frac{\tau}{\tau_0}\right)^{3/8} \tag{1.11}$$

Heating of the material by the laser is mainly applied in the laser heat treatment technique. Usually, under the heat treatment condition, the laser beam applies to the material surface in the form of scanning, and the laser beam in this case is regarded as the moving heat source. The temperature field generating the moving heat source is the quasi-steady state, equivalent to the static temperature field dragging in the material at the laser scanning speed v, in which case, it's necessary to define an equivalent laser action time $\tau^* = Cr/v$. To approximately describe the temperature change of the moving laser surface using the static laser surface temperature, the constant C can be properly adjusted, usually equal to 1.25.

When the laser equivalent action time $\tau^* = 1.25r/v$ is substituted to Eq. (1.7), the maximum temperature at the center of the light spot can be obtained.

$$\Delta T = \frac{AP}{k\pi^{3/2}r}\arctan\sqrt{\frac{5\alpha}{vr}} \tag{1.12}$$

The characteristic velocity $v_O = 5\alpha/r$ is introduced. Generally, in the actual lasering heat treatment, the laser scanning speed is near v_O, in which case, Eq. (1.12) can be rewritten as:

$$\Delta T = \frac{AP}{k\pi^{3/2}r}\arctan\sqrt{\frac{v_0}{v}} \tag{1.13}$$

As it can be seen from Eq. (1.13), the material surface temperature is directly proportional to the square root of the heating time. For the laser heating pulse at the given energy, when the power density increases, the pulse duration (i.e., heating time) will definitely be shortened and the temperature on the material surface will raise; that is to say, the laser pulse with high peak power and short duration can heat the material surface more effectively.

1.1.3 Material Melting Under Laser Action

Physical issues more closely related to the laser processing is melting and vaporization of the material caused by the laser. When the temperature of the material surface heated by the laser is up to the melting temperature and vaporization temperature, Eq. (1.7) is no longer true. The material has to absorb latent heat during melting and vaporization. After melting and vaporization, the heat conductivity of the material changes greatly and the heat conduction gets very complex.

When the laser with a certain strength beats down on the material surface, and the material surface temperature is up to the melting point T_m, the isothermal level (prior to the melting wave $T = T_m$) will transfer into the material at certain speed, with the propagation velocity depending on the laser power density and thermodynamic parameters of the material in both solid and liquid phases. The melting without vaporization is usually defined to be shallow melting. In case of shallow melting, the light spot is larger than the weld pool in diameter, and the effect of transverse thermal diffusion can be ignored. Maximum depth at the shallow melting zone can be represented as:

$$Z\frac{1.2k}{AP}{}_{v}\left(\frac{T_v}{T_m} - 1\right)_{m,\max} \tag{1.14}$$

where T_m—Melting temperature of the material;

T_v—Vaporization temperature of the material.

In terms of the material characteristics, the larger the ratio of the thermal conductivity coefficient k of the material to T_v/T_m is, the greater the melting depth $Z_{\max}$ will be. For the laser characteristics, to improve the melting depth, the relatively low laser power density shall be adopted because it can take relatively long time to heat the material surface to T_v. Therefore, at the given laser pulse, the laser power density shall be adjusted so that the material surface can just reach the vaporization temperature at the end of laser pulse, in order to obtain the maximum melting depth. Under normal circumstances, the maximum depth of the pure melting of the material under laser irradiation is about several microns to several hundred microns.

What the shallow melting corresponds to is deep melting which is defined to be situation where the melting depth is no less than the radius of the light spot. For deep melting, only equilibrium vaporization occurs at the gas–liquid phase surface, the vapor is transparent to the laser. The threshold value of the laser power density causing deep melting of the material under continuous Gaussian beam irradiation is about:

$$I_{md} = 2kT_m/(\sqrt{\pi}\alpha A) \tag{1.15}$$

During deep melting, the diameter of the weld pool is larger than that of the light spot. A stable vaporization wellblock with the diameter less than that of the light spot appears in the center of the weld pool. The vapor in the wellblock has low density, basically transparent to the laser, so that the laser can directly enter and irradiate on the gas–liquid interface at the bottom of the wellblock, and then be absorbed. The absorbed laser energy is used for heat dissipation of the side wall and vaporization at the bottom. Assumed that the radius of both laser spot and vaporization wellblock is R_s, the depth of the deep melting can be expressed as:

$$Z_v = \frac{R_s^2 A I_0}{k_q T_v}\left[1 - \exp\left(-\frac{k_q T_v}{R_s^2 L_v \rho_q}t\right)\right] \tag{1.16}$$

In the deep melting, the length-diameter ratio of the vaporization wellblock Z_v/R_s cannot be too large (generally $0.5 \le Z_v/R_s \le 40$), or unstable movement may occur to vapor and solution.

If the laser intensity is relatively high, vaporization at the gas–liquid interface in the wellblock will exacerbate, and the vapor pressure will rise. The vapor pressure and vapor reaction can overcome the tension on the solution surface and the solution static pressure, resulting in appearance of keyholes and transfer of liquid mass. The keyhole is like black body, which is in favor of absorption of the beam energy, indicating the "wall focusing effect". Since the laser beam after focusing is not parallel beam, it will make certain angle of incidence with the keyhole wall. After radiating on the keyhole wall, the laser beam reaches the bottom of the keyhole after multiple reflection, and finally be completely absorbed.

The high-pressure steam in the keyhole pushes the melt to splash out along the edge of the weld pool or the wall of the vapor well, causing liquid mass transfer. Liquid mass transfer of this kind will greatly improve the efficiency of some laser processing, such as laser drilling or laser cutting. The squeezing effect of material vapor is an important cause of liquid mass transfer. Assuming that the vapor is produced from equilibrium vaporization at the gas–liquid section, ignoring the momentum and energy of the vapor, and only taking into consideration the vapor pressure P_v, when the vapor pushes the nonviscous incompressible solution layer of a certain thickness, the work of the vapor will be completely converted into the kinetic energy of the solution, and the mass of the liquid material transferred from a unit area of laser beam spot within a unit time at the ambient pressure of P_0 is the liquid mass transfer rate $\dot{m}_q$ of liquid mass, which is

$$\dot{m}_q = \left[\frac{2\lambda_q}{R_s}\ln\left(\frac{T_v}{T_m}\right)\right]^{1/2} \rho_q^{3/4}[2(P_v - P_0)]^{1/4} \tag{1.17}$$

1.1.4 Evaporation of Materials Under Laser Action

It can be seen from the melting of material under laser action, the melting process is usually accompanied by evaporation of the material. The evaporation mechanism is closely related to laser power density.

The time from the start of laser irradiation to the material surface reaching the evaporating temperature T_v is called the evaporation starting time t_v, which can be estimated as

$$t_v = \frac{\pi}{4\alpha}\left(\frac{kT_v}{AI_0}\right)^2 \tag{1.18}$$

For the metal with a low evaporating temperature, when AI_0 *is* 10^4~10^6 W/cm^2, t_v is about a few milliseconds to microseconds; for the metal with a high evaporating temperature, when AI_0 is 10^5~10^7 W/cm^2, t_v is about a few milliseconds to several hundred nanoseconds; If t_v is much less than the width of the laser pulse, the evaporation of the material surface is considered to start immediately once the laser irradiates.

When the laser power density is not very high, the evaporation of material is not drastic, the saturated vapor pressure is balanced with the ambient pressure, and the velocity distribution of the vapor particle is isotropic, in the Maxwell Distribution of translational equilibrium. Myriad laser equilibrium evaporation models have been established by scholars across the world. At present, the theoretical model of laser evaporation has only qualitative significance, and the quantitative understanding mainly relies on experiments. In all evaporation pressure formulas obtained with laser equilibrium evaporation models, the vapor pressure is considered to be proportional to the laser power density.

When the laser power density is high, the material evaporation rate increases, and the vapor pressure increases. When it is significantly higher than the ambient pressure, the number of particles in the vapor that return to the solution decreases, and the velocity distribution will deviate from the balanced Maxwell Distribution. The vapor particles leaving the liquid surface must pass a distance and collide with each other to reestablish a translational equilibrium. The thin layer above the liquid surface where the vapor is in the transition from a non-equilibrium state to an equilibrium state is called the Knudsen layer. In the extreme of strong evaporation, the vapor on the outer surface of the Knudsen layer flows at the sound velocity, and the vapor flowing outward is noticeably colder and thinner than the saturated vapor on the surface of the solution.

Under continuous laser irradiation, the velocity of evaporation front (namely the receding velocity) rises rapidly from zero to an approximately constant value, and the evaporation enters a steady state. The establishment time t_{sv} of a steady state is related to the steady receding velocity U_v. Usually t_{sv} is dozens of times greater than t_v. For example, when AI_0 is 10^7 W/cm^2, the receding speed of aluminum is about 2.34 m/s, and the corresponding velocity $t_{sv} \approx 15$ μs, while t_v is less than 0.3 μs.

Due to the action of the vapor pressure, t_v in Eq. (1.18) should be greater than the normal evaporating temperature under standard atmospheric pressure. When the laser power density is between 10^6 and 10^7 W/cm^2, the temperature of the evaporation front of the metal is slightly greater than the normal evaporating temperature. When the laser power density is between 10^6 and 10^{10} W/cm^2, the temperature of the former evaporation front is several times to dozens of times that of the latter. Since T_v increases with the laser power density, it leads to the receding velocity U_v in the range above 10^8 W/cm^2 decreasing with the increase of the laser power density, and the decrease is more obvious in the range above 10^0 W/cm^2.

The gaseous mass transfer rate $\dot{m}_v$ depends on the receding velocity, $\dot{m}_v = \rho U_v$, and the total gaseous transfer mass is $\rho \int_{t_v}^{t} U_v dt$. When the laser power density is low, the heat lost by thermal diffusion has an obvious influence on U_v, and the rate of gaseous mass transfer is also low. The calculated values of the receding velocity and gaseous mass transfer rate are usually noticeably different from the experimental results, which is caused by two reasons. For one thing, it is because the reflectivity, absorptivity and thermal physical characteristics are all related to temperature and the actual laser waveform is highly irregular. For the other, it is because that mass transfer is actually a comprehensive result of various mechanisms, in which the main part of the transferred mass is composed of the splashed droplets by vapor pressure.

1.1.5 Laser-Induced Plasma and Its Effects

When laser acts on the surface of the material, vapor is induced. The vapor continues to absorb laser energy, causing an increase in the temperature and eventually forming a high-temperature and high-density plasma. A plasma is a mass composed of large quantities of charged particles (electrons and ions), atoms and molecules, which as a whole is electrically neutral. There are three main mechanisms of plasma generation by high power density laser: photoionization, thermal electric ionization and collision ionization. Photoionization refers to the phenomenon that when electrons in atoms are irradiated by laser and absorb sufficient photon energy, ionization occurs due to the photoelectric effect or the multi-photon effect. Photoionization is mainly applicable to the generation process of initial charge carriers in a relatively cold medium, while the laser plasma is in a completely ionized state, and therefore photoionization is not the main mechanism for its formation. Thermal electric ionization refers to that when the temperature of the vapor under laser action is high enough, the atoms with high thermal velocities at the high temperature collide with each other, making their electrons excited, and the energy of some of the electrons exceeds the ionization potential, which leads to the ionization of atoms. Collision ionization refers to that charged particles in vapor accelerate under the action of electric field and collide with neutral atoms, resulting in energy exchange, which enables the electrons in the atom to acquire sufficient energy to ionize.

The ionization degree of vapor in the thermodynamic equilibrium state is completely determined by the density and temperature of vapor. In a partially ionized gas, the energy of the incident laser is absorbed by the thermally excited atoms through the binding-free mechanism and by the particles through the inverse bremsstrahlung. Vapor absorbs laser energy and heats up, leading to further increase in the ionization degree and absorption coefficient. This kind of positive feedback contributes to the formation of plasma in the vapor.

The plasma absorbs the energy of the laser beam propagating in it through various mechanisms, which increases its temperature and ionization degree. Inverse bremsstrahlung absorption is the main mechanism of plasma absorbing laser energy. Inverse bremsstrahlung absorption refers to that the electrons in a laser electric field are excited to oscillate at high frequency, and collide with particles with a certain probability to transfer energy to heavier particles (ions and atoms), thus heating up the plasma.

The vapor of a metallic material is a monatomic gas, which has a very high evaporating temperature T_v but a very low ionization potential, so the evaporating rate of the material is not high. When the vapor pressure is only slightly higher than the ambient pressure, there are also a large number of free electrons in the weakly ionized vapor, which can lead to effective inverse bremsstrahlung absorption of laser. In other words, free electrons in an ion Coulomb field absorb laser energy, and accelerate or change the direction of motion, thus transferring the laser energy to ions and heating up the vapor.

When the light intensity is high and the energy absorption rate of the vapor exceeds its various losses, the vapor ionizes and the number of free electrons increases exponentially with time, making the vapor completely ionized and opaque to the laser. As the size of the plasma increases by absorbing energy, the power density of the laser reaching the surface of the material decreases, and therefore the evaporation effect weakens, resulting in a drop in the density and temperature of the plasma, as well as in its absorptivity; conversely, as the absorptivity weakens, the laser power density reaching the surface of the material increases, and therefore the evaporation effect is enhanced, resulting in an increase in the density and temperature of the plasma, as well as in its absorptivity.

The absorption of the laser by the plasma prevents the laser from reaching the surface of the material, cutting off the energy coupling between the laser and the material. This effect is called plasma shielding. The ratio of energy absorbed by plasma to the energy of the incident laser is called plasma shielding coefficient. The plasma shielding coefficient is related to the laser wavelength. The plasma shielding effect of a long wavelength laser is stronger than that of a short wavelength laser, and appears earlier. In laser welding technology, the absorption and scattering of plasma affect the transmission efficiency of laser and reduce the laser energy reaching the work piece. Meanwhile, the negative lens effect (refraction) of plasma expands the area of action of laser energy on the work piece, thus reducing welding quality.

1.2 Principles and Characteristics of Laser Welding

Laser welding is a special fusion welding method in which a focused laser beam with high energy density (10^6–10^{12} W/cm^2) is used as a heat source to heat and melt the work piece. It is a fusion welding based on the above-mentioned photothermal effect of the interaction between the laser and the material. Its premise is that the laser is absorbed by the material and converted into the thermal energy required for welding. In general, the physical phenomena resulting from laser action on the material surface vary, including the increase of surface temperature, melting, evaporation, formation of keyhole, and generation of laser induced plasma (see Fig. 1.2).

These physical phenomena determine the thermal action mechanism of the welding process, which leads to two laser welding modes: heat conduction welding and deep penetration welding. The transition of the two modes depends primarily

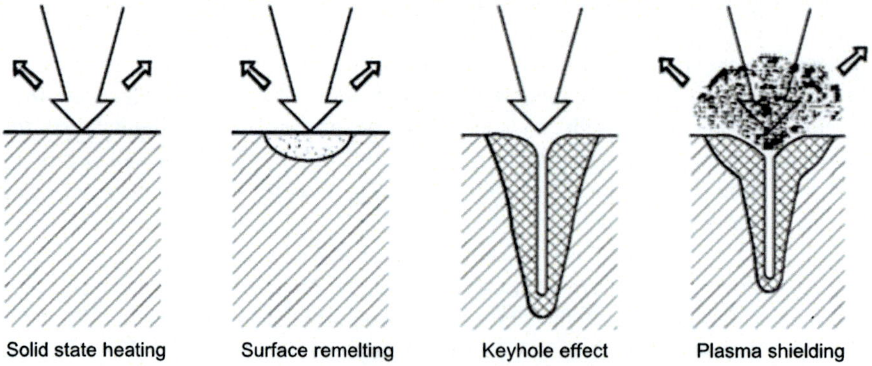

Fig. 1.2 Physical processes of laser of different intensities acting on metal surface

on the power density of the laser spot acting on the material. For some particular materials, there is a specific threshold of power density. When the power density of the laser acting on the material is lower than this threshold value, the laser energy is absorbed by the surface of the material and transferred to the interior of the material quickly, forming a heat conduction weld with a relatively large width and depth. However, when the power density of the laser acting on the material is higher than this threshold value, the laser energy makes the material surface heat up, melt and evaporate rapidly before the work piece surface can transfer the heat into the material, and with the continuous input of laser energy, a keyhole in the direction of penetrating the thickness is formed. The keyhole, surrounded by a weld pool of liquid metal, is filled with high-temperature metal vapor and plasma. The expansion force of the high-temperature metal vapor and plasma acts together with the gravity and surface tension of the liquid metal around the keyhole to maintain a stable existence of the keyhole (see Fig. 1.3).

The keyhole moves along the welding direction, and the weld pool behind it quickly cools and solidifies, forming a deep penetration weld with a large width and depth, as shown in Fig. 1.4. Therefore, the laser welding mode is closely related to the laser power density and welding heat input which determine the thermal mechanism.

Laser welding is an extremely complex physical and chemical process, containing phenomena such as rapid heating, melting, evaporation, ionization, rapid cooling and non-equilibrium solidification of the material, and involving the formation and of different states of matter, such as solid, liquid, gaseous and plasma forms and the complex interaction between them. The welding area is in a dynamic supernormal thermophysical state, in which the behavior of heat and mass transfer and the process of non-equilibrium solidification and structure evolution are more complicated than that of conventional heat source welding. The dynamic behavior of these complex phenomena has an important influence on the welding quality and the mechanical properties of the joint. Furthermore, the key components to be welded often have complex three-dimensional shapes, and dynamic working conditions, such as

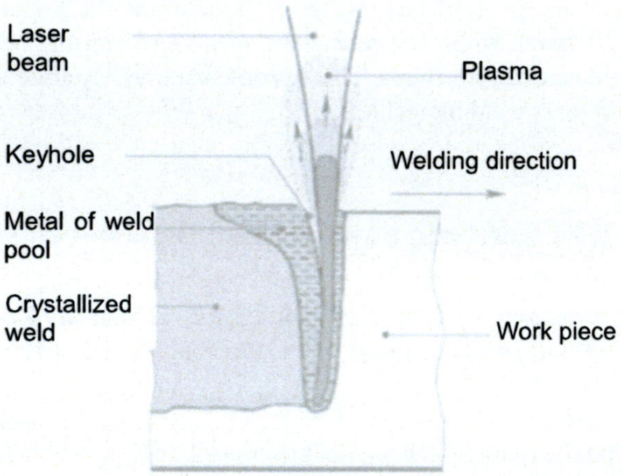

Fig. 1.3 Diagram of keyhole formation in laser welding

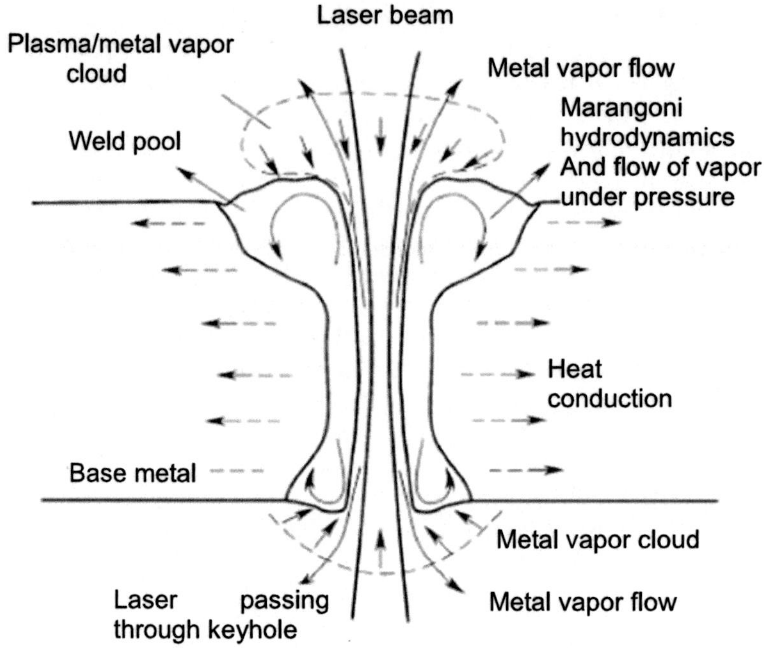

Fig. 1.4 Diagram of laser penetration welding principle

weld layout and slight thermal deformation caused by welding process, inevitably affect the stability of welding process. Therefore, various complicated theoretical and technical problems are exposed in the welding process. The first one is the poor welding process stability and repeatability, such as frequent interruptions in the welding process, which make it impossible to maintain a stable welding in the whole process. The second one is that the generation mechanism of welding defects is unclear, and the control technology is not well developed, which lead to negligible advancement of laser welding. In addition, the performance of laser welding joint is uneven with poor consistency, which makes the welded structure fail to meet the demand for functions. These problems seriously affect the application and promotion of laser welding technology.

1.3 Research on Weld Pool Behavior in Laser Welding

Compared with the conventional arc welding method, the deep penetration laser welding, which is based on the theoretical basis of keyhole effect, has attracted wide attention for its high energy density, low heat input, high depth-to-width ratio of weld, small heat affected zone (HAZ), small deformation, high welding quality, high production efficiency, flexible control, etc.

But in the welding process, due to reasons such as multiple process parameters and complex working conditions, welding quality problems often occur. Especially in the process of welding thick section structures and large size panels with long weld joints, it is easy to cause poor welding forming, serious welding defects, or unstable welding process with frequent interruptions, which affect the qualified rate of the weldment.

As studies have shown that in any welding process, the main factor closely related to the quality of the weld joint is the stable existence of a weld pool, and for the deep penetration laser welding process, the main manifestation is the stable existence of the keyhole in the weld pool. Therefore, a stable welding process for deep penetration laser welding is a prerequisite of ensuring the quality of the weld joint, and the basis for a stable welding process is the stable existence of a keyhole in the welding process. By observing the deep penetration laser welding process and by dissecting the weld joints, it is found that the defect of welding quality is often caused by the instability of the weld pool keyhole in the welding process. Researches have demonstrated that the main factor closely related to the quality of a weld joint during the welding process is the stable existence of a weld pool. Based on experiments and researches on the deep penetration laser welding process, it is revealed that, in the deep penetration laser welding process, many parameters such as laser welding power, focus position, welding speed, assembly gap, etc. in existence of the keyhole and in formation and disappearance of the keyhole affect directly existence state and morphology of the keyhole. So, the influence of different process parameters on the behavior of the keyhole under deep penetration laser welding conditions is studied and analyzed to identify the pattern by which different welding process parameters exert influence on the welding quality, to create favorable conditions for further realization of the control over the welding process and quality.

In coping with the issue that the keyhole simulation and the calculation of convective heat transfer of the weld pool in deep penetration laser welding cannot be organically combined or that the coupling relationship is extremely complicated, and with the view of practical application, the research team of the author employs a method of combining experimental research with computational simulation, on the basis of works of current mathematical simulation research on deep penetration laser welding and the energy conservation theory (see Fig. 1.5), in an attempt to build a heat source model suitable for the finite volume method for deep penetration laser welding process simulation, and uses FLUENT, a commercial computational fluid dynamics software, to simulate the basic features of the keyhole formation and the flow of the weld pool in the laser welding process. On this basis, the interacting flow field between the side blowing auxiliary airflow and the keyhole jet flow is calculated, which provides a basic theoretical foundation for improving protection of the welding area and the quality of weld joints. Applying the basic process theories on deep penetration laser welding, the hydrodynamics of the weld pool and the stability of the keyhole are studied jointly from the perspective of weld pool hydrodynamics, with the expectation of improving the quality of laser welding. The main research works and achievement are as follows:

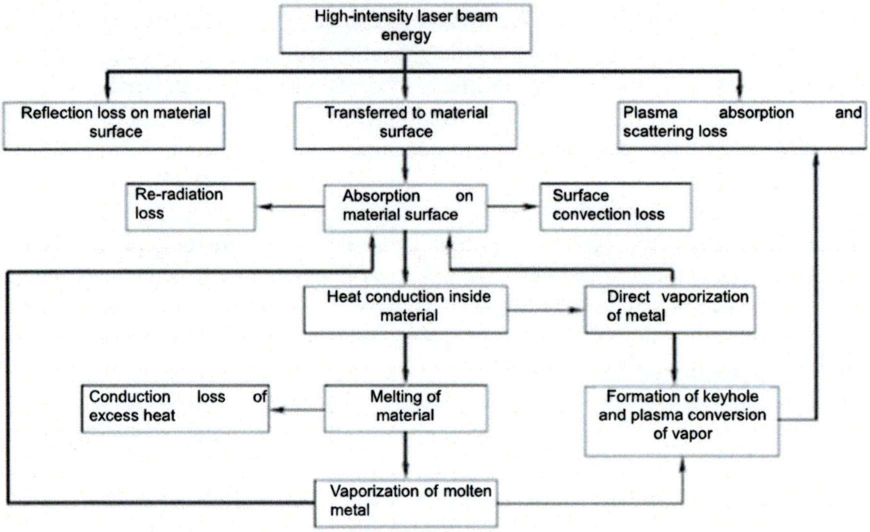

Fig. 1.5 Diagram of energy conservation in deep penetration laser welding process

(1) Establishment of a mathematical model of volume heat source in deep penetration laser welding;
(2) Applying basic equations of fluid dynamics to establish a three-dimensional mathematical model for melting and solidification of weld pool in deep penetration laser welding under action of a moving heat source to simulate the shape of the keyhole formed by the titanium alloy under different welding heat input;
(3) Based on software FLUENT, applying the finite volume method to study temperature distribution and velocity distribution characteristics of weld pool in deep penetration laser welding and verifying the correctness of the mathematical model through deep penetration laser welding experiment;
(4) Simulating the effective position and size of the velocity field and protected area under the interaction of the auxiliary side airflow and the keyhole jet near the keyhole and the weld pool.

Figure 1.6 shows the main contents and their relationship in three aspects of fluid dynamics research on deep penetration laser welding. The study concludes as follows:

(1) It is suitable to employ a combined heat source model consisting of the rotating Gauss heat source and double-ellipsoid heat source to simulate the formation of the keyhole in deep penetration laser welding of titanium alloy and the flow velocity field of the weld pool. The body heat source model can reflect the basic physical process in deep penetration laser welding, and also embodies the simulation characteristics of the control volume method.
(2) In the process the deep penetration laser welding with full penetration, the diameter of the keyhole is not very sensitive to the laser power, but the inclined

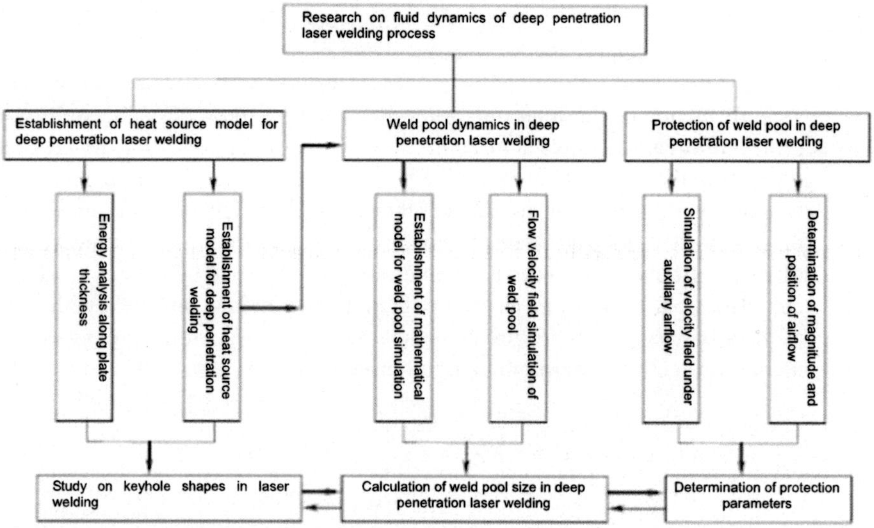

Fig. 1.6 Contents of research on fluid flow dynamic behavior in deep penetration laser welding process

angle of the keyhole changes obviously with the increase of the laser welding speed. The surface tension pressure caused by the curvature of the free surface of the keyhole has a strong relationship with the welding speed under condition of constant airflow pressure in the keyhole. With the increase of the welding speed, the stability of the keyhole decreases gradually.

(3) The flow velocity above the surface of the weld pool in deep penetration laser welding is the largest, and Marangoni force plays a dominating role in convective heat transfer on the surface of the weld pool. Under the action of the moving heat source, the Marangoni flow evolves from symmetrical shape around the laser heating center to tadpole shape where the long axis of the weld pool is parallel to the welding direction. In a velocity "virtual" circle, on the surface of the weld pool, with half of the width of the weld pool as the radius, the flow velocity of the weld pool is relatively large, and outside the "virtual" circle, the flow velocity of fluid in the weld pool decreases gradually. Inside the weld pool, the fluid flow velocity is significantly lower than that on the surface of the weld pool, but much larger than the welding speed. The flow velocity value of fluid on the back side of the weld pool is larger than the flow speed value of metal inside the weld pool.

(4) The shape and size of the weld pool correspond well with the size and position of the flow velocity vortex in the weld pool. The weld pool vortex is the result of the combined action of the counter forces of Marangoni flow and the solid–liquid interface, with buoyancy and gravity only playing an auxiliary role. The existence of the fluid flow vortex in the weld pool greatly enhances the

convective heat transfer between the high temperature metal fluid and the cold fluid, and directly affects the size and shape of the weld pool.

(5) The parameter of side blowing auxiliary airflow in deep penetration laser welding is one of the main factors in determining the size of the weld pool protected area and in reducing emission of plasma plume from the keyhole. Increasing the flow velocity near the plasma plume is conducive to convective heat dissipation, and in turn enhances the probability of recombination between plasma positive ions and negative ions, therefore facilitating the improvement of welding quality. There is a relatively large difference between the component flow field with helium as the auxiliary airflow and the flow field with argon as the auxiliary airflow. From the perspective of the size of the protected area, helium protection is better than argon protection.

Chapter 2
Model of Quasi-Steady Weld Pool Dynamics and Numerical Simulation

2.1 Introduction

Laser welding is a joining process in which a focused laser beam is coupled to the surface of a material to be welded to generate heat and melt the material. The welding process can be optimized in three aspects: determining a set of process parameters which make the welding process basically stable and are reproducible; monitoring the changes of various parameters in the welding process to ensure the quality of the joint; real-time control of the parameters of the laser welding process.

The instability of laser welding is its intrinsic characteristic. The instability is caused by fluctuation of parameters in four aspects during the laser welding process, including the fluctuation of laser output parameters, the effect of changing plasma flow at the exit of the keyhole on the incident laser beam, the dynamic instability of the weld pool motion, and the instability of the vapor flow in the keyhole. The interaction of these parameters makes the laser welding process extremely complicated. Hence, the basic point of departure is to simplify some complex parameters, ignoring the influence of some minor factors, to study the influence of the laser welding parameters on the temperature field and velocity field of the weld pool in quasi-steady state by numerical simulation.

The steady state refers to a state in which the weld pool does not change with time under certain conditions. The quasi-steady state means that the temperature field and velocity field of the laser welding pool do not change with time, but the instantaneous quantities are changing at all times. The "Quasi" means that although the degree is not completely enough, it can still be regarded as a certain kind of thing, that is, it is close to a stable state but still unstable.

In any welding process, the main factor closely related to the quality of the weld joint is the stable existence of a weld pool, and for the deep penetration laser welding process, the main manifestation is the stable existence of the keyhole in the weld pool. Experiments and researches on the deep penetration laser welding process have shown that, in the deep penetration laser welding process, many parameters such as laser welding power, focus position, welding speed, assembly gap, etc. in

S. Gong et al., *Weld Pool Dynamics in Deep Penetration Laser Welding*,
https://doi.org/10.1007/978-981-16-0788-2_2

existence of the keyhole and in process of formation and disappearance of the keyhole affect directly the existence state and morphology of the keyhole.

This chapter takes titanium alloy laser welding as the research subject, focusing on research of the temperature field model in the quasi-steady state in laser keyhole welding, the fluid dynamics model of the weld pool in quasi-steady state and the solving techniques, distribution of the welding temperature field and of the velocity in quasi steady state.

2.2 Basic Models of Quasi-Steady Laser Welding

When the energy density of the focused laser beam is greater than 10^6 W/cm^2, the metal is melted rapidly and evaporated and gasified. The strong evaporative recoil force pushes out the melted metal around it, forming a keyhole in a very short time. The keyhole is surrounded by molten liquid metal. Under the action of pressure gradient, the metal fluid in the front of the keyhole flows from both sides of the keyhole to the back of the weld pool, as shown in Fig. 2.1. After the welding heat source has left its original position, the metal in the weld pool is cooled to form a weld, which connects the metals together. The flow of metal in weld pool will affect directly the quality of welded joints. It is very important to simulate the energy input of laser welding and the heat transfer of welding pool for accurately predicting the transient temperature field and further studying the microstructure, properties, stress and strain of the joint.

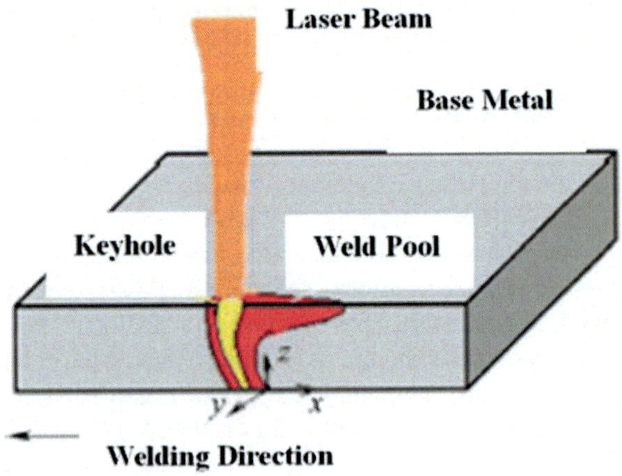

Fig. 2.1 Schematic of deep penetration laser welding

2.2.1 *Governing Equations of the Flow and Heat Transfer in Quasi-Steady Laser Welding*

The governing equations of the main variables (velocity, temperature, etc.) that need to be solved in the flow and heat transfer problems can be expressed in the following general forms:

$$\frac{\partial(\rho\varphi)}{\partial t} + \mathrm{div}\left(\rho \vec{U} \varphi\right) = \mathrm{div}\left(\Gamma_\varphi \mathrm{grad}\varphi\right) + S_\varphi \tag{2.1}$$

where: φ—General variable;

$\vec{U}$—Cartesian coordinate velocity component u in the x direction, v in the y direction, and w in the z direction;

Γ_φ—General diffusivity;

S_φ—General source term, including Darcy source term of the mushy zone of momentum conservation equations, Boussinesq buoyancy source term, internal heat source term and latent heat source term in the energy equation;

ρ—Density of material.

Suppose m_l represents the mass component of chemical composition l. If we have speed $\vec{U}$, the conservation equation of m_l can be expressed as follows:

$$\frac{\partial(\rho m_l)}{\partial t} + \mathrm{div}(\rho m_l \vec{U}) = \mathrm{div}(\Gamma_l \mathrm{grad} m_l) + R_l \tag{2.2}$$

where: R_l—Productivity of composition l per unit volume, (kg/s m^3);

Γ_φ—Diffusion coefficient of composition l.

In Eq. (2.1), the expressions of Darcy source term and buoyancy source term are as follows:

$$S_\varphi = -\frac{\mu}{K}\vec{U} + \rho_0 \beta g(T - T_0) \tag{2.3}$$

where: K—Permeability, calculated by Karman-Kozeny equation;

ρ_0—Density of material at reference temperature;

β—Thermal expansion coefficient of representative material;

g—Gravitational acceleration;

T_0—Reference temperature.

Karman-Kozeny equation is expressed as follows:

$$\mathrm{K} = \frac{(f_l)^3}{D_l(1 - f_l)^2} \tag{2.4}$$

where: f_l—Liquid fraction;

D_l—the Darcy constant associated with the liquid fraction.

In liquid phase zone, $f_l = 1$; in solid phase zone, $f_l < 1$; in mushy zone, $0 < f_l < 1$.

The expression of latent heat source term in Eq. (2.2) is as follows:

$$S_T = -\rho \frac{\partial (f_l L_m)}{\partial t} - \rho \nabla \left(\vec{U} f_l L_m \right) \tag{2.5}$$

where: L_m—Latent heat of fusion of the material;

$\vec{U}$—Velocity vector.

2.2.2 3D Mathematical Model of Deep Penetration Laser Welding Under Moving Heat Source

2.2.2.1 Governing Equations of Quasi-Steady Deep Penetration Laser Welding Under Moving Heat Source

Continuity equation:

$$\frac{\partial (\rho u)}{\partial x} + \frac{\partial (\rho v)}{\partial y} + \frac{\partial (\rho w)}{\partial z} = 0 \tag{2.6}$$

where: *u, v,* and *w*—Cartesian coordinate velocity component in the x direction, in the y direction, and *w* in the z direction in that order for the flow of heat or fluid a weld pool;

ρ—Density of weld pool or solid metal.

Energy equation:

$$\begin{aligned} \rho C_p \left((u - v_w) \frac{\partial T}{\partial x} + v \frac{\partial T}{\partial y} + w \frac{\partial T}{\partial z} \right) &= \frac{\partial}{\partial x} \left(\lambda \frac{\partial T}{\partial x} \right) + \frac{\partial}{\partial y} \left(\lambda \frac{\partial T}{\partial y} \right) \\ &+ \frac{\partial}{\partial z} \left(\lambda \frac{\partial T}{\partial z} \right) + S(x, y, z) \\ &- \frac{\partial}{\partial x} (\rho u \Delta H) - \frac{\partial}{\partial y} (\rho v \Delta H) - \frac{\partial}{\partial z} (\rho w \Delta H) \end{aligned} \tag{2.7}$$

where: v_w = Welding velocity;

T—Temperature;

λ—Thermal conductivity of materials;

$S(x, y, z)$—The internal heat source term in welding;
ΔH—Latent heat during phase transition.
Momentum equations in the x direction

$$\frac{\partial(\rho uu)}{\partial x}+\frac{\partial(\rho uv)}{\partial y}+\frac{\partial(\rho uw)}{\partial z}=-\frac{\partial p}{\partial x}+\frac{\partial}{\partial x}\left(\mu\frac{\partial u}{\partial x}\right)+\frac{\partial}{\partial y}\left(\mu\frac{\partial u}{\partial y}\right)+\frac{\partial}{\partial z}\left(\mu\frac{\partial u}{\partial z}\right)-\frac{\mu}{K}(u-u_w) \tag{2.8}$$

Momentum equations in the y direction

$$\frac{\partial(\rho vu)}{\partial x}+\frac{\partial(\rho vv)}{\partial y}+\frac{\partial(\rho vw)}{\partial z}=-\frac{\partial p}{\partial y}+\frac{\partial}{\partial x}\left(\mu\frac{\partial v}{\partial x}\right)+\frac{\partial}{\partial y}\left(\mu\frac{\partial v}{\partial y}\right)+\frac{\partial}{\partial z}\left(\mu\frac{\partial v}{\partial z}\right)-\frac{\mu}{K}v \tag{2.9}$$

Momentum equations in the z direction

$$\frac{\partial(\rho wu)}{\partial x}+\frac{\partial(\rho wv)}{\partial y}+\frac{\partial(\rho ww)}{\partial z}=-\frac{\partial p}{\partial z}+\frac{\partial}{\partial x}\left(\mu\frac{\partial w}{\partial x}\right)+\frac{\partial}{\partial y}\left(\mu\frac{\partial w}{\partial y}\right)+\frac{\partial}{\partial z}\left(\mu\frac{\partial w}{\partial z}\right)-\frac{\mu}{K}w+\rho g\beta(T-T_{ref}) \tag{2.10}$$

In Eqs. (2.8)–(2.10):
p—pressure;
μ—liquid viscosity;
K—permeability;
β—the thermal expansion coefficient of material;
g—gravitational acceleration;
T_{ref}—reference temperature (300 K).

2.2.2.2 Boundary Conditions

The computational domain of laser welding is symmetrical along the welding centerline. The symmetrical plane coincides with the center line of the weld. The pressure inlet and outlet of the fluid reflect the characteristics of the main flow direction, which is mainly affected by the welding direction. The upper and lower surfaces of weld pools are free surfaces, and surface tension plays an important role. In the quasi-steady state, the center subject to the laser heat source is located at the origin

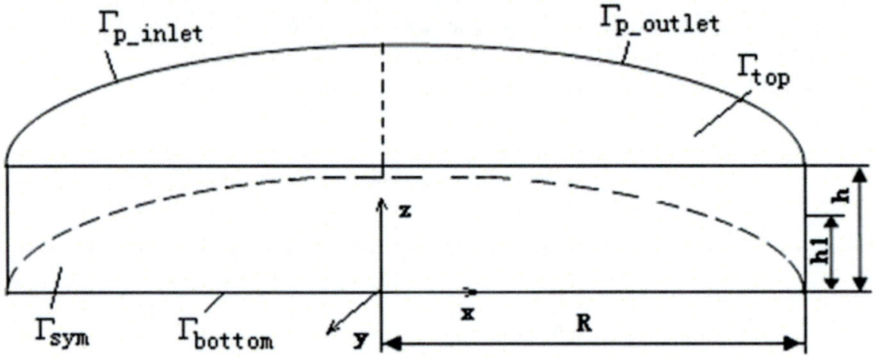

Fig. 2.2 Computational domain for deep penetration laser welding

of the coordinate. h1 is the demarcation line of the combined heat source. h is the maximum depth of the keyhole and also the thickness of the welded material when the laser keyhole welding is fully penetrated, as shown in Fig. 2.2. In order to save computing time, half of the sheets is selected for numerical simulation, given the symmetry between the temperature field and flow field along the welding center line. In Fig. 2.2, R is the computational domain radius, Γ_{p_out}, Γ_{p_inlet} are the fluid outflow and inflow boundaries, respectively, Γ_{sym} is the symmetric boundary, Γ_{top} is the upper surface boundary, and Γ_{bottom} is the lower surface boundary.

The convective heat loss per unit area of the upper surface of the workpiece is expressed by the following formula:

$$q_c = h_c(T - T_{ref}) \tag{2.11}$$

In this equation:

q_c—Convective heat flux;

h_c—Convective heat transfer coefficient.

To determine the convective heat transfer coefficient h_c, it is necessary to define the feature length L of the workpiece.

$$L = \frac{A}{P} \tag{2.12}$$

where: A—the surface area of workpiece;

P—the perimeter.

The corresponding Nusselt number is as follows:

$$N_u = 0.27 \cdot {R_a}^{0.25} \tag{2.13}$$

where Ra is the Rayleigh number. It can be expressed as:

$$Ra = \frac{g\beta}{\lambda\eta} L^3 \left(T - T_{ref}\right) \tag{2.14}$$

where: g—Gravitational acceleration;

β—Thermal expansion coefficient of material;

λ—Thermal diffusivity of material;

η—The kinematic viscosity of gas flow around workpiece.

For the definition of other parameters, refer to Eqs. (2.7), (2.10), and (2.12).

The convective heat transfer coefficient h_c on the workpiece surface is calculated by the following formula:

$$h_c = N_u \frac{k_{air}}{L} \tag{2.15}$$

In this equation:

k_{air}—The heat conductivity of gas flow around workpiece.

The heat loss flow q_r per unit area caused by radiation is calculated as follows:

$$q_r = \varepsilon\sigma\left(T^4 - T_{amb}^4\right) \tag{2.16}$$

where: ε—The emissivity of materials on surface;

σ—The Stefan-Boltzmann constant, 5.6697×10^{-8}.

The upper surface of weld pools:

$$\mu \frac{\partial u}{\partial z} = -\frac{\partial \gamma}{\partial T} \frac{\partial T}{\partial x} \tag{2.17}$$

In this equation:

$\frac{\partial \gamma}{\partial T}$—The temperature coefficient of surface tension of weld pool.

$$\mu \frac{\partial v}{\partial z} = -\frac{\partial \gamma}{\partial T} \frac{\partial T}{\partial y}, \, w = 0 \tag{2.18}$$

In solid zone:

$$u = v_w, \, v = 0, \, w = 0 \tag{2.19}$$

The lower surface of weld pools:

$$\mu \frac{\partial u}{\partial z} = -\frac{\partial \gamma}{\partial T} \frac{\partial T}{\partial x} \tag{2.20}$$

$$\mu \frac{\partial v}{\partial z} = -\frac{\partial \gamma}{\partial T} \frac{\partial T}{\partial y}, \, w = 0 \tag{2.21}$$

In the symmetry plane:

$$\frac{\partial u}{\partial y} = \frac{\partial w}{\partial y} = 0,\ v = 0,\ \frac{\partial T}{\partial y} = 0 \tag{2.22}$$

2.2.3 Heat Source Model in Deep Penetration Laser Welding

The Gauss heat source model is suitable for most finite element simulations that do not combine the convective heat transfer of weld pool with pore simulation organically. In these simulations, however, the model describes surface heat source and body heat source quite differently. Surface heat source only acts on the surface element of the workpiece, while body heat source is applied to some elements inside the workpiece. Considering the gradual attenuation of laser beam energy in the direction of plate thickness during deep penetration laser welding, Wu Su et al. [27] of Tsinghua University proposed a Gauss rotating heat source model:

$$q_{rotary}(x, y, z) = q(0, 0)exp\left[\frac{-3c_s}{log10\left(\frac{h}{z}\right)}\left(x^2 + y^2\right)\right] \tag{2.23}$$

In this equation, $q(0, 0)$ is the maximum heat flow intensity at the exit of rotating body heat source. It is expressed as follows:

$$q(0, 0) = \frac{3(1 - \chi)c_s\eta Q_{total}}{\pi h\left(1 - \frac{1}{e^3}\right)} \tag{2.24}$$

where: c_s—The concentration coefficient of heat flow distribution on this section $(1/m^2)$;

h—The height of body heat source (m);

Q_{total}—The power of the heat source;

η—The effective absorption coefficient of laser beams;

χ—The distribution coefficient of the heat flow of body heat source.

The disadvantage of the above body heat source model is that the energy attenuation rate in the thickness direction of the plate is too fast to fully reflect the energy distribution in the whole process of deep penetration laser welding. The double ellipsoid model proposed by Goldak from Canada takes into account the influence of welding speeds on heat flux distribution. The energy distribution in the direction of plate thickness can reflect the attenuation of laser beam energy [28], that is

$$q_{double-ellipsoidal}(x, y, z) = \frac{6\sqrt{3}\eta\chi Q_{total}}{ach\pi\sqrt{\pi}}\exp\left(-3\frac{x^2}{a^2}\right)\exp\left(-3\frac{y^2}{b^2}\right) \tag{2.25}$$

where: *a, b*—The half-axis lengths of the double ellipsoid model in the x and y directions, respectively. The focal radius of laser beam is used.

c—The half-axis length of the double ellipsoid model in the z direction. The effective depth of heat source is used.

Therefore, the study proposes the rotating Gauss heat source in the upper part of the workpiece and the double ellipsoid heat source model in the lower part of the workpiece. The body heat source mode means that the heat flow acts directly on the metal inside the material, enabling the keyhole to form rapidly. These keyhole effects are considerably different from those produced when heat source acts only on the surface of the metal surface. As far as the physical nature of heat transfer in deep penetration laser welding is concerned, the combination heat source is the progress and development of the combination heat source composed of surface heat source and columnar heat source.

2.3 Implementation of Numerical Solution

2.3.1 Upwind Scheme

The upwind scheme appeared in the 1950s and has been developing ever since. It was initially proposed by Courant, Issacson and Rees, and later on, was re-expounded by Gentry, Martin and Daly, Barakat and Clark, Runchal and Wolfsh-tein, etc. It gives full consideration to the influence of the flow direction on the derivative difference calculation formula and the function value-taking method on the interface.

(1) Definition of Taylor expansion method

As shown in Fig. 2.3, in the direction of flow, the first derivative of point *P* is always the backward difference in that direction. That means the information necessary to

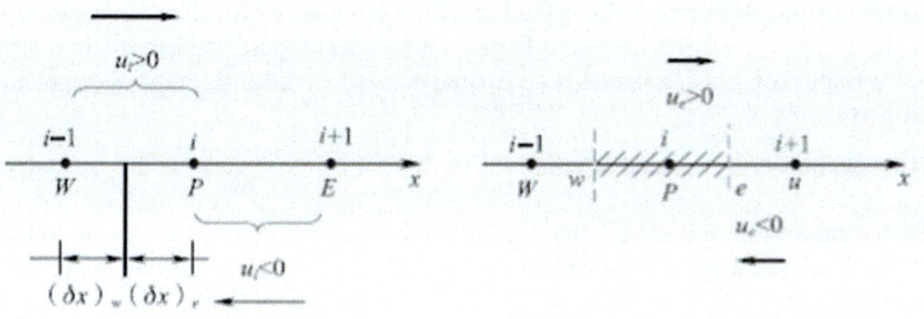

(a) Construction of upwind scheme based on Taylor expansion method

(b) Construction of upwind scheme based on control volume integral method

Fig. 2.3 First-order upwind difference scheme

form the first derivative is always obtained from the upstream, the formula expression is

$$\left[\frac{d\varphi}{dx}\right]_i = \frac{\varphi_i - \varphi_{i-1}}{\delta x}, u_i > 0 \tag{2.26}$$

$$\left[\frac{d\varphi}{dx}\right]_i = \frac{\varphi_{i+1} - \varphi_i}{\delta x}, u_i < 0 \tag{2.27}$$

(2) Definition of control volume integral method

The value of variable Φ on the control volume interface is defined as follows (see Fig. 2.3b):

On interface e: $u_e > 0, \varphi = \varphi_P; u_e < 0, \varphi = \varphi_E$ (3.28a).

On interface w: $u_w > 0, \varphi = \varphi_w; u_w > 0, \varphi = \varphi_P$ (3.28b).

That is to say, the unknown quantities on the interface always take the values at the upstream node while the central difference takes the arithmetic mean of those at the upstream and downstream nodes, and this is the basic difference between the two schemes.

(3) Meaning of false diffusion

The false diffusion is defined as a phenomenon of relatively large numerical calculation error for the reason that the truncation error of the discrete scheme obtained by the first derivative term of the convection–diffusion equation is smaller than the second order.

The partial difference of the first derivative term is the cause of the false diffusion. From the characteristics of the physical process itself, the effect of diffusion always reduces the change rate of the physical quantity, which in turn makes the whole field uniform. In a discrete scheme, the presence of false diffusion exacerbates the extent to which the results of numerical solutions deviate from the true solution. The greater the degree of false diffusion, the more serious the deviation. Where the first term of the truncation error of a discrete equation is an even-order spatial derivative, the error of the numerical calculation result features diffusion property; where the first term of the truncation error is an odd-order spatial derivative, the error features dispersion property.

(4) Second-order upwind scheme

The second-order upwind scheme can be employed to overcome the shortcomings of the first-order upwind scheme while taking advantage of the benefit of the upwind scheme.

For the evenly divided grid shown in Fig. 2.4, the second-order upwind scheme is defined as

$$u\frac{\partial\varphi}{\partial x} \approx \frac{u_i}{2\Delta x}(3\varphi_i - 4\varphi_{i-1} + \varphi_{i-2}), u_i > 0 \tag{2.29a}$$

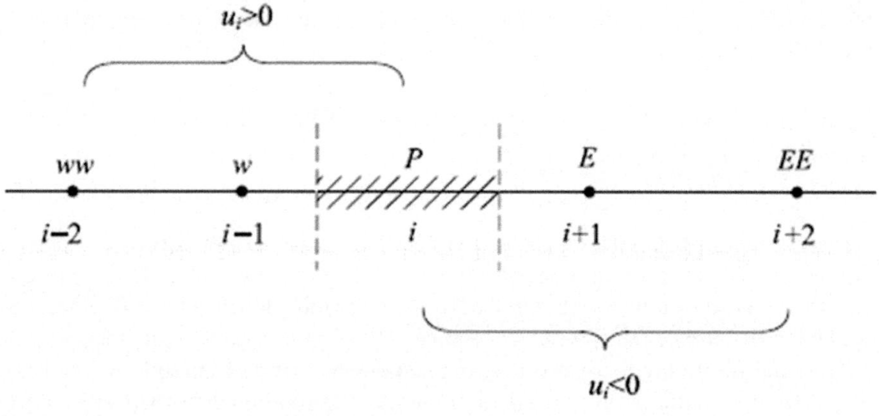

Fig. 2.4 Illustration of second-order upwind scheme

$$u\frac{\partial \varphi}{\partial x} \approx \frac{u_i}{2\Delta x}(-3\varphi_i + 4\varphi_{i+1} - \varphi_{i+2}),\ u_i < 0 \tag{2.29b}$$

The interface interpolation of the second order upwind is defined as follows (taking φ_w for example)

$$\begin{aligned}\varphi_w &= 1.5\varphi_W - 0.5\varphi_{WW},\ u_W > 0\\ \varphi_w &= 1.5\varphi_P - 0.5\varphi_E,\ u_W < 0\end{aligned} \tag{2.30}$$

When the convection term adopts the second-order upwind scheme and the diffusion term adopts the central difference scheme, the obtained discrete equation has a second-order precision truncation error.

(5) QUICK scheme

The QUICK scheme improves the truncation error of the scheme by increasing the interpolation function on the interface.

A curvature correction rate correction method introduced by Leonard based on piecewise linear interpolation for the quadratic upwind scheme is

$$\varphi_e = \frac{\varphi_P + \varphi_E}{2} - \frac{1}{8}Curv \tag{2.31}$$

where symbol *Curv* represents the correction of curvature.

QUICK is the English abbreviation for "Quadratic Upwind Interpolation of Convective Kinematics". "Quadratic" is relative to linear interpolation ("primary"); "upward" means that the curvature correction *Curv* is always composed of two points on both sides of the curved interface and another point in the upward direction. The QUICK scheme of the convection term has a third-order precision truncation error, but the diffusion term still uses the central difference scheme of the second-order truncation error.

2.3.2 Staggered Grid Method

The methods for establishing discrete equations include Taylor expansion method, polynomial fitting method, control volume integral method, balance method, etc.; and the three difference methods that are often used include forward difference, backward difference, and center difference. The Taylor expansion method and the polynomial fitting lay more stress on mathematical derivation, mainly by replacing the derivatives in the governing equations with corresponding difference expressions to form discrete equations. The control volume integral method and the balance method lay particular stress on the physical point of view, and the physical meaning is clear. Currently, popular commercial software such as PHOENICS, FLUENT, STAR-CD, CFX, and FLOW-3D are developed based on the finite volume method.

There are two key problems in solving the nonlinear Navier-Stokers equation: one is that the discrete form of momentum equation may fail to detect the unreasonable pressure field when the pressure gradient term is discretized by using the conventional grid and central difference method; the other is how to construct an equation for calculating the pressure improvement value when the relationship between pressure and velocity is implicit in the continuity equation. The two key problems mentioned above are related to the dispersion of the pressure gradient and the solution of the pressure, collectively referred to as the "coupling problem of pressure and velocity". If a waveform pressure field is obtained by a numerical solution, it is called decoupling between pressure and velocity. To overcome the decoupling between pressure and velocity, a staggered grid may be used.

The staggered grid means different grids are used for different variables. There are three types of grids in the two-dimensional SIMPLE algorithm (semi-implicit method for pressure-linked equations): the main control body grid, the control body grid of x-direction velocity u and the control body grid of y-direction velocity v. The main control body grid is suitable for pressure and scalar equations including temperature and concentration. The control body for equations of velocity u and v is dislocated from the main control body in spatial position.

Such a staggered grid for velocity component was first adopted by Harlow and Welch in their MAC methods, and then applied to other methods developed by Harlow and his colleagues, forming the basis of the SIVA program of Caretto, Curr and Spalding and the basis of the SIMPLE program of Patankar and Spalding.

To solve the incompressible Navier-Stokers equation, the technique of the staggered grid is needed when the central difference method is used to discretize partial differential equations.

A staggered grid has two important advantages: the staggered grid can avoid the situation that the wave velocity field satisfies the continuity equation. The pressure difference between two adjacent grid nodes becomes the natural driving force of the velocity component between two grid points, and the establishment of the relationship between pressure gradient and velocity creates conditions for the iterative solution.

A staggered control volume of the x-direction momentum equation is shown in Figure 2.5a. The difference p_P-p_E can be used to calculate the pressure resultant force acting on the control volume of velocity u.

The discretization equation for u_n is as follows:

$$a_n u_n = \sum a_{nb} u_{nb} + b + (p_P - p_E) A_e \tag{2.32}$$

The momentum equations in other directions can be processed in a similar way. Figure 2.5b shows the control volume of the y-direction momentum equation. It is misplaced in the y direction. The discretization equation for v_n can be written as:

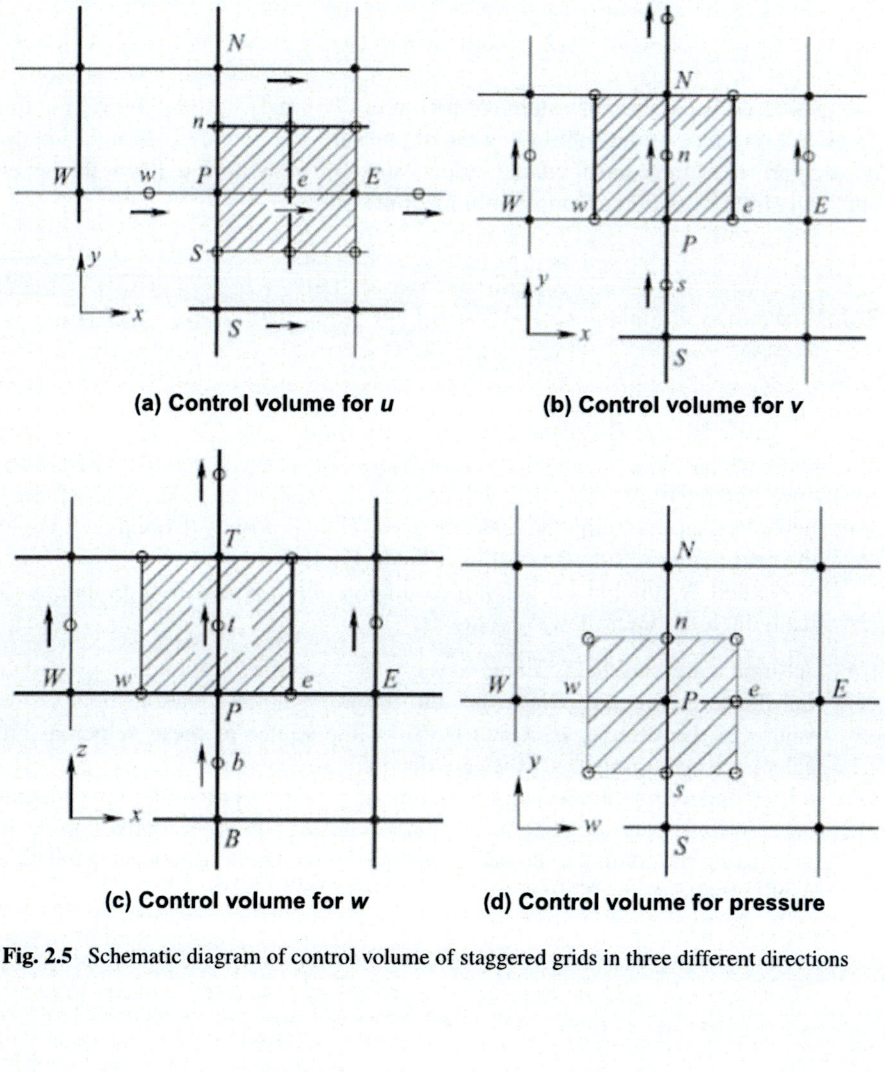

(a) Control volume for u (b) Control volume for v

(c) Control volume for w (d) Control volume for pressure

Fig. 2.5 Schematic diagram of control volume of staggered grids in three different directions

$$a_n v_n = \sum a_{nb} v_{nb} + b + (p_P - p_N) A_n \tag{2.33}$$

where: $(p_P - p_N)\, A_n$—corresponding pressure resultant force.

The shadows in the figure indicate the control volume, the short arrows indicate the positions of *u*, *v* and *w*, and the main grid points are indicated by small black dots. Relative to the main grid point, the position of *u* is only misplaced in the *x* direction. In other words, whether the position of *u* is exactly at the midpoint between two grids is related to the method of defining the control volume. The position of *u* must fall on the surface of the control volume, independent of whether the surface of the control volume is at the midpoint between the two grids.

Figure 2.5c shows the control volume of the *z*-direction momentum equation. It has a half step displacement in the *z* direction. The discretization equation for velocity component *w* is as follows:

$$a_n w_n = \sum a_{nb} w_{nb} + b + (p_P - p_T) A_t \tag{2.34}$$

Figure 2.5d is the control volume of pressure. Obviously, velocity *u* exists at the east–west interface of the control volume of pressure, velocity *v* exists at the south-north interface of the control volume of pressure, and velocity *w* exists at the upper and lower interfaces of the control volume of pressure.

2.3.3 Pressure Correction Method for Solving Navier-Stokers Equation

2.3.3.1 Basic Idea of Pressure Correction Method

In essence, the pressure correction technique is an iterative method, in which only some new physical reasoning methods are used. The purpose is to build a bridge to solve the next iteration from the results of the last iteration.

This method is suitable for solving incompressible viscous flow problems. It specifically includes the following contents:

(1) Assume a pressure field, denote it as p^*, and start the iteration.
(2) Substitute the already known p^* into the momentum equation to solve the values of the velocity *u*, *v*, and *w*. Since the values of these velocities are calculated according to p^*, they are denoted as u^*, v^*, and w^*.
(3) u^*, v^*, and w^* are obtained according to the predicted pressure, so they do not necessarily satisfy the continuity equation. Form a pressure correction value p' by using the continuity equation. $(p^* + p')$ may be more consistent with the continuity equation. The pressure correction formula is:

$$p = p^* + p' \tag{2.35}$$

The corresponding velocity correction formula is:

$$u = u^* + u' \tag{2.36}$$

$$v = v^* + v' \tag{2.37}$$

$$w = w^* + w' \tag{2.38}$$

where, the velocity correction equations are respectively:

$$u'_e = \left(\frac{A_e}{a_e}\right)\left(p'_P - p'_E\right) = d_e\left(p'_P - p'_E\right) \tag{2.39}$$

$$v'_n = \left(\frac{A_n}{a_n}\right)\left(p'_P - p'_N\right) = d_n\left(p'_P - p'_N\right) \tag{2.40}$$

$$w'_t = \left(\frac{A_t}{a_t}\right)\left(p'_P - p'_T\right) = d_n\left(p'_P - p'_T\right) \tag{2.41}$$

Therefore, the improved velocities are:

$$u_e = u_e^* + d_e\left(p'_P - p'_E\right) \tag{2.42}$$

$$v_n = v_n^* + d_n\left(p'_P - p'_N\right) \tag{2.43}$$

$$w_t = w_t^* + d_t\left(p'_P - p'_T\right) \tag{2.44}$$

(4) Give a new pressure value to the formula (2.35), return to step 2, and repeat each step until the velocity field satisfying the continuity equation is obtained. When both the continuity equation and the momentum equation are satisfied, the obtained flow field is the correct result.

2.3.3.2 Pressure Correction Equation

The pressure correction value is determined by the pressure correction equation. The condition that the pressure correction value shall meet is that the velocity field improved according to the pressure correction value can satisfy the continuity equation. The improved velocity is substituted into the continuity equation, and the following pressure correction algebraic equation is obtained

$$a_P p_P^{'} = a_E p_E^{'} + a_W p_W^{'} + a_N p_N^{'} + a_S p_S^{'} + a_T p_T^{'} + a_B p_B^{'} + b \tag{2.45}$$

where

$$a_E = \rho d_e \Delta y \Delta z \tag{2.46a}$$

$$a_W = \rho d_w \Delta y \Delta z \tag{2.46b}$$

$$a_N = \rho d_n \Delta z \Delta x \tag{2.46c}$$

$$a_S = \rho d_s \Delta z \Delta x \tag{2.46d}$$

$$a_T = \rho d_t \Delta x \Delta y \tag{2.46e}$$

$$a_B = \rho d_b \Delta x \Delta y \tag{2.46f}$$

$$a_P = a_E + a_W + a_N + a_S + a_T + a_B \tag{2.46g}$$

$$\begin{aligned} b = {} & \left[\left(\rho u^*\right)_w - \left(\rho u^*\right)_e\right]\Delta y \Delta z + \left[\left(\rho v^*\right)_s - \left(\rho v^*\right)_n\right]\Delta z \Delta x \\ & + \left[\left(\rho w^*\right)_b - \left(\rho w^*\right)_t\right]\Delta y \Delta x \end{aligned} \tag{2.46h}$$

2.3.4 *SIMPLE Algorithm*

Patankar proposed a SIMPLE algorithm to solve the fluid flow and heat transfer problem, namely the semi-implicit method for pressure-linked equations.

The calculation steps for the SIMPLE algorithm are as follows:

(1) The staggered grid is shown in Fig. 2.6. The real points of the main grid are the nodes of the predicted pressure value $(p)^n$, and the virtual points represent the grid points of the velocity values $(\rho u*)^{nn}$ and $(\rho v*)^n$.

(2) The $(\rho u*)^{n+1}$ and $(\rho v*)^{n+1}$ are respectively solved from the following equations

$$\left(\rho u^*\right)_{i+1/2,j}^{n+1} = \left(\rho u^*\right)_{i+1/2,j}^{n} + A^* \Delta t - \frac{\Delta t}{\Delta x}\left(p_{i+1,j}^*\right) \tag{2.47}$$

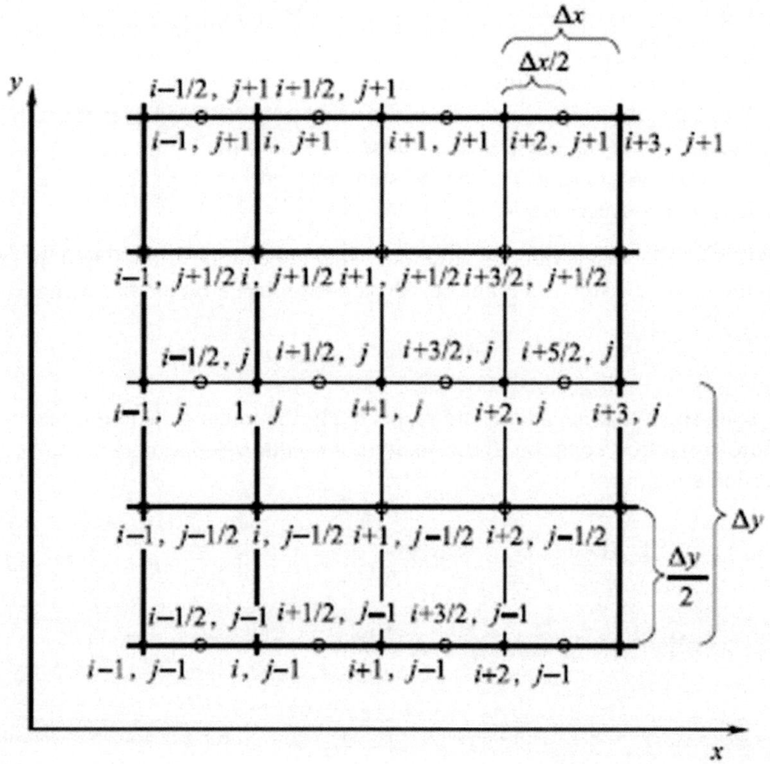

Fig. 2.6 Staggered grid

$$(\rho v^*)^{n+1}_{i,j+1/2} = (\rho v^*)^n_{i,j+1/2} + B^*\Delta t - \frac{\Delta t}{\Delta y}\left(p^*_{i,j+1} - p^*_{i,j}\right) \tag{2.48}$$

where

$$A = -\left[\frac{(\rho u^2)^n_{i+3/2,j} - (\rho u^2)^n_{i-1/2,j}}{2\Delta x} + \frac{(\rho uv)^n_{i+1/2,j+1} - (\rho uv)^n_{i+1/2,j-1}}{2\Delta y}\right]$$
$$+ \mu\left[\frac{u^n_{i+3/2,j} - 2u^n_{i+1/2,j} + u^n_{i-1/2,j}}{(\Delta x)^2} + \frac{u^n_{i+1/2,j+1} - 2u^n_{i+1/2,j} + u^n_{i+1/2,j-1}}{(\Delta y)^2}\right]$$

$$B = -\left[\frac{(\rho v^2)^n_{i,j+3/2} - (\rho v^2)^n_{i,j-1/2}}{2\Delta y} + \frac{(\rho uv)^n_{i+1,j+1/2} - (\rho uv)^n_{i-1,j+1/2}}{2\Delta x}\right]$$
$$+ \mu\left[\frac{v^n_{i+1,j+1/2} - 2v^n_{i,j+1/2} + v^n_{i-1,j+1/2}}{(\Delta x)^2} + \frac{v^n_{i,j+3/2} - 2v^n_{i,j+1/2} + v^n_{i,j-1/2}}{(\Delta y)^2}\right]$$

(3) Substitute the obtained $(\rho u*)^{n+1}$ and $(\rho v*)^{n+1}$ into the formula (2.45) to solve the pressure p'.
(4) Calculate ρ^{n+1} according to formula (2.36).
(5) Solve the momentum equation again using ρ^{n+1} obtained from the fourth step, use the ρ^{n+1} as a new $(\rho)^n$ and substitute it into the formula (2.42)—formula (2.44), then return to the second step, and repeat the second to the fifth step until the convergence.

To avoid the dispersion of solutions, the under-relaxation iteration is often performed, and the formula in the fourth step is replaced by the following formula:

$$p^{n+1} = (p^*)^n + \alpha_p p^{'} \tag{2.49}$$

Among them, a_p, a_u, and a_v are respectively the under-relaxation factors of the pressure correction equation, the velocity u correction equation, and the velocity v correction equation.

$$\left(\rho u^*\right)_{i+1/2,j}^{n+1} = \alpha_u\left(\rho u^*\right)_{i+1/2,j}^{n} + A^*\Delta t - \frac{\Delta t}{\Delta x}\left(p_{i+1,j}^*\right) \tag{2.50}$$

$$\left(\rho v^*\right)_{i,j+1/2}^{n+1} = \alpha_v\left(\rho v^*\right)_{i,j+1/2}^{n} + B^*\Delta t - \frac{\Delta t}{\Delta y}\left(p_{i,j+1}^* - p_{i,j}^*\right) \tag{2.51}$$

2.3.5 *Programming and Solving*

The UDF (User Defined Function) is written and compiled by the user of FLUENT. The original program of FLUENT is inserted into the UDF, and the use function of FLUENT is added to the UDF to meet users' need of special problem simulation. UDF is often used to customize boundary conditions, define the relationship between thermophysical properties and temperature of materials, and add the source terms of momentum equation or energy equation.

When the material characteristic parameters are updated, different heat source intensities are given in subsection by using the body heat source mode, that is, along the incident direction of the laser beam, the energy of laser beam is gradually attenuated by combined action of Fresnel absorption and inverse bremsstrahlung absorption when the laser enters the keyhole. The attenuation law is described by two heat source modes: In the upper part of the workpiece thickness ($h \geq h_1$), the body heat source is in the rotating Gauss mode, and in the lower part of the workpiece ($h < h1$), the body heat source is in double-ellipsoid heat source mode. The requirements of UDF program are written according to FLUENT, and the calculation program of heat source model is compiled in C language. Considering the complexity of the relationship between the thermophysical property parameters of titanium alloy and the temperature, when writing the program, the changes of the fluid viscosity of weld pool and the thermal conductivity of the material during deep penetration laser welding are neglected.

After inputting material parameters and determining the melting/solidification model, the calculation program of the body heat source is added to the energy equation through the user-defined function. After defining the boundary conditions, a solution method is selected for iterative calculation.

The separation solution algorithm in the separation solver means that the equations are independent of each other in the process of sequentially solving each equation in the governing equations. Because all the equations in the governing equations are nonlinear, several cycles of loop iteration are needed to make the solution converge. Each iteration loop consists of the steps shown in Fig. 2.7. The specific steps are as follows:

(1) Given the initial conditions of calculation, update the material parameters.
(2) By using the current pressure values and mass flow on each surface of each calculation unit, calculate the momentum equations of u, v, and w by turns to obtain new values of the velocities, i.e. values of u, v, and w, which meet the current conditions.
(3) Since the values of all velocities obtained in the first step may not satisfy the continuity equation, it is necessary to derive the Poisson type pressure correction equation from the continuity equation and the linearized momentum

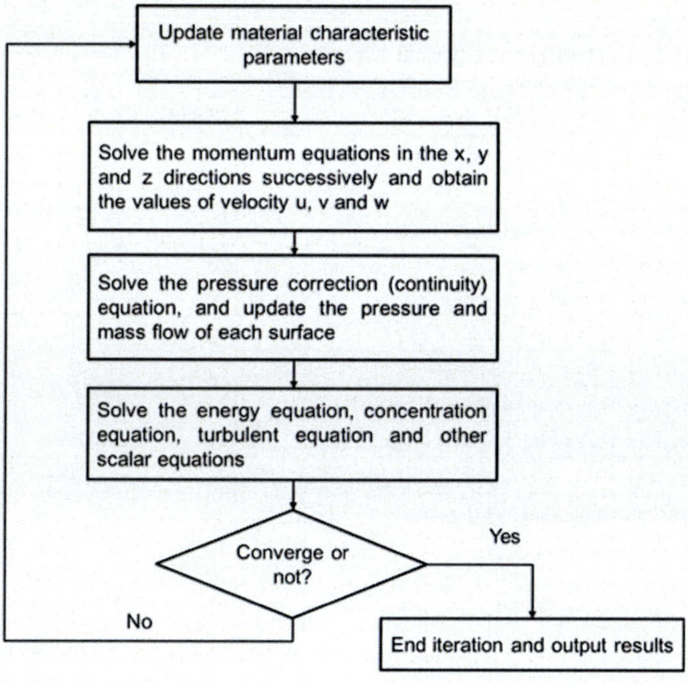

Fig. 2.7 General steps of problem solving

equations. Solve the pressure correction equation and obtain the necessary pressure correction values so that the velocity field and the mass flow on each face satisfy the continuity equation.

2.4 Physical Properties and Calculation Parameters

The established three-dimensional mathematical model of laser welding keyhole formation and weld pool flow process is used for numerical simulation of the welded workpiece of TC4 titanium alloy with a radius of 0.015 m and thickness of 0.025 m. The thermophysical property parameters of TC4 alloy are shown in Table 2.1. The CO_2 laser is used for welding, the focal radius of the laser is 0.2 mm, and the focal plane is on the upper surface of the workpiece. According to the characteristics that welding heat transfer is from the heat source center to the periphery, the selected computational domain is circular. The size of the computational domain is determined according to the range of thermal effect. Considering the symmetry of the welding temperature field and the flow field, only a semicircle with a radius of 0.015 m is taken in the calculation. For the mesh generation of the computational domain, the change of welding heat flow direction and the temperature gradient is considered. After discretization, the number of hexahedral elements is 6×10^4, and the partial enlarged drawing of the grid is shown in Fig. 2.8.

Table 2.1 Thermophysical property parameters of material TC4 used for heat flow calculation of deep penetration laser welding

Name	Symbol	Numerical Value
Liquidus temperature/K	T_L	1974
Solidus temperature/K	T_S	1966
Vaporization phase line temperature/K	T_V	3591
Density of solid phase/(kg·m^{-3})	ρ_S	4500
Density of liquid phase/(kg·m^{-3})	ρ_L	4100
Specific heat capacity of solid phase/(J·kg^{-1}·K^{-1})	c_{pS}	520
Specific heat capacity of liquid phase/(J·kg^{-1}·K^{-1})	c_{pL}	743
Thermal conductivity of solid phase (W·m^{-1}·K^{-1})	k_S	32.5
Thermal conductivity of liquid phase (W·m^{-1}·K^{-1})	k_L	32.5
Viscosity/(Pa·s)	η	3.4E−3
Surface tension temperature coefficient/(N·m^{-1}·K^{-1})	$\frac{d\sigma}{dT}$	−0.28E−3
Latent heat of melting/(J·kg^{-1})	L_m	0.389E+5
Thermal expansion coefficient/(K^{-1})	β	1.1E−5
Ambient temperature/K	T_{sur}	300
Surface tension/(N·m^{-1})	σ	1.6

Fig. 2.8 Temperature field numerical simulation grid for deep penetration laser welding process

Table 2.1 shows the basic parameters of the heat source model during simulation of the keyhole and weld pool in the process of deep penetration laser welding.

2.5 Effect of Welding Speed on Keyhole Profile

2.5.1 Temperature Distribution in Deep Penetration Laser Welding of Titanium Alloys

2.5.1.1 Computational Hypothesis

(1) The free surface of weld pool is flat.
(2) The weld pool fluid is incompressible Newtonian fluid.
(3) The flow pattern of molten metal is laminar flow.
(4) Buoyancy calculation is based on the Businesk hypothesis.
(5) The node above the material boiling point is regarded as the gasification equivalent point, and the region above the boiling point is regarded as the gasification evaporation zone. The material gasified can be keyhole in the weld pools.
(6) Quasi-steady state condition is met.
(7) The influence of the upper and lower surface airflow on the temperature field is not considered.

2.5.1.2 Computational Results and Analysis

The TC4 titanium alloy with radius 0.015 m was calculated by using the established three-dimensional laser welding process dynamics model. Because the thermophysical properties in the high temperature region cannot be obtained, and the data obtained by extrapolation method cannot accurately reflect the essential characteristics of materials, the thermophysical parameters in the high temperature region of numerical experiments are all taken as constants. The laser processing parameters used in numerical experiments are as follows: the laser power is selected between 1600 and 3000 W, and the welding speed is selected between 0.025 and 0.070 m/s.

Figures 2.9 and 2.10 show the temperature distribution on the upper surface of the workpiece and the temperature gradients before and behind the welding heat source. In Fig. 2.9, the positive direction of x axis is the moving direction of the workpiece. As can be seen from Fig. 2.9, the temperature near the center of the heat source is 3510 K, the shape of the temperature profile is like an ellipse, and the long axis of the ellipse is parallel to the welding direction. It is easy to infer that the outlet of the keyhole has the same characteristics. From the distribution characteristics of the isotherms, the isotherm in front of the keyhole has a considerably higher density than the isotherm at the back of the keyhole. In Fig. 2.10, the temperature

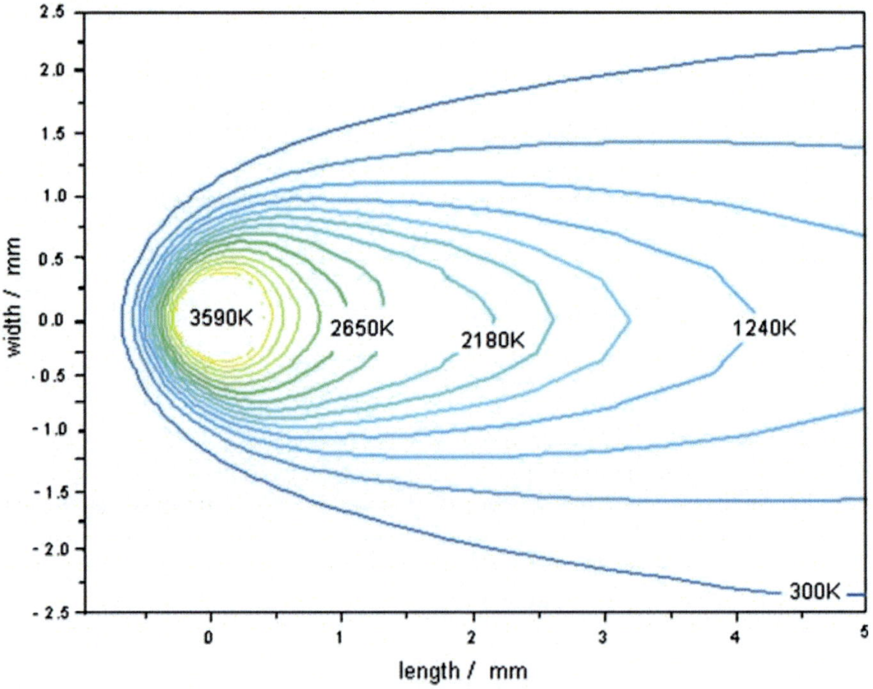

Fig. 2.9 Temperature distributions in the XOY plane at upper surface of the piecework during deep penetration laser welding

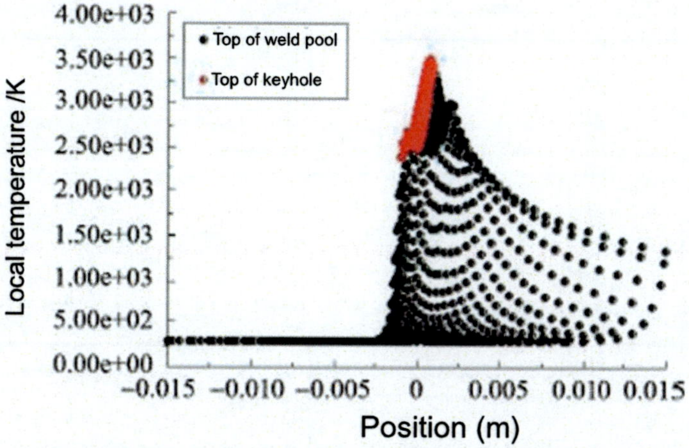

Fig. 2.10 Temperature distribution around the heat source at top surface

distribution illustrates the characteristics of temperature gradient before and behind the keyhole in deep penetration laser welding. Most of the isotherms are elliptical, which indicates that the heat flow has obvious directivity. Figure 2.10 shows that in the front of the heat source, the temperature rises from 300 to 3500 K at the heating stage. The difference of heat flux density between the front and the back of the keyhole is the driving force of heat flow from the front of the keyhole to the back of the keyhole. Heat flow has obvious directional characteristics, which will directly affect the crystallization direction of weld metal and the shape and size of weld cross section.

Figure 2.10 shows that on the left side of the origin, within the range of -0.0015–0.00 m, the maximum temperature variation is 3200 K, and the maximum temperature gradient is about 1280 K/mm. Each small dot in the figure represents a position on the surface of the workpiece. A lower temperature point indicates a greater distance of the dot towards the heat source. The temperature changes from the location of the highest temperature to the position of 0.015 m are about 2300 K, and the corresponding temperature gradient is 153.3 K/mm. Within this temperature range, it belongs to the solidification and crystallization stage of the weld pool.

Figure 2.11 shows the temperature field of deep penetration laser welding of XOZ section and the shape and size of keyhole. In deep penetration laser welding, the keyhole is asymmetrical, and the diameter of keyhole decreases gradually from top to bottom. The temperature field near the keyhole reflects the characteristics of higher isotherm density in front and looser isotherm density behind keyhole. After comparing the temperature field range of thickness direction, it is found that the temperature field range of the upper surface of the workpiece is obviously larger than that of the bottom of the workpiece, while the isotherm of the waist is more concentrated.

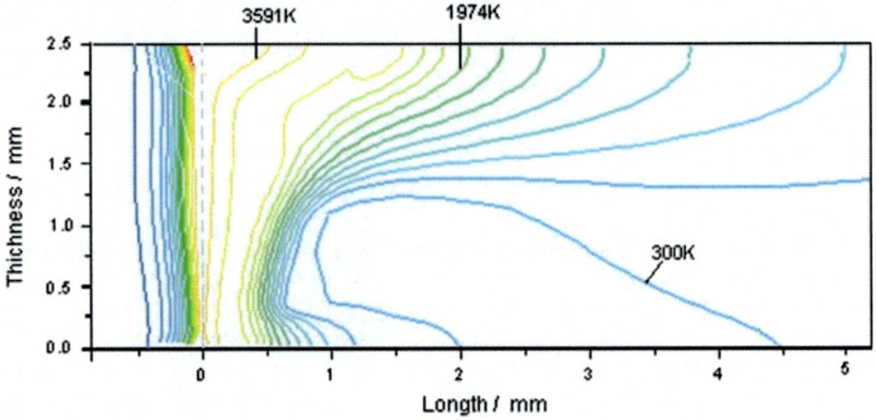

Fig. 2.11 Temperature distribution and the keyhole shape and its size in XOZ plane

Figure 2.12 shows the temperature field of laser welding of TC4 titanium alloy cross section and the geometry of laser keyhole. It can be seen that the keyhole is bilaterally symmetric, but in the direction of thickness, the size of the keyhole gradually decreases. The keyhole resembles a trumpet in appearance. Its outlet diameter is about 1.4 mm on the upper surface of the workpiece and about 0.3 mm on the lower surface. The size of the keyhole is affected mainly by the power of the incident laser beam, the welding speed and the convection velocity of the weld pool. It can be seen from the experimental observation that the power of incident laser beam is affected by the plasma flow emitted by the keyhole, and the size and position of the keyhole

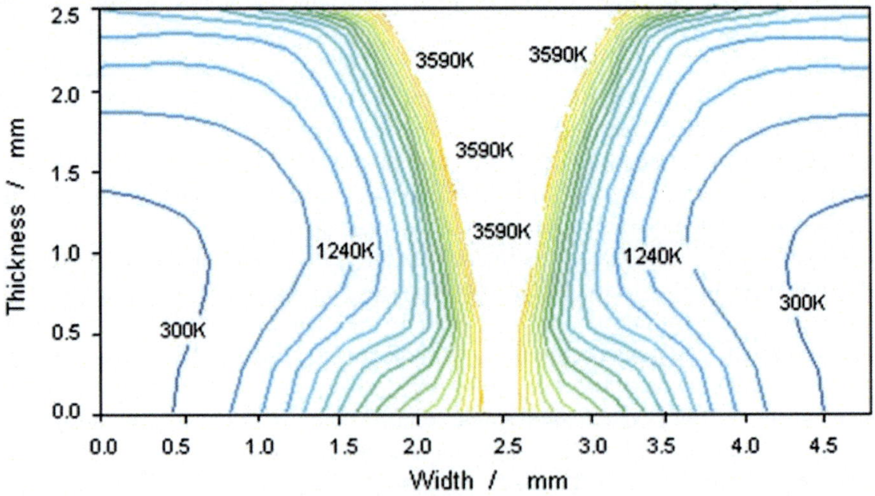

Fig. 2.12 Temperature field and corresponding keyhole dimensions in the YOZ plane of the laser beam centerline

are constantly changing. Similarly, in the thickness direction of the workpiece, the energy of the laser beam is gradually attenuated due to the interaction of the Fresnel absorption of the keyhole wall and the inverse bremsstrahlung absorption of the plasma in the keyhole. This energy attenuation is the direct reason for the gradual reduction of the keyhole size in the direction of workpiece thickness. In the case of full penetration, the upper and lower surfaces of the weld pool are free surfaces, and Marangoni convection strengthens the convection of the weld pool. Therefore, the range of heat flow action at the upper and lower surfaces of the workpiece shown in Fig. 2.12 is larger than that at the middle of the workpiece. The asymmetry of the action range of heat flow in Figs. 2.11 and 2.12 are the final results of the direct influence of welding speed on the direction of flow and heat transfer in weld pool.

2.5.2 *Effect of Welding Speed on Keyhole Size*

Laser keyhole welding involves many technological parameters, such as focus position, laser power, laser beam quality, welding speed, and auxiliary gas flow.

Figure 2.13 shows the temperature field in the XOZ plane. The figure shows the upper and lower outlets of the keyhole. Obviously, the laser power density at this time has been able to ensure the penetration through the titanium alloy plate, that is, the keyhole is a penetrating keyhole. From the position of the upper and lower outlets of the keyhole, the keyhole has a certain inclination, and the lower outlet of the keyhole is offset to a certain size backward.

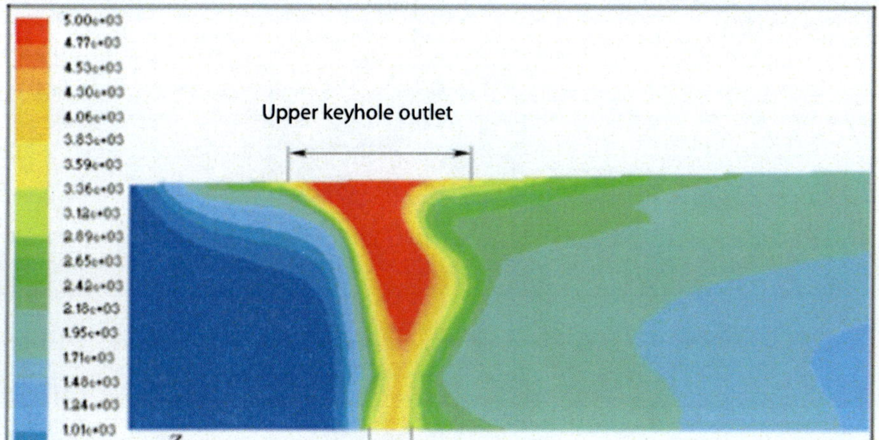

Fig. 2.13 Keyhole dimensions calculated under high power laser density

It can be seen that in the front of the keyhole, the temperature changes greatly within a small distance, and the temperature gradient is greater at the position below 2/3 of the plate thickness. We note that the temperature gradient at the front of the keyhole is about 8334.7 K/mm on the lower surface of the workpiece. This temperature gradient is much larger than the temperature gradient on the upper surface of the workpiece. In contrast, the temperature gradient at the back of the keyhole is significantly lower than that at the front of the keyhole. The weld pool appears like an inverted horseshoe, which is closely related to the heat transfer mechanism in the laser welding.

In order to discuss the relationship between keyhole geometric parameters and welding parameters, Fig. 2.14 defines keyhole geometric parameters φ_t 、 φ_b 、 Δ 、 β 、 R_1 、 R_2 and R_3. $\varphi_t$is the diameter of the upper keyhole outlet, φ_b is the diameter of the lower keyhole outlet, Δ is the deviation between the central position of the upper outlet and the lower outlet of the keyhole, β is the inclination angle of the front wall of the upper keyhole outlet, R_1 is the curvature radius of the front wall of the keyhole, R_2 and R_3 are the outer radius and inner radius of the curvature of the back wall of the keyhole. It should be noted that the curvature radius of the front wall and the back wall of the keyhole directly determines the curvature of the keyhole wall. The curvature of the keyhole wall is calculated as follows:

$$\kappa = -\left[\nabla \cdot \left(\frac{\vec{n}}{|\vec{n}|}\right)\right] = \frac{1}{|\vec{n}|}\left[\left(\frac{\vec{n}}{|\vec{n}|}\right) \cdot \nabla|\vec{n}|\right] - (\nabla \cdot \vec{n}) \tag{2.52}$$

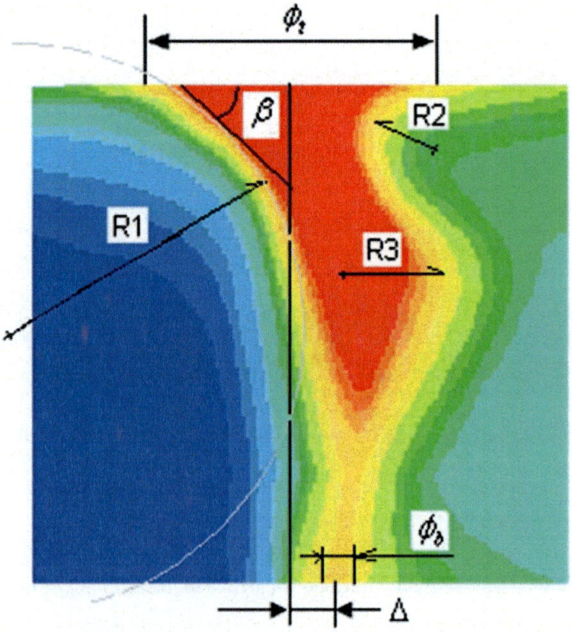

Fig. 2.14 The defined keyhole dimensions induced by laser energy

Table 2.2 Comparison of the keyhole geometry parameters at different welding speeds

Welding speed (m/s)	β(°)	φ_t (mm)	φ_b (mm)	Δ (mm)	P_γ N/m^2
0.025	61.9	1.22	0.39	0.056	200
0.04	58.1	1.06	0.28	0.333	230
0.05	52.0	1.01	0.20	0.366	270
0.06	43.4	0.966	0.11	0.398	310

$$P_\gamma = \kappa\gamma \tag{2.53}$$

where: P_γ—surface tension pressure;
k —keyhole wall curvature;
γ—surface tension coefficient of keyholes;
$\vec{n}$ —normal direction of the keyhole surface.

Obviously, Fig. 2.14 simplifies the complex variation of the keyhole geometry with three parameters R_1, R_2, and, R_3. Formula (2.7) can be rewritten as follows:

$$\kappa = \gamma\left(\frac{1}{R_1} + \frac{1}{R_2} + \frac{1}{R_3}\right) \tag{2.54}$$

Table 2.2 shows the calculation results of the keyhole diameter and the center position offset of the upper and lower outlets at different welding speeds when the laser power is 3000 W. In addition, the.

keyhole inclination angle β and the keyhole surface tension pressure P_γ corresponding to the welding speeds are also given. Figure 2.15 shows the variation of keyhole parameters with the increase of welding speed during deep penetration laser welding. It can be seen that the diameter of keyhole outlet decreases with the increase of welding speed, and the change of upper keyhole outlet is greater than that of lower keyhole outlet. In other words, at constant laser power, the diameter of the lower and upper keyhole outlets changes by 71.8% and 20.8%, respectively. The lower keyhole outlet is more sensitive to the change of welding speed. In Fig. 3.15, it can be seen that with the increase of welding speed, the offset of the center position of the upper and lower outlets of the keyhole increases continuously, but between 0.025 m/s and 0.04 m/s, the variation of the offset is greater. The increase of the offset of the center position of the upper and lower outlets of the keyhole means that the probability of interaction between laser beam and material is increased. In this way, the material welded can receive more laser energy when the welding speed is higher, so as to compensate for the decrease of the interaction time between laser beam and the material welded due to the acceleration of welding speed.

Figure 2.16 shows that the inclination angle of the front wall and the surface tension pressure of the keyhole vary with the increase of welding speed when the laser power is 3000 W. Obviously, the inclination of the front wall of the keyhole decreases with the increase of welding speed, while the change pattern of the surface tension

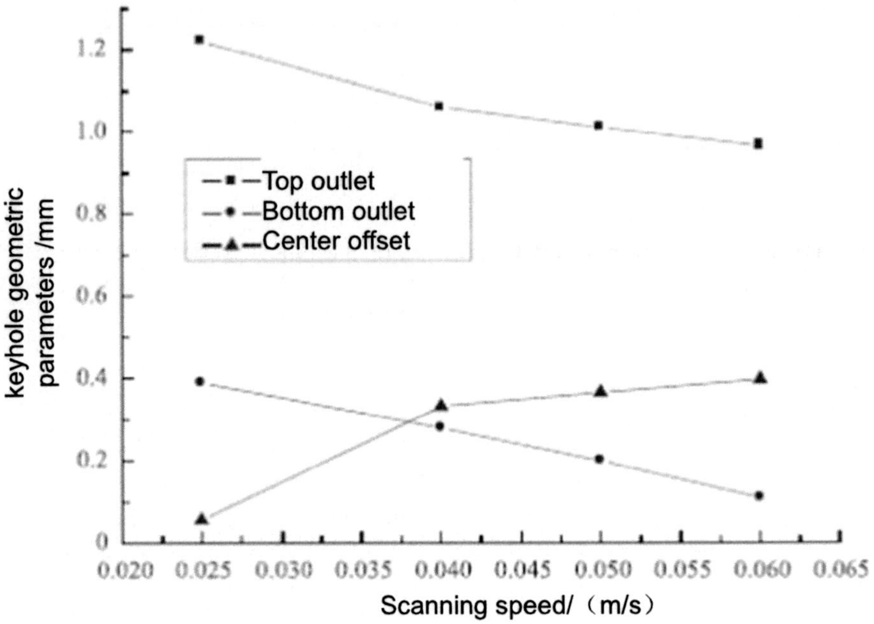

Fig. 2.15 The keyhole diameter and its center offset at the upper surface and lower surface of workpiece varied with welding speed

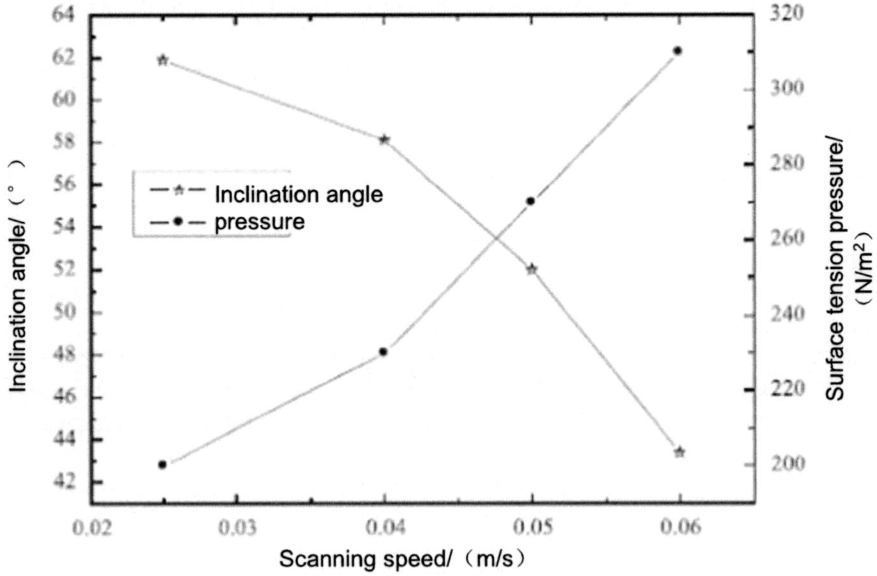

Fig. 2.16 The inclination angle of keyhole and its surface tension pressure varied with welding speed of workpiece

and pressure of the keyhole is just the opposite. The decrease of the inclination of the front wall of the keyhole is beneficial to the absorption of laser beam by the keyhole wall. In deep penetration laser welding, the stability of keyhole is the result of the combined effect of fluid flow pressure in keyhole and surface tension pressure, static pressure and dynamic pressure of weld pool fluid. On the premise that other pressures remain unchanged, the increase of the surface tension pressure of the keyhole means the decrease of the stability of the keyhole.

2.5.3 Effect of Laser Power on Keyhole Size

Deep penetration laser welding is divided into partial deep penetration welding and full deep penetration welding. The former refers to the generation of laser keyholes without penetrating the base metal, while the latter refers to full penetration of keyhole through the plate. In partial deep penetration welding, it is easy to produce process pore, or porosity. Such porosity can be caused by many factors; a major one is the instability of the keyhole. To keep the keyhole wall open, two basic conditions must be met: energy balance and the pressure balance. The following describes the impact of laser power changes on keyhole size from the perspective of energy balance.

In laser welding, the existence of plasma considerably affects the efficiency of energy coupling between laser and material. The periodic plasma flow from the keyhole directly takes the energy of the air flow inside keyhole. Based on energy balance, the stability of the keyhole wall depends on the heat absorption and dissipation. According to the available data, laser energy is absorbed by the air flow and laser beam inside keyhole through the Fresnel absorption mechanism of the keyhole wall and the reverse bremsstrahlung absorption mechanism. The heat absorbed is used to overcome the resistance of forcing the keyhole closed, such as the surface tension, the static pressure of the weld pool fluid and the dynamic pressure of the flow.

Figure 2.17 shows the temperature field when the welding speed is 0.025 m/s. In the picture, the temperature in the red area is 5000 K, and the temperature in the dark yellow area is above 3590 K, which is also in the evaporation temperature range. This temperature region is the equivalent pore. It can be seen that the keyhole is in the shape of an upright wine cup, showing typical asymmetry in appearance. The double horn shape of weld pool is also very obvious. The incident laser beam enters from the upper part of the workpiece. If the plasma density in the keyhole is not high, part of the laser energy passes through the vapor in the keyhole and interacts directly with the welded base metal. According to the Fresnel absorption mechanism, the energy of the laser beam is converted into the energy required for welding. In fact, the vapor pressure in the keyhole changes in a quasi-periodic manner, so does the amount of plasma. In this changing process, the Fresnel absorption mechanism has been playing a role, but the absorption of laser beam energy by this mechanism has also undergone a quasi-periodic change. Seto et al. from Osaka University observed the quasi-periodic eruption process of keyhole plasma through high-speed photography.

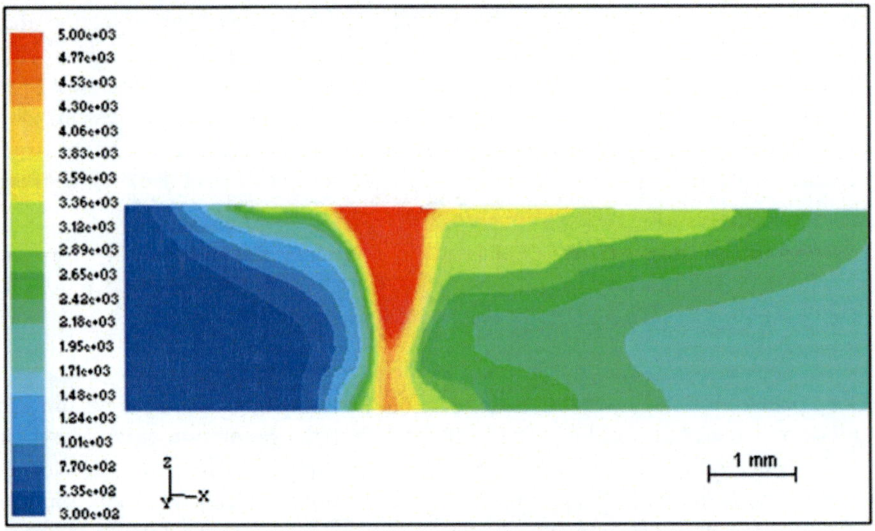

Fig. 2.17 Keyhole dimensions at welding speeds of 0.025 m/s during laser welding

In addition, the flow of laser beams into the keyhole transmits the laser energy to the workpiece by convection and heat conduction of the air flow in the keyhole through the inverse bremsstrahlung absorption mechanism. The size of keyholes changes from large to small from top to bottom, which indicates the gradual attenuation of laser energy.

Figure 2.18 shows considerable changes of keyhole shape at laser power of 3000 W and constant input welding energy when the welding speed increases. When the welding speed increases to 0.06 m/s, the shape of the keyhole changes remarkably. The change of welding speed means the change of interaction time between laser beam and welded material. It also indirectly reflects the change of energy absorbed by laser keyhole. In order to further study the effect of laser energy on keyhole size, the variation of keyhole size with the increase of laser power at welding speed of 0.025 m/s is given in Fig. 2.19. It can be seen from Fig. 2.19 that the size of the keyhole changes little with the increase of laser power. This is related to the facts that air pressure inside keyhole is close to atmospheric pressure, and that the absorption of incident laser by the keyhole is realized by indirect heat transfer from plasma in the keyhole to the keyhole wall.

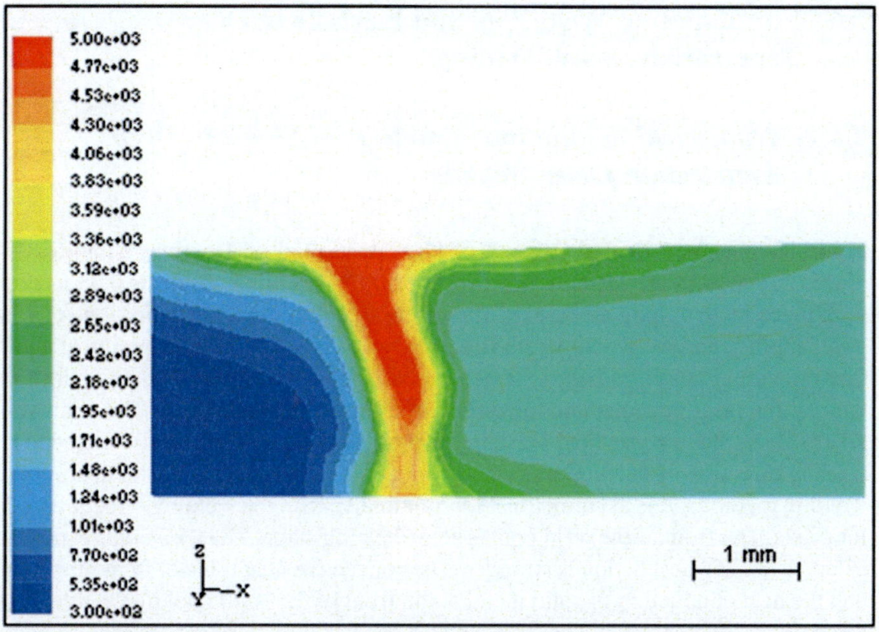

Fig. 2.18 Keyhole geometry calculated at welding speed of 0.075 m/s

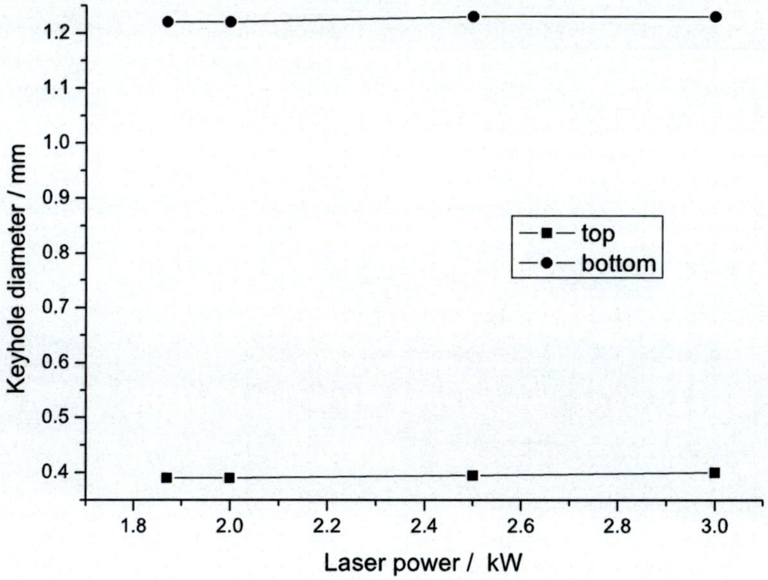

Fig. 2.19 Keyhole diameter varied with laser power at 0.025 m/s welding speed

2.6 Simulation of Weld Pool and Keyhole in Deep Penetration Laser Welding

2.6.1 Numerical Simulation of Weld Pool Shape in Deep Penetration Laser Welding

Figure 2.20 is the calculation result of laser weld pool of Titanium alloy in XOY plane (z = 0.0025 m) when laser power is 3000 W, welding speeds is 0.03 m/s and workpiece moves along the X-axis. For discussion purposes, the definitions of weld pool length L and weld pool width B are shown in Fig. 2.20. The experiment shows that the temperature gradient is larger in the front of the laser keyhole than in the back of the laser keyhole, and the direction of heat flow is obvious. The simulation results show that the principal axis direction of the weld pool is closely related to the moving direction of the workpiece. The shape of isotherm is similar to that in other welding methods. It is asymmetrical longitudinally with the welding direction as the long axis. The front of the weld pool is at the heating stage. The temperature gradient is larger and the heat action is stronger. The convective heat transfer of molten metal and the heat conduction of solid metal at the front of the weld pool play a role at the same time. After carefully observing the position of the laser keyhole center and the position of the widest weld pool, it is found that the position of the widest weld pool

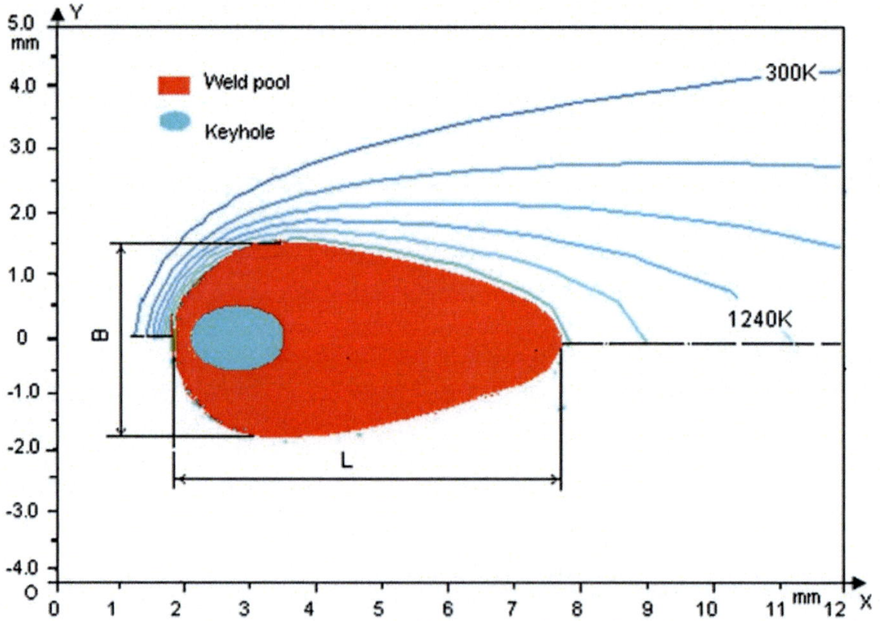

Fig. 2.20 The shape and size of weld pool on upper surface of the workpiece in XOY plane during deep penetration laser welding

is slightly farther than that of the laser beam center. In other words, the width of the weld pool increases slightly after the heat source leaves, and there is a small lag at the position of its maximum value. This shows that there is a lag time for thermal action, and the latent heat of crystallization from liquid to solid phase transition is also an important factor.

Figure 2.21 shows the calculation result of weld pool in XOZ plane at laser welding speed of 0.03 m/s, laser power of 3000 W. Obviously, the front of the weld pool has a certain inclination, and the asymmetric structure of the weld pool is very obvious.

Figure 2.22 shows the geometric shape and size of the weld pool in YOZ plane when laser power is 3000 W and welding speed is 0.03 m/s. It can be seen that the weld pool is symmetrical in the OZ direction but asymmetric in the thickness

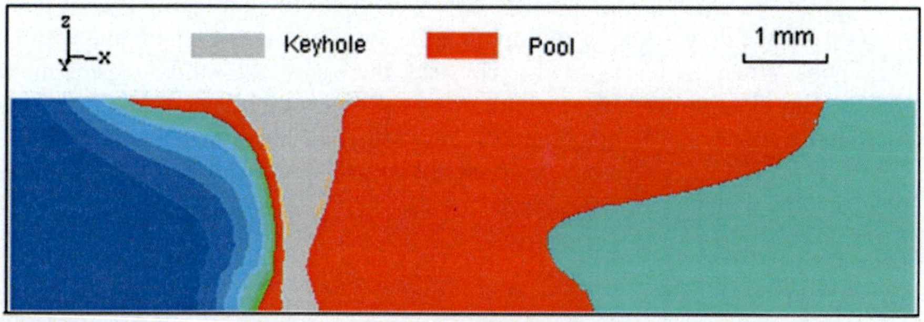

Fig. 2.21 Weld pool shape and size in XOZ plane during deep penetration laser welding

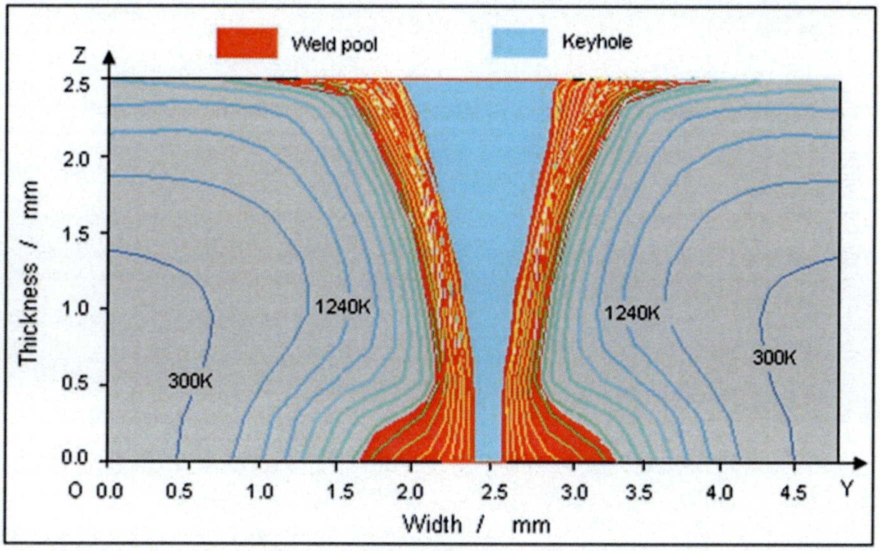

Fig. 2.22 Shape and size of the cross section of weld pool in deep penetration laser welding

direction. The shape of both the upper and lower surfaces of weld pool is open. This indicates that the surface tension gradient of free surface of weld pool plays an important role in heat transfer of weld pool.

2.6.2 Effect of Welding Speed and Laser Power on Weld Pool Size

Under the assumption that the laser beam is CO_2 laser, the radius of focal plane is 0.2 mm and the defocus is 0, the relationship between laser weld pool, welding speed and laser input power is simulated. Figure 2.23 shows the relationship between weld width and welding speed during laser welding of titanium alloy at laser power of 3000 W. Welding speed is an important factor affecting the length and width of weld pool. When the laser power is constant, increasing the welding speed means that the energy of the welding body decreases. On the other hand, with the increase of welding speed, the interaction time between laser beam and titanium alloy is relatively shortened. Therefore, the laser energy of input material decreases in unit time, and the width of weld pool is decreases.

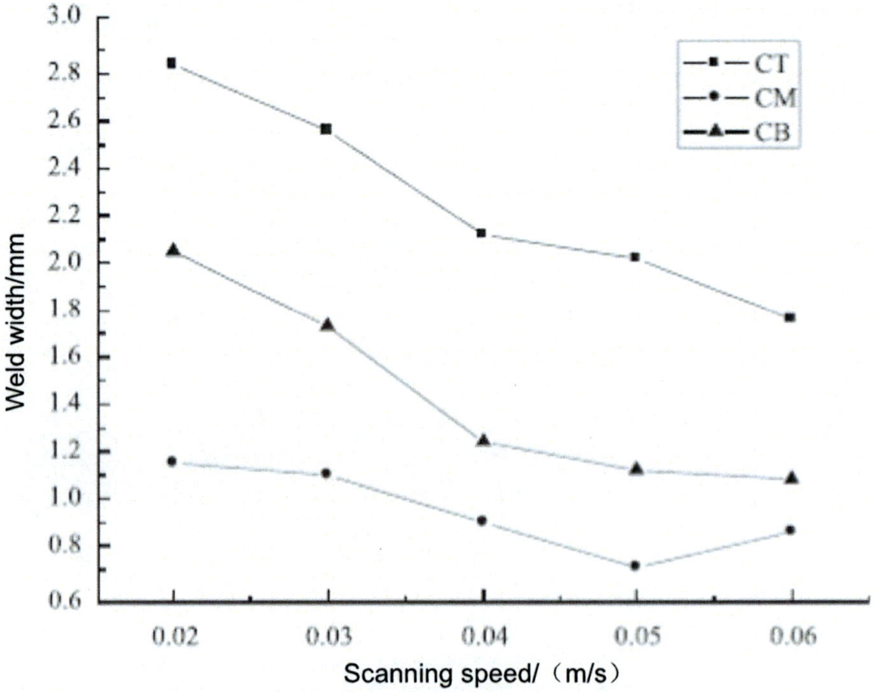

Fig. 2.23 Weld pool width dependence on traveling speeds of workpiece

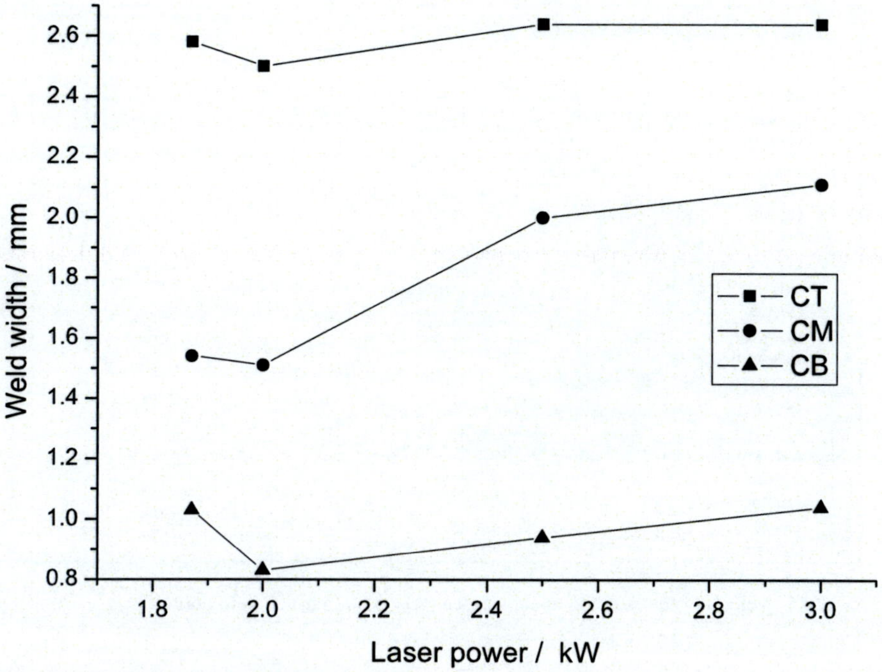

Fig. 2.24 Weld width varied with laser power

Figure 2.24 shows the variation of weld pool width during the increase of laser power. With the increase of laser power, the width of the weld increases, but the width of the middle part of the weld changes more obviously.

2.7 Numerical Simulation of Fluid Flow in Weld Pool During Deep Penetration Laser Welding

2.7.1 Characteristics of Fluid Flow in Weld Pool

Figure 2.25 shows the velocity vector diagram of the upper surface of the weld pool ($z = 0.0025$ m) when the laser power is 3000 W and the welding speed is 0.05 m/s. It can be seen that the velocity distribution on the upper surface of weld pool is not uniform. The velocity vectors with a velocity value of about 1.05 m/s form a distinct circle. These velocity vectors originate from the center of the heating source and diverge from one heating center to all sides. That is to say, the melted liquid metal flows from the heating center to the surrounding area, and the flow velocity is approximately the same. This convection has a higher velocity value, which indicates that the convection heat transfer is stronger.

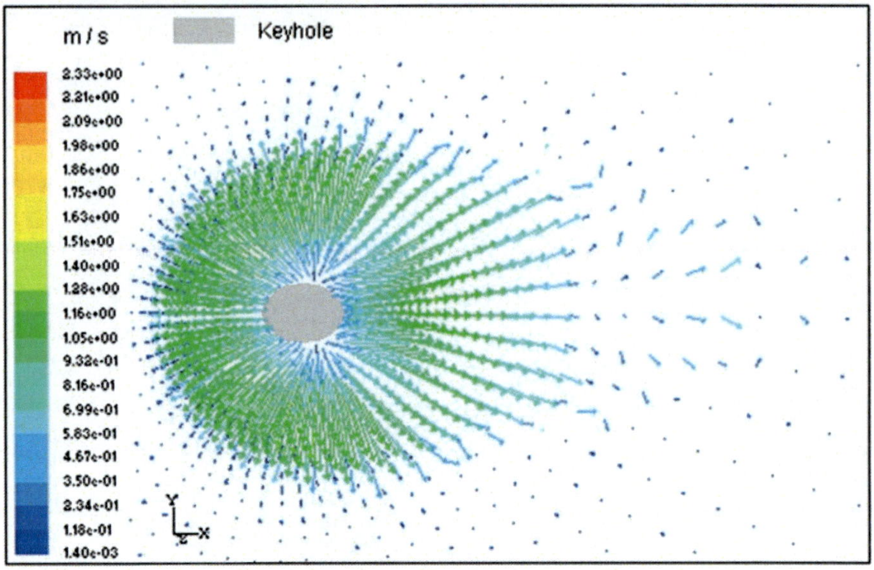

Fig. 2.25 Velocity vectors in the upper surface (Z = 0.0025 m) of weld pool

In fact, the driving force of this convection is the surface tension gradient, the so-called Marangoni convection. This driving force is also called Marangoni force. In the figure, the velocity values of each point farther from the heating center are very low and the same, i.e., 0.05 m/s. This is because the calculation assumes that the laser beam is stationary and the workpiece moves at a speed along the positive X-axis. Compared with the moving speed of workpiece, the flow velocity of weld pool is much larger. The maximum flow velocity on the surface of weld pool is 1.16 m/s. It is 23.2 times faster than the welding speed.

For the cold metal in front of the heat source, the convective heat will melt the solid metal, and the melted metal will become a part of the weld pool. For the rear part of the heat source, Marangoni flow will bring the metal from high-temperature areas to the weld pool metal which is still in liquid state. This not only supplements the heat loss of the metal from the back part of the pool to the surrounding cold metal, but also maintains the convective inertia for a period of time, but the flow speed is obviously lower than the convective speed in the invisible circle. With the increase of laser power, the intensity of heat source increases, the Marangoni convection strengthens and the weld pool width increases naturally. During deep penetration laser welding, keyholes are produced, and their size increases with laser power density. If the intensity of Marangoni flow remains unchanged and the size of keyhole increases, the effective range of heat flow will naturally increase. In this way, the length and width of weld pool increase with laser power. At the same time, with the increase of laser power, the attenuation of laser beam by plasma above the keyhole will become greater, and the effective heat input of weld pool will be reduced. Therefore, the rate of increase in the size of weld pool decreases as the laser power increases.

Figure 2.26 shows the velocity vector diagram of weld pool fluid flow in XOZ plane when laser power is 3000 W and welding speed is 0.06 m/s. The weld pool has an asymmetric horseshoe shape. As can be seen in Fig. 2.26, there is an obvious eddy on the upper surface of the weld pool and at a little distance from the keyhole. The weld pool is horseshoe-shaped, and there is an obvious eddy on the shoulder of the weld pool. The maximum speed of the weld pool surface reaches 1.4 m/s. The velocity of fluid flow in weld pool is small, but the direction of flow is obviously affected by the fluid flow on the surface of weld pool. Careful observation shows that the flow velocity on the surface layer of the weld pool is significantly higher than that on the subsurface layer.

From the point of view of Newtonian fluid, the velocity difference between surface fluid and subsurface fluid is caused by the shear stress of the fluid. Newton shear stress is calculated using the following formula:

$$\tau = \eta \frac{dw}{dz} \tag{2.55}$$

Substitute η of material viscosity and the obtained velocity in the equation. The viscous shear stress between the surface and subsurface is about 61.2 Pa. The results show that the flow direction of Z = 0.0023 m plane fluid is opposite to that of surface

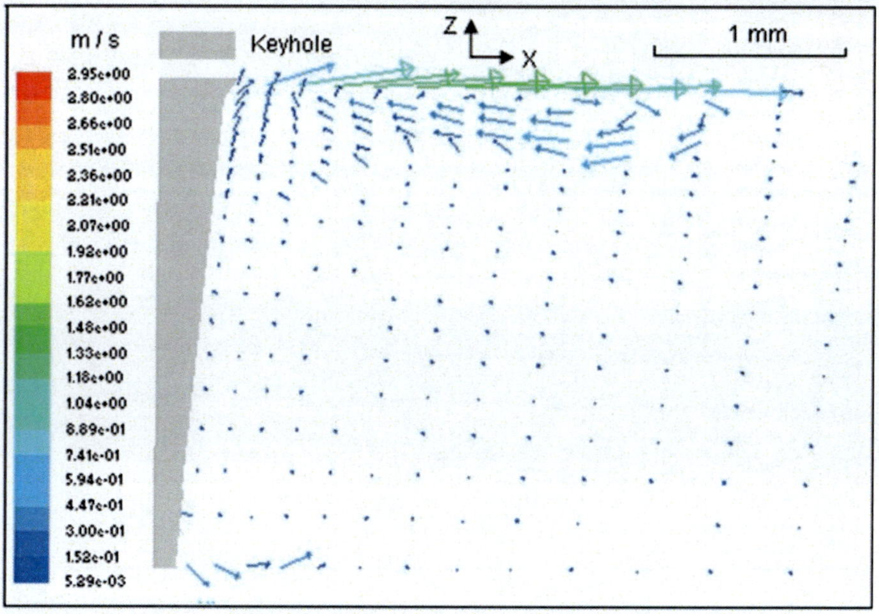

Fig. 2.26 Velocity vector distribution of weld pool in XOZ plane in deep penetration laser welding

and subsurface fluid. That is to say, the flow and heat transfer of metal on the surface of weld pool is strong, but Marangoni flow mainly affects the surface, but has little effect on the metal flow under the subsurface layer. The generation of eddy is a strong proof of convective heat transfer. The fluid on the surface of the weld pool runs so fast that the heat carried by the fluid is transferred rapidly to the surrounding areas. However, due to the limited melting area of the weld pool, the rapidly flowing fluid may show two flow patterns. One pattern is that the hot metal on the surface of the weld pool center is pushed to the edge of the weld pool to form a protuberance. After the laser heat source moves away, the temperature of the small laser weld pool drops rapidly, the viscosity increases rapidly, and finally welds with an extra height are formed. In calculation, it is assumed that the weld pool is plane, which limits the deformation of the weld pool surface. The second pattern is that the surface metal will encounter significant resistance when it flows to the edge of the weld pool. After the metal collides with the solid metal, the backlash force of the solid wall forces the flow direction velocity to change, resulting in eddy. From the two cases, it can be inferred that the actual eddy may not be as significant as those calculated.

Figure 2.27 shows the velocity field of weld pool in the YOZ plane when laser power is 3000 W and welding speed is 0.03 m/s. It can be seen that there are two distinct eddies near the surface of the weld pool. The existence of eddy indicates

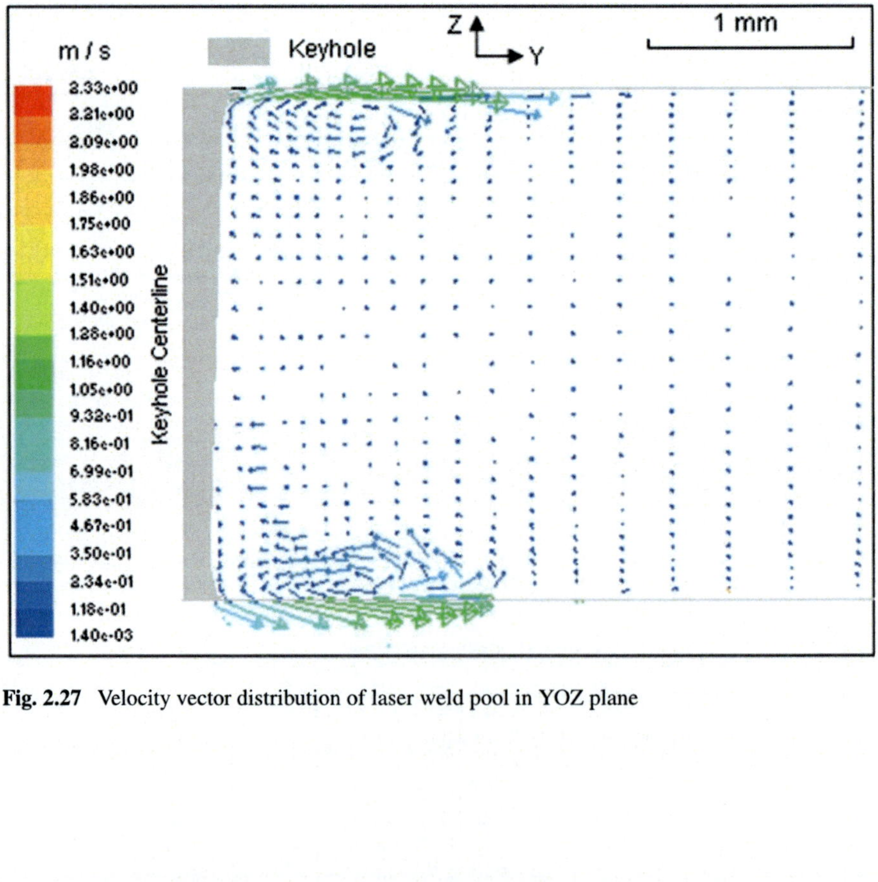

Fig. 2.27 Velocity vector distribution of laser weld pool in YOZ plane

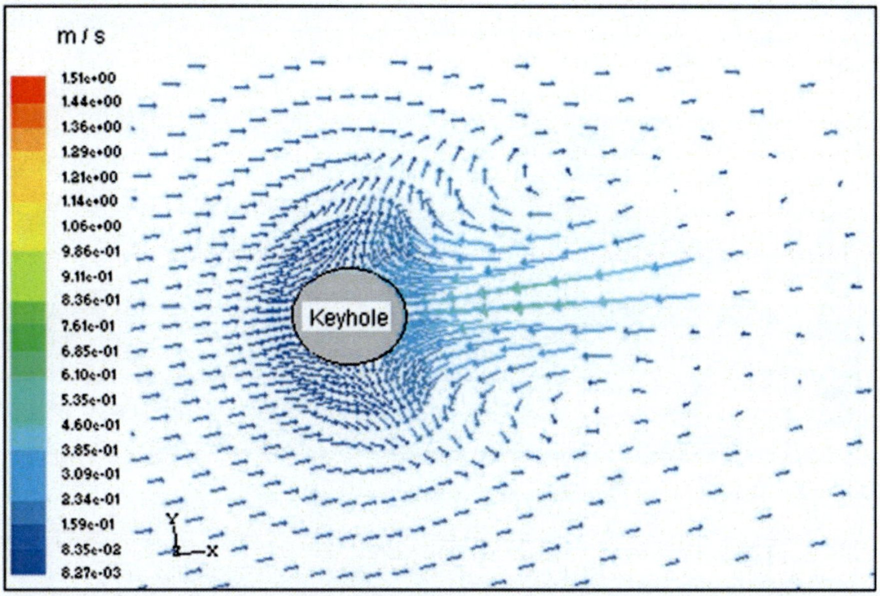

Fig. 2.28 Velocity field in z = 0.0023 m plane in laser welding

strong convective heat transfer. Eddy fluids bring heat from heat sources around the weld pool, which will inevitably affect the shape and size of the weld pool.

Figure 2.28 is the velocity vector diagram of the fluid flow in the weld pool on the sub-surface and z = 0.0023 m XOY plane when the laser power is 3000 W and the welding speed is 0.05 m/s. The recirculating eddies of the weld pool fluid are very obvious in the figure. This flow makes the cold metal at the edge of the weld pool migrate to the heating center and cooperates with the Marangoni flow on the surface of the weld pool, which promotes the flow and heat transfer in the weld pool.

Figure 2.29 shows the velocity field in z = 0.0024 m plane. It can be seen that the direction of fluid flow velocity is similar to that on the surface of weld pool, but the velocity value is much lower. The representative fluid flow is at 0.3 m/s.

2.7.2 Distribution of Flow Velocity in Weld Pool

Figure 2.30 shows the distribution of the weld pool flow velocity in the direction of workpiece thickness when laser power is 3000 W and the welding speed is 0.05 m/s, 0.075 m/s, and 0.10 m/s, respectively. In the case of full penetration in deep penetration laser welding, the upper and lower surfaces of the weld pool are free surfaces, and the surface tension gradient drives the weld pool fluid from the center of the heat source to low-temperature areas. The existence of eddies greatly accelerates the flow rate of metal near the surface of the weld pool, but their sphere of influence is

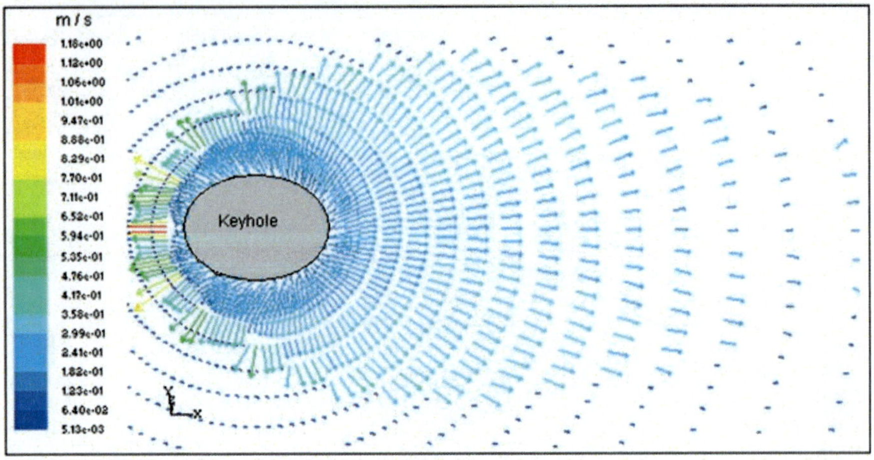

Fig. 2.29 Velocity vector distribution of the subsurface of weld pool

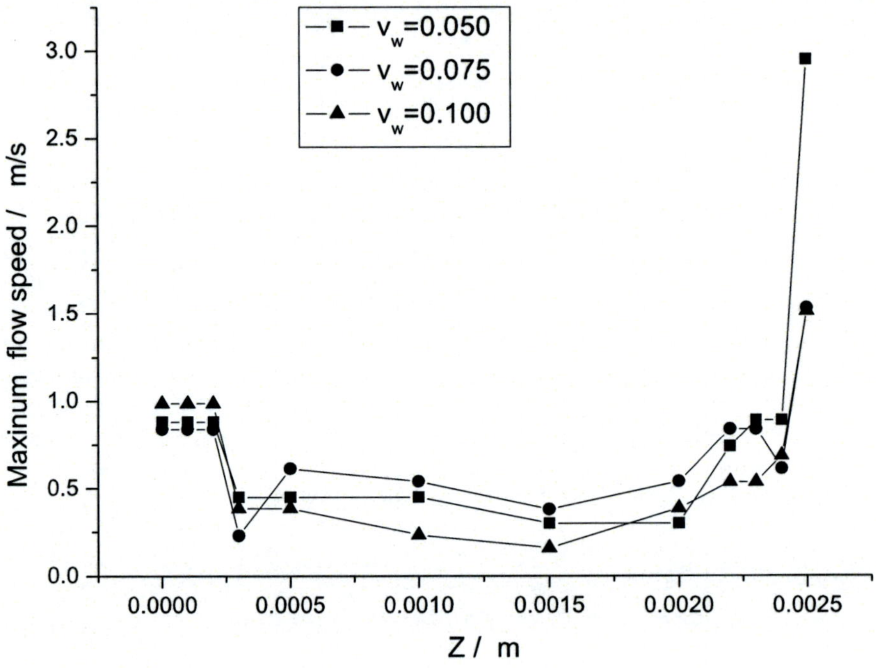

Fig. 2.30 Velocity distribution of weld pool at different welding speeds in the thickness direction of the workpiece

affected by the size of the eddies. The flow rate of weld pool fluid decreases rapidly with the increase of the distance from the free surface. In terms of the driving force of laser welding pool flow, there are Marangoni force, recoil force, buoyancy force, gravity force and friction force of irregular eruption of keyhole gas flow. Because this section focuses on the flow of laser welding pool in the case of full penetration, deep penetration laser welding is different from laser spot welding or partial deep penetration laser welding whose weld pool flow is driven by the recoil force generated by vapor evaporation. In the weld pool, if there is no auxiliary effect of eddy flow, the flow driving force is only the buoyancy of the weld pool, so the flow velocity is much lower than that of the weld pool surface. On the back of the weld pool, the weld pool fluid is still the working surface, but because of the difference of the temperature gradient between the positive and negative sides of the workpiece, the surface tension gradient of the weld pool is smaller on the back side than on the front side, and the size of the weld pool is also significantly smaller than that on the front side. Therefore, the maximum flow velocity on the surface of the weld pool is much lower than that on the front side, due to the Marangoni force.

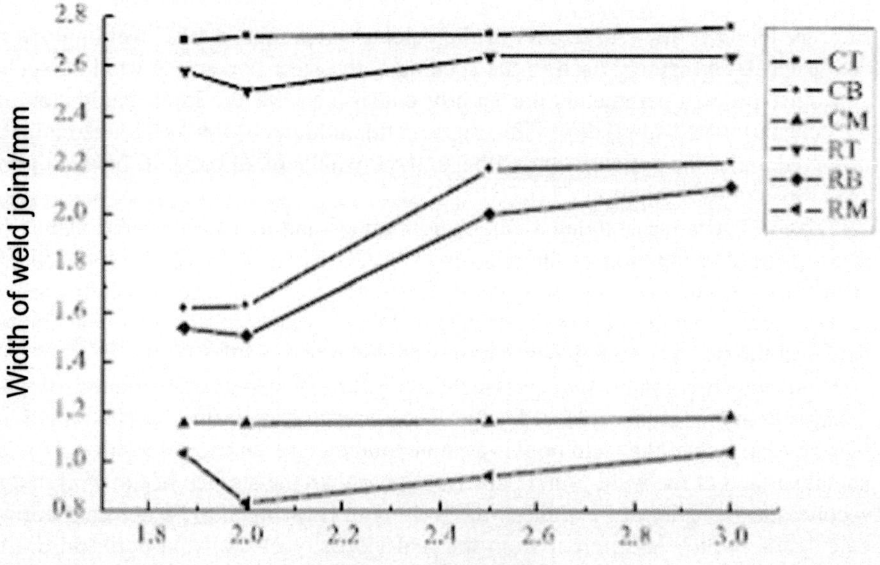

Fig. 2.31 Effect of laser power on sizes of the cross section of TC4 titanium alloy CO_2 laser weld joint

Table 2.3 Comparison of calculated value and actual measured value of the cross section of the weld joint

Power/kW	1.87		2. 00		2.50		3.00	
Size of weld joint/mm	Simulated value	Actual value	Simulated value	Actual value	Simulated value	Actual value	Simulated value	Actual value
Width of the upper part	2.70	2. 58	2.62	2. 50	2.73	2. 64	2.76	2.64
Width of the lower part	1.62	1.54	1.63	1.51	2.18	2. 00	2.11	1.89
Width of the middle part	1.15	1.03	0.86	0. 83	1. 15	0.94	1. 14	1.04

2.7.3 Verification

In order to verify the correctness of the calculation results, a laser welding test is conducted. The material used in the specimen, the laser power, the welding speed and other process parameters are strictly consistent with the input parameters in the calculation and simulation. The cross-sectional morphology of the weld joint is observed under the optical microscope, and the widths of all parts of the weld joint are measured by a microscope for comparison.

Table 2.3 lists the simulated calculation values and actual measured values of sizes of the cross section of the weld joint of CO_2 laser welding of TC4 titanium alloy. These values are obtained under different power conditions at a welding speed of 0.025 m/s. It can be seen from the data in Table 2.3 that the simulated calculation values of the widths of the upper, middle and lower parts of the cross section of the.

weld joint are similar to the actual measured values. In general, the calculated value is larger than the actual measured value. There are several reasons for this situation: First, the surface of the weld pool is assumed to be plane during calculation, but the actual surface of the weld joint is always higher than the surface of the workpiece; second, the viscosity of the fluid in the weld pool is assumed to be constant during calculation, which is different from the real viscosity of the weld pool; third, the flow state of the weld pool is assumed to be laminar flow during calculation, and this cannot fully reflect the actual flow state of the weld pool. Figure 2.31 shows the effect of laser power on sizes of the cross section of TC4 titanium alloy CO_2 laser weld joint. RT is the actual measured value of the size of the upper part of the weld joint, RB is the measured value of the size of the lower part of the weld joint, RM is the measured value of the size of the middle part of the weld joint, CT is the calculated value of the size of the upper part of the weld joint, CB is the calculated value of the size of the lower part of the weld joint, and CM is the calculated value of the size of the middle part of the weld joint. As can be seen from the figure, when the laser power increases from 1.87 kW to 2.00 kW, sizes of the cross section of the weld joint decrease. This relationship is related to the gravity of fluid in the weld pool in the initial stage of full penetration and the fact that the recoil pressure of the air flow in the keyhole forces the molten metal to fall. When the laser power continues to increase, sizes of the cross section of the weld joint are increased. In this steady increase phase, the welding speed remains constant. Despite greater fall in the weld pool, the fall rate is far less than the rate of increase in the overall size of the weld pool, and as a result, the size of the cross section of the weld joint increases continuously.

Figure 2.32 shows the effect of welding speed on widths of all parts of the cross section of TC4 titanium alloy CO_2 laser weld joint when the laser power is 3000 W. As the welding speed increases, test values and simulated values of widths of all parts of the weld joint decrease, and most of the simulated values are slightly higher than the test values. For the width of the middle part of the weld joint at a welding speed of 0.05 m/s and the width of the upper part of the weld joint at a welding speed of 0.04 m/s, the simulated value and the actual measured value are in good agreement. Figure 2.33 compares the simulation results of the geometric shape and the size of the weld joint at the CO_2 laser power of 3000 W and the welding speed of 0.05 m/s and the test results obtained after TC4 titanium alloy is welded under the

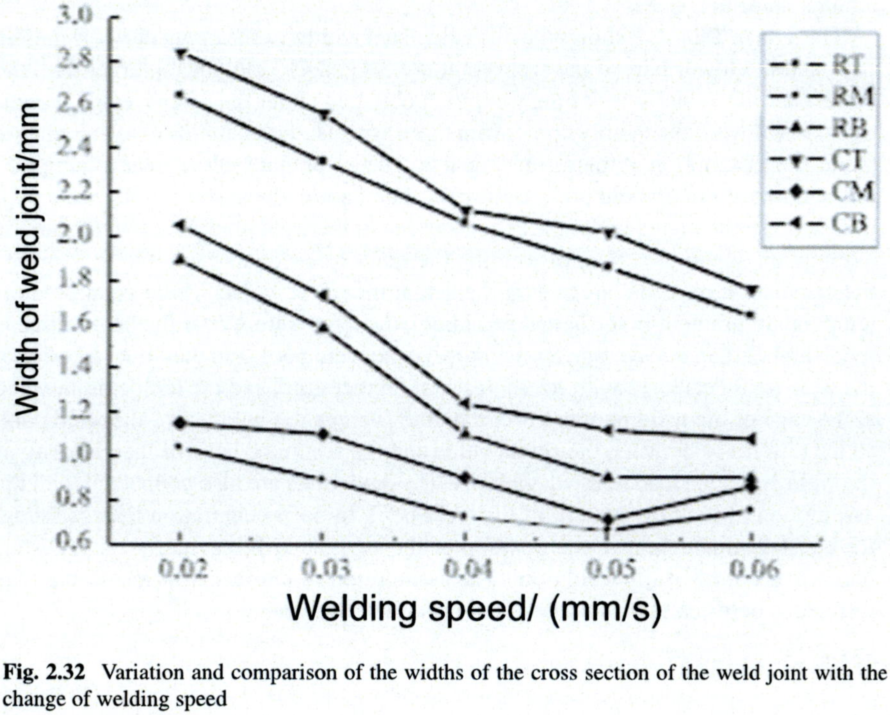

Fig. 2.32 Variation and comparison of the widths of the cross section of the weld joint with the change of welding speed

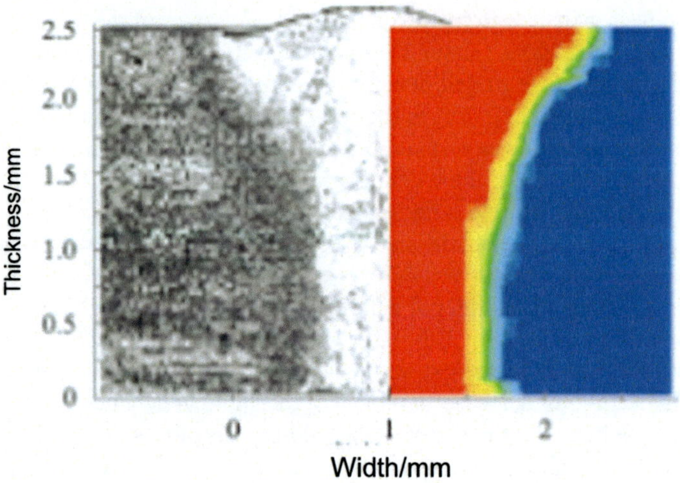

Fig. 2.33 Comparison of the calculated shape and size of the weld joint and the actual shape and size of the weld joint

same conditions. The results show that the simulated and actual weld joints share a similar shape and size.

The data in Table 2.3 shows that the calculated widths of the upper part, the middle part, and the lower part of the weld joint are 3.41, 8.41, and 9.30% larger than the corresponding actual sizes of the weld joint obtained under the same welding conditions. The deviations mainly come from three aspects: the assumption of calculation plane, the assumption of thermo-physical properties of materials and the assumption of the flow pattern of weld pool. Before the calculation, the surface of the weld pool is assumed to be plane, while the actual surface of the weld pool is deviated from the assumption due to the effect of welding thermal expansion and gravity. In this way, the calculation result is larger than the actual measured value. There is an obvious temperature gradient in the actual weld pool, the temperature near the heat source is much higher than the average temperature of the weld pool, and the flow viscosity of the weld pool decreases with the increase of temperature, and the temperature value at the edge of the weld pool is lower than the average temperature of the weld pool, so the difference between the actual value and the assumed value of the viscosity of the weld pool also lead to the deviations. The deviations are also connected with the assumption of the flow pattern of the weld pool. In deep penetration laser welding, the constant fluctuation of the position of the keyhole will inevitably cause disturbance to the flow of the weld pool. The assumption of laminar flow also causes the difference between the calculated value and the actual value.

2.8 Summary

On the basis of mass conservation equation, momentum conservation equation and energy conservation equation in fluid mechanics, the component conservation equation and the κ–ε equation are added according to different simulation objectives. Given boundary conditions of deep penetration laser welding, the control volume method is used to discretize nonlinear partial differential equations, and with the CFD commercial software FLUENT solver and SIMPLE algorithm, the keyhole and the flow velocity field of the weld pool, and the mass fraction of components under the interaction of inert auxiliary airflow, the keyhole erupting airflow and the upper part of the keyhole during the deep penetration laser welding are simulated. The calculation results are compared with the changes in the shape and size of the weld of titanium alloy laser welding, or the changes of the plasma smoke photo of high-speed photography. It is found that the simulation results are basically consistent with the experimental results.

Through the above study, the following conclusions are drawn:

(1) It is suitable to employ a combined heat source model consisting of the rotating Gauss heat source and double-ellipsoid heat source to simulate the formation of the keyhole in deep penetration laser welding of titanium alloy and the flow velocity field of the weld pool. The body heat source model can reflect the basic physical process in deep penetration laser welding, and also embodies the simulation characteristics of the control volume method.

(2) In the process the deep penetration laser welding with full penetration, the diameter of the keyhole is not very sensitive to the laser power, but the inclined angle of the keyhole changes obviously with the increase of the laser welding speed. The surface tension pressure caused by the curvature of the free surface of the keyhole has a strong relationship with the welding speed under the condition of constant airflow pressure in the keyhole. With the increase of the welding speed, the stability of the keyhole gradually decreases.

(3) The flow velocity above the surface of the weld pool for deep penetration laser welding is the largest, and Marangoni force plays a leading role in the convective heat transfer on the surface of the weld pool. Under the action of moving heat source, the Marangoni flow evolves from a symmetrical shape around the laser heating center to a tadpole shape in which the direction of the long axis is parallel to the welding direction of the weld pool. In a velocity "virtual" circle, on the surface of the weld pool, with half of the width of the weld pool as the radius, the flow velocity of the weld pool is relatively large, and outside the "virtual" circle, the flow velocity of the fluid in the weld pool decreases gradually. Inside the weld pool, the flow velocity of the fluid is significantly lower than that on the surface of the weld pool, but much larger than the welding speed. The flow velocity value of fluid on the back side of the weld pool is larger than the flow velocity value of metal inside the weld pool.

(4) The shape and size of the weld pool have a good correspondence with the size and position of the flow velocity vortex in the weld pool. The vortex in the weld

pool is the result of the combined action of the recoil pressure of Marangoni flow and the solid–liquid interface, while buoyancy and gravity only play an auxiliary role. The existence of the flow vortex of fluid in the weld pool greatly enhances the convective heat transfer between the high-temperature metal fluid and the cold fluid, thereby directly affecting the shape and size of the weld pool.

Chapter 3
Coupling Model and Numerical Computation Method of Keyhole and Weld Pool

Abstract This chapter introduces a self-consistent model of coupled keyhole and weld pool dynamics and its numerical implementation for keyhole mode laser welding. A self-consistent heat source model dependent on the transient three-dimensional keyhole profiles is presented by using the ray tracing algorithm. Governing equations for three-dimensional transient heat transfer and fluid flow of weld pool and keyhole free surface tracking are given in detail. The discontinuous boundary conditions of thermocapillary force, surface tension and recoil pressure on the interface between keyhole and weld pool are deduced based on incompressible viscous flow assumptions. Numerical implementations of this self-consistent model are also introduced.

3.1 Introduction

In the process of deep penetration laser welding, the laser heat source makes the material rapidly melted and vaporized to form weld pool, metal vapor/plasma and dynamic keyhole. With the movement of the heat source, the keyhole and the weld pool continue to advance, and at the same time, complex free-surface motions occur in the millimeter-scale weld pool, together with various behaviors such as solid–liquid–gas conversion, heat transfer and mobility, throughout the entire welding process. The stability of the deep penetration laser welding process and the joint quality after welding rely heavily on the coupling of the dynamic keyhole and the moving weld pool. Therefore, it is necessary to establish a coupling model for the laser welding keyhole and weld pool and use an accurate and efficient numerical computation method to provide theoretical tools for the quantitative simulation analysis of the complex multi-phase coupling dynamics behavior and mechanism of the keyhole and weld pool in the welding process.

This chapter introduces in detail how to establish a laser welding multi-phase coupling model. A hybrid model is used to describe the real-time changing solid–liquid surface, assuming that the molten metal in the weld pool is an incompressible Newtonian thermal fluid. The heat transfer and flow in the weld pool are expressed by the conservation of energy equation and the Navier–Stokes equation, respectively.

S. Gong et al., *Weld Pool Dynamics in Deep Penetration Laser Welding*,
https://doi.org/10.1007/978-981-16-0788-2_3

As for heat source model of the keyhole laser-absorbing wall, the analytical keyhole linear heat source model commonly used in the early stage of laser welding simulation, and the dynamic keyhole multiple Fresnel absorption model are introduced. As for tracking method of the real-time changing gas–liquid free surface during the welding process, the commonly used VOF (volume of fluid) method and Level Set method are mainly introduced. The precise treatment of the gas–liquid surface boundary conditions is a guarantee essential to the calculation accuracy of mesoscale surface flow in laser welding. The gas–liquid sharp surface method proposed by the author can accurately treat the surface boundary conditions, avoiding the numerical error caused by the CSF (continue surface force) model, and greatly improving the accuracy of the simulation calculation. Therefore, the multi-phase coupling model of the laser keyhole and weld pool established by this method is also known as laser welding sharp surface model.

This chapter also gives a detailed introduction of the numerical solution method and process corresponding to the multi-phase coupling model, including the algorithm based on linear heat source model, the ray tracing algorithm based on multiple Fresnel absorption heat source, the high precision finite difference solution format of Level Set method, the fast solving algorithm of weld pool heat transfer and flow, the OpenMP parallel solving language, and the iterative solving process in the coupled solution. The multi-phase coupling model introduced in this chapter has more comprehensive considerations than the quasi-steady-state model as provided in Chap. 2. It can describe the dynamics features of the keyhole and moving weld pool during the entire welding process from the initial welding to the quasi-steady state. But it is more complicated and time-consuming in calculation.

3.2 Governing Equations of Coupled Model

3.2.1 Heat Transfer and Fluid Flow Equations

It is assumed that the molten metal in the weld pool during laser welding is an incompressible fluid, and the density of the molten metal does not change during the solid–liquid phase transition. Therefore, the mass, momentum and energy conservation is expressed as follows

$$\nabla \cdot \vec{U} = 0 \tag{3.1}$$

$$\rho\left(\frac{\partial \vec{U}}{\partial t} + \left(\vec{U} \cdot \nabla\right)\vec{U}\right) = \nabla \cdot \left(\mu_l \nabla \vec{U}\right) - \nabla p - \frac{\mu_l}{K}\vec{U} - \frac{C_\rho}{\sqrt{K}}|\vec{U}|\vec{U} + \rho\vec{g}\beta\left(T - T_{ref}\right) \tag{3.2}$$

$$\rho C_p\left(\frac{\partial T}{\partial t} + \left(\vec{U}\cdot\nabla\right)T\right) = \nabla\cdot(k\nabla T) \tag{3.3}$$

where:

$\vec{U}$	Three-dimensional velocity vector;
μ_l	Dynamic viscosity of fluid;
ρ	Density;
p	Pressure;
$\vec{g}$	Three-dimensional gravity acceleration vector;
β	Coefficient of thermal expansion;
C_p	Specific heat capacity;
k	Heat conductivity coefficient;
T	Temperature;
T_{ref}	Reference temperature;
K	Carman-Kozeny coefficient in mixed phase model, i.e., transmission coefficient, which is closely related to the liquid mass fraction f_l of current grid cell.

K can be determined by the following equation.

$$K = \frac{f_l^2 d^2}{180(1 - f_l)^2} \tag{3.4}$$

where: d is closely related to the size of dendritic arms, which is usually a constant. For laser welding, its size is above 10^{-2} cm. In addition, C in the Eq. (3.2) is a nonreactivity coefficient related to liquid mass fraction. It is calculated by the following equation.

$$C = 0.13 f_l^{-3/2} \tag{3.5}$$

In order to improve the value calculation speed, the liquid mass fraction may be deemed to have a linear relation with the temperature, thus:

$$f_l = \begin{cases} 1 & T > T_l \\ \frac{T - T_s}{T_l - T_s} & T_l >= T >= T_s \\ 0 & T < T_s \end{cases} \tag{3.6}$$

where: T_l and T_s—Liquidus temperature and solidus temperature of welded alloy material.

3.2.2 Heat Source Model

During the deep penetration welding process, the laser beam is irradiated onto the surface of workpiece to lead to rapid melting and evaporation, thus forming keyhole in the weld pool. For this reason, the heat source based on the change of the keyhole morphology fits more the physical facts in the laser welding simulation calculation. Earlier, Kaplan et al. proposed a linear heat source model based on the analytical keyhole model in the 1990s. With the practice of keyhole surface real-time tracking and the breakthrough of ray tracing technology in laser welding heat source application, the multiple Fresnel absorption heat source based on keyhole free surface has been greatly concerned and widely applied in simulation applications, especially in numerical simulation for accurate calculation of keyhole and weld pool dynamics.

3.2.2.1 Linear Heat Source Model

Kaplan proposed a calculation model for the keyhole shape of deep penetration laser welding based on the idea of point-by-point energy balance of keyhole wall. Since the laser welding heat source is assumed as a fast moving linear heat source, the model is suitable for laser rapid welding and sheet welding. Since the laser heat source is assumed as a linear heat source in the model, the temperature field can be expressed by Rosenthal's fast moving linear heat source equation.

$$T(r, \varphi) = T_{\partial} + \frac{P'}{2\pi \lambda_{th}} K_0\left(P_e' r\right) e^{-P_e' r \cos \varphi} \tag{3.7}$$

where:

T_a Ambient temperature, K;
P' Average intensity of the heat source along the thickness of the workpiece;
λ_{th} Thermal conductivity coefficient, W· m^{-1} ·K^{-1};
K_0 Modified second-class zero-order Bessel function;
r Distance of point (r, ϕ) from the heat source, m;
ϕ Angle of point (r, ϕ) with the heat source line and the welding direction;
Pe' Corrected Peclet number.

The corrected Peclet number can be expressed by the following equation.

$$P_e' = \frac{\rho \cdot C \cdot v}{r_0 \lambda_{th}} = \frac{r_0 v}{k} \tag{3.8}$$

where:

k Thermal diffusion coefficient;
v Welding velocity, m/s;

r_0 Radius of heat source, m.

Pe' is proportional to the velocity v under given diffusion coefficient and radius of heat source. The temperature of any point in the weld pool (including any point on the inner wall of the keyhole) can be expressed by $T(r, \phi)$ theoretically. The temperature of the keyhole inner wall is assumed as the metal evaporation temperature T_v; If the left side (3.6) of equation is equal to T_v, the heat source intensity P' at each point on the inner wall of the keyhole can be expressed as follows.

$$P'(r, \varphi) = (T_v - T_\partial)\frac{2\pi\lambda_{th}}{K_0(P_e'r)}e^{P_e'r\cos\varphi} \tag{3.9}$$

3.2.2.2 Heat Source Model Based on Fresnel Absorption

The approximate distribution of laser energy density is assumed to be a Gaussian distribution, and then the distribution function equation of laser energy can be expressed as follows.

$$I_0(r, z) = 3Q/(\pi R^2)\exp(-3(r^2)/R^2) \tag{3.10}$$

where:

R Light spot radius;
Q Laser power density.

In addition, let's assume that the laser energy is not attenuated along the depth direction. As the laser head moves, the Gaussian heat source must move with it, so as to consider the effect of welding velocity.

Considering the Fresnel absorption effect, the laser energy density q absorbed at any position on the keyhole wall can be expressed as follows.

$$q = I_0(r, z)\left(\vec{I}_0 \cdot \vec{n}_0\right)\alpha_{Fr}(\theta_0) + \sum_{m=1}^{N} I_m(r, z)\left(\vec{I}_m \cdot \vec{n}_m\right)\alpha_{Fr}(\theta_m) \tag{3.11}$$

$$\alpha_{Fr}(\theta) = 1 - \frac{1}{2}\left(\frac{1 + (1 - \varepsilon\cos\theta)^2}{1 + (1 + \varepsilon\cos\theta)^2} + \frac{\varepsilon^2 - 2\varepsilon\cos\theta + 2\cos^2\theta}{\varepsilon^2 + 2\varepsilon\cos\theta + 2\cos^2\theta}\right) \tag{3.12}$$

where:

θ Angle between the incident laser beam and the normal vector of the keyhole wall;
$\alpha_{Fr}(\theta)$ Fresnel absorption coefficient;
N Number of laser beam incidence considering multiple reflection;

$\vec{I}$ Normalized laser beam direction;
$\vec{n}$ Normalized normal vector of the keyhole wall;
ε Constant related to the laser device and material;
$I_0(r, z)$ Energy distribution function of laser beam;
$I_m(r, z)$ Remaining energy density of the laser beam after the m reflection.

The constant ε is theoretically determined by the following equation.

$$\varepsilon^2 = \frac{2\varepsilon_2}{\varepsilon_1 + \left[\varepsilon_1^2 + (\sigma_{st}/\omega\varepsilon_0)^2\right]^{1/2}} \tag{3.13}$$

where:

σ_{st} Electrical conductivity per unit depth of metallic material;
$\varepsilon_1, \varepsilon_2$ Real component of the metal and plasma dielectric constants;
ε_0 Vacuum transmittance;
ω Laser angular frequency.

As a matter of fact, ε value is normally determined by trial-and-error method, which is larger than the theoretical value in general. There are main two reasons for using the trial-and-error method: ① The keyhole wall has an extreme high temperature, close to boiling point, which is likely to affect the Fresnel absorption effect; ② The approximation error can be compensated by a large ε value, because the formula used for calculating the Fresnel absorption is an approximation by itself. For CO_2 laser welding, ε is normally equal to 0.08; for YAG laser, ε is normally equal to 0.2–0.25. Considering the fiber laser has a better beam quality than YAG in general, this study proposes that for the fiber laser welding, ε can be determined as 0.2–0.5, which varies with the laser power.

3.3 Free-Surface Tracking Method

The keyhole normally has a changing transient morphology in the laser welding process, and its shape may even be highly topologically deformed, such as metal spatter and porosity. In addition, the normal vector and the curvature at any position of the keyhole wall should be calculated precisely in order to accurately consider the influence of Fresnel effect and surface tension on the transient behavior of keyhole. The aforesaid two problems cause great difficulty in choosing a reasonable numerical method to track the moving surface of transient keyhole.

It is very difficult to use Lagrangian-based surface tracking methods such as MAC (Mark and Cell) and Front Tracking to deal with the topological deformation of complex surface. Now, two main surface tracking methods are VOF (Volume of Fluid) method and Level Set method, with the advantages of high calculation accuracy, short calculation time and less memory. So, this chapter mainly introduces the Level Set and VOF methods to track the keyhole free-surface movement.

3.3.1 Level Set Method

The Level Set method uses an implicit scalar function to describe the free-surface position. The scalar function is normally a signed distance field function. It is defined as the shortest distance from the keyhole interface in respect of laser welding. For the Level Set function $\phi(\vec{x})$ in this document, $R^3 \rightarrow R$ is defined as follows.

$$\varphi(\vec{x}) = \begin{cases} -d(\vec{x}, \Omega) & \vec{x} \in \Omega_{in} \\ 0 & \vec{x} \in \Omega \\ -d(\vec{x}, \Omega) & \vec{x} \in \{R^3 - \Omega - \Omega_{in}\} \end{cases} \tag{3.14}$$

where:

Ω	Keyhole surface;
Ω_{in}	Area occupied by molten metal or workpiece;
$\{R^3 - \Omega - \Omega_{in}\}$	Area of metal vapor/plasma.

Based on Level Set method, the kinematic equation describing the transient keyhole surface in laser welding can be expressed as follows.

$$\frac{\partial \varphi}{\partial t} + \vec{U} \cdot \nabla \varphi = 0 \tag{3.15}$$

where: $\vec{U}$—Kinematic velocity of keyhole surface. In this study, $\vec{U}$ is determined as the kinematic velocity of the molten metal in the weld pool closest to the current position.

If the Level Set method is used to track the moving surface, the current Level Set function should be re-initialized after solving of the Eq. (3.15), so as to ensure that the Level Set function in each time step is a signed distance field. The re-initialization equation can be described as follows.

$$|\nabla \varphi| = 1 \tag{3.16}$$

Using the Level Set method to describe the free surface of the keyhole can facilitate the computation of the exact normal vector and curvature value at any position on the keyhole surface. Based on the distance field function, the calculation formulas of normal vector and curvature k are as follows.

$$\vec{n} = \frac{\nabla \varphi}{|\nabla \varphi|} \tag{3.17}$$

$$\kappa = \nabla \cdot \frac{\nabla \varphi}{|\nabla \varphi|} \tag{3.18}$$

Large numerical dissipation is the biggest problem of the Level Set method in tracking the free surface, meaning that it will cause serious surface quality loss. The numerical dissipation problem can be addressed by the Particle Level Set method. Its principles are as follows. Some virtual particles are arranged artificially on both sides of the surface described by the Level Set to make the particles move passively with the fluid. The position information of the particles is used to correct the numerical dissipation of the Level Set. The motion equation of particles can be determined as follows.

$$\frac{d\vec{x}_p}{d_t} = \vec{U}(\vec{x}_p) \tag{3.19}$$

where:

$\vec{x}_p$ Current position vector of particle;
$\vec{U}(\vec{x}_p)$ Velocity vector of the fluid at particle location.

3.3.2 VOF Method

The VOF method, proposed by Hirt and Nichols in 1981, is used to handle arbitrary free surface problems. Its basic principle is to determine the position and shape of the free surface by the ratio function F (also known as volume function F) between the changing volume of fluid in the computational regional grid cell and the volume of the grid cell itself. The VOF method, which is evolved based on the MAC method, requires less computing time, less memory and has easier boundary conditions compared with the MAC method.

The free surface tracked by the VOF method is determined by the volume fraction F of fluid in the grid cell at each moment. For instance, if the volume fraction F of fluid in a grid cell is equal to 1 at a certain moment, the grid cell is filled with fluid. If the volume fraction F of fluid in another grid cell is equal to 0, the grid cell is filled with another fluid. If the volume fraction F of fluid is more than 0 and less than 1 in a grid cell, the grid cell is a surface containing two-phase substance, so that the position and shape of the free surface can be determined according to the volume fraction F of fluid in grid cell. The volume function is shown by the equation below.

$$\frac{\partial F}{\partial t} + \frac{\partial (F_u)}{\partial t} + \frac{\partial (F_v)}{\partial t} + \frac{\partial (F_w)}{\partial t} = 0 \tag{3.20}$$

The free surface is assumed as a horizontal plane based on the determination of the surface normal vector. For the grid cell (i, j, k), the adjacent grid cells in the same horizontal plane are represented as F_R, F_L, F_T, F_B, so the vector is expressed as follows.

$$\overrightarrow{F_L F_R} = (2d_x, 0, F_R^2 - F_L^2) \tag{3.21}$$

$$\overrightarrow{F_B F_T} = (0, 2d_y, F_T^Z - F_B^Z) \tag{3.22}$$

where:

d_x, d_y	Grid step size of x and y equation;
$F_L^Z\ F_R^Z\ F_B^Z\ F_T^Z$	z coordinate value of the corresponding grid.

Thus, the normal vector of the free surface in the horizontal direction can be expressed as follows.

$$\overrightarrow{F_L F_R} \times \overrightarrow{F_B F_T} = (-2(F_R^Z - F_L^Z)d_y - 2(F_T^Z - F_B^Z)dx, 2dxdy) \tag{3.23}$$

The free-surface normal vector of positive plane and side plane can be obtained similarly.

During computation, the material property parameters of the control body can be determined according to the proportion of each phase in the control body. For instance, for a system containing n phases, the density of each control body can be determined by the following equation.

$$\rho = \sum_{l=1}^{n} \alpha_i \rho_i \tag{3.24}$$

where:

α_i	Volume fraction of phase *i* in the control body;
ρ_i	Density of phase *i*, kg· m^{-3}.

The density and viscosity are determined by the volume fraction of each phase in the control body, so other parameters (such as viscosity) can be determined in the same way.

3.4 Boundary Conditions of the Coupling Model

Now, relevant studies show that serious parasitic flow problem may arise if the CSF model is used to handle the recoil pressure, surface tension and thermal capillarity force caused during deep penetration laser welding. Theoretically, the existence of surface tension will result in mathematical discontinuity of the pressure of gas–liquid two-phase fluid across both sides of the keyhole. Besides, the existence of thermal capillary force and friction force will cause mathematical discontinuity of the gas–liquid two-phase viscosity stress tensor. In numerical simulation, precisely considering these mathematical discontinuities is an essential prerequisite for obtaining accurate solutions consistent with physical processes. Therefore, discontinuous numerical method is used to handle these important boundary conditions.

The theoretical derivation of discontinuous conditions with the presence of thermal capillary force and surface tension is lacked at home and abroad. In this section, a detailed theoretical derivation is made for the discontinuous boundary conditions on the free surface of the keyhole during deep penetration laser welding based on the fundamental principles of viscous fluid mechanics.

3.4.1 Basic Agreements

First, the vector is defined as $\vec{u} = u_x\vec{i} + u_y\vec{j} + u_z\vec{k}$, where $\vec{i}$, $\vec{j}$, $\vec{k}$ are the unit vector in the direction of x, y, and z, respectively. It is agreed that the vector is expressed in the form of a column vector matrix.

$$\vec{u} = \begin{pmatrix} u_x \\ u_y \\ u_z \end{pmatrix} \tag{3.25}$$

In addition, it is also agreed that the second-order stress tensor is expressed in the form of a 3 × 3 matrix.

$$\sigma = \left(\sigma_{ij}\right) = \begin{pmatrix} \sigma_{xx} & \sigma_{xy} & \sigma_{xz} \\ \sigma_{yx} & \sigma_{yy} & \sigma_{yz} \\ \sigma_{zx} & \sigma_{zy} & \sigma_{zz} \end{pmatrix} \tag{3.26}$$

where:

σ_{ij} Stress on each surface;

i and j The stress is applied on a plane perpendicular to the x_i axis in the same direction as the x_i axis.

Finally, the second-order tensor of vector left dot product and right dot product is expressed as follows respectively.

$$\text{Left dot product: } \vec{u} \cdot \sigma = \begin{pmatrix} u_x & u_y & u_z \end{pmatrix} \begin{pmatrix} \sigma_{xx} & \sigma_{xy} & \sigma_{xz} \\ \sigma_{yx} & \sigma_{yy} & \sigma_{yz} \\ \sigma_{zx} & \sigma_{zy} & \sigma_{zz} \end{pmatrix}$$

$$\text{Right dot product: } \sigma \cdot \vec{u} = \begin{pmatrix} \sigma_{xx} & \sigma_{xy} & \sigma_{xz} \\ \sigma_{yx} & \sigma_{yy} & \sigma_{yz} \\ \sigma_{zx} & \sigma_{zy} & \sigma_{zz} \end{pmatrix} \begin{pmatrix} u_x \\ u_y \\ u_z \end{pmatrix} \tag{3.27}$$

The normal vector of any tiny curved surface is assumed as $\vec{n}$, so the stress vector $T^{(\vec{n})}$ on the surface is defined as follows.

$$T^{(\vec{n})} = \vec{n} \cdot T = \begin{pmatrix} n_x & n_y & n_z \end{pmatrix} \cdot \begin{pmatrix} \sigma_{xx} & \sigma_{xy} & \sigma_{xz} \\ \sigma_{yx} & \sigma_{yy} & \sigma_{yz} \\ \sigma_{zx} & \sigma_{zy} & \sigma_{zz} \end{pmatrix} \tag{3.28}$$

3.4.2 *Discontinuous Boundary Conditions in Gas–Liquid Two-Phase Flow*

In the presence of surface tension and thermal capillary force at the gas–liquid two-phase fluid surface, the free surface (S) tension discontinuity can be easily deduced according to the principle of fluid mechanics.

$$\int_s \left(T_G^{(\vec{n})} - T_L^{(\vec{n})}\right) dS = \int_S (\sigma \vec{n} \nabla \cdot \vec{n} - \nabla_s \sigma) dS \tag{3.29}$$

It can also be expressed as follows.

$$[\vec{n} \cdot T] = \vec{n} \cdot T_G - \vec{n} T_L^{(\vec{n})} = T_G^{(\vec{n})} - T_L^{(\vec{n})} = \sigma \vec{n} \nabla \cdot \vec{n} - \nabla_S \sigma = \sigma \kappa \vec{n} - \nabla_S \sigma \tag{3.30}$$

where, the subscripts L and G represent liquid and gas, σ is surface tension coefficient and k is curvature. The arbitrary discontinuity A on the free surface S is $[A] = A_G - A_L$. $\nabla_s = (I - \vec{n}\vec{n})$, ∇ is surface tangential derivative, and I is unit matrix.

Through careful observation of Eqs. (3.29) and (3.30), it is apparent that $\sigma \kappa \vec{n}$ is normal component, and $\nabla_S \sigma$ is tangent component. The right dot product in the Eq. (3.30) is multiplied by the normal vector and the tangent vector, respectively, leading to:

$$\begin{cases} (\vec{n} \cdot T_G - \vec{n} \cdot T_L) \cdot \vec{n} = \vec{n} \cdot T_G \cdot \vec{n} - \vec{n} \cdot T_L \cdot \vec{n} = \vec{n} \cdot [T] \cdot \vec{n} = \sigma \kappa \\ (\vec{n} \cdot T_G - \vec{n} \cdot T_L) \cdot \vec{t} = \vec{n} \cdot T_G \cdot \vec{t} - \vec{n} \cdot T_L \cdot \vec{t} = \vec{n} \cdot [T] \cdot \vec{t} = -\nabla_s \sigma \cdot \vec{t} \end{cases} \tag{3.31}$$

According to the Navier–Stokes equation, the stress tensor of fluid can be expressed as:

$$\sigma = -\begin{pmatrix} p & 0 & 0 \\ 0 & p & 0 \\ 0 & 0 & p \end{pmatrix} + \begin{pmatrix} \sigma_{xx} + p & \sigma_{xy} & \sigma_{xz} \\ \sigma_{yx} & \sigma_{yy} + p & \sigma_{yz} \\ \sigma_{zx} & \sigma_{zy} & \sigma_{zz} + p \end{pmatrix} \tag{3.32}$$

where:

p Fluid pressure, $p = -\frac{1}{3}(\sigma_{xx} + \sigma_{yy} + \sigma_{zz})$;
τ Stress bias tensor.

The fluid in the weld pool is assumed as Newtonian fluid during the laser welding process. As for Newtonian fluid, there are three assumptions: ① The stress tensor is a linear function of the strain rate; ② Fluid is isotropic; ③ When the fluid is static, $\nabla \cdot \tau$ is equal to 0. Thus, we can obtain:

$$\tau_{ij} = \mu\left(\frac{\partial u_i}{\partial x_j} + \frac{\partial u_j}{\partial x_i}\right) + \delta_{ij}\lambda\nabla \cdot \vec{U} \tag{3.33}$$

where:

δ_{ij} Dirac symbol;
$\vec{U} = (u\ v\ w)^T$ Velocity vector.

In the process of laser welding, the metal liquid in the weld pool is generally considered to be incompressible, i.e. $\nabla \cdot \vec{U} = 0$. Therefore, the deviatoric tensor of stress can be simplified as

$$\tau = \mu\left((\nabla\vec{U}) + (\nabla\vec{U})^T\right) \tag{3.34}$$

By substituting Eq. (3.33) into Eq. (3.30), it can be deduced that the discontinuity caused by surface tension and thermal capillary force in gas–liquid two-phase flow is

$$\begin{aligned} \sigma\kappa &= [\vec{n} \cdot T \cdot \vec{n}] = \left[\vec{n} \cdot \left(-pI + \mu\left(\left(\nabla\vec{U}\right) + \left(\nabla\vec{U}\right)^T\right)\right) \cdot \vec{n}\right] \\ &= \left[-p + 2\mu\vec{n} \cdot \nabla\vec{U} \cdot \vec{n}\right] = [-p] + \left[2\mu\vec{n} \cdot \nabla\vec{U} \cdot \vec{n}\right] \\ -\nabla_s\sigma \cdot \vec{t} &= \left[\vec{n} \cdot T \cdot \vec{n}\right] = \left[\vec{n} \cdot \left(-pI + \mu\left(\left(\nabla\vec{U}\right) + \left(\nabla\vec{U}\right)^T\right)\right) \cdot \vec{t}\right] \\ &= \left[\mu\vec{n} \cdot \nabla\vec{U} \cdot \vec{t} + \mu\vec{t} \cdot \nabla\vec{U} \cdot \vec{n}\right] = \left[\mu\vec{n} \cdot \nabla\vec{U} \cdot \vec{t}\right] + \left[\mu\vec{t} \cdot \nabla\vec{U} \cdot \vec{n}\right] \end{aligned} \tag{3.35}$$

3.4.3 Discontinuous Boundary Conditions for Surface Tension, Thermal Capillary Force and Recoil Pressure

Next, Eq. (3.35) is simplified to deduce the discontinuous boundary conditions used to calculate the capturing method of boundary conditions. Several identities on viscous incompressible fluid are given first before the discontinuous boundary conditions are deduced.

Since the metal liquid and the metal vapor/plasma in the weld pool are all viscous fluids, their velocities on the free interface of the keyhole are continuous. Otherwise, the free interface will be torn or overlapped non-physically. That is, for viscous fluids, the following equation is satisfied:

$$\left[\vec{U}\right] = \left[u\ v\ w\right] = \vec{0} \tag{3.36}$$

At the same time, because the tangential speed of two viscous fluids on the interface is also continuous, the following equation is satisfied:

$$\left[\vec{t} \cdot \nabla\vec{U}\right] = (0\,0\,0) \tag{3.37}$$

Equation (3.37) is converted into the components in three directions:

$$\left[\vec{t} \cdot \nabla u\right] = \left[\vec{t} \cdot \nabla v\right] = \left[\vec{t} \cdot \nabla w\right] = 0 \tag{3.38}$$

$$\left[\nabla u \cdot \vec{t}\right] = \left[\nabla v \cdot \vec{t}\right] = \left[\nabla w \cdot \vec{t}\right] = 0 \tag{3.39}$$

By transforming the continuity equation $\nabla \cdot \vec{U} = 0$ describing the incompressible fluid in coordinate system $\vec{i}, \vec{j}, \vec{k}$ into the coordinate system $\vec{n}, \vec{t}_1, \vec{t}_2$, we can obtain the following identity:

$$\vec{n} \cdot \nabla\vec{U} + \vec{t}_1 \cdot \nabla\vec{U} \cdot \vec{t}_1 + \vec{t}_2 \cdot \nabla\vec{U} \cdot \vec{t}_2 = 0 \tag{3.40}$$

Based on Eq. (3.40), the following discontinuity equation can be directly obtained:

$$\left[\vec{n} \cdot \nabla\vec{U} \cdot \vec{n}\right] = -\left[\vec{t}_1 \cdot \nabla\vec{U} \cdot \vec{t}_1\right] - \left[\vec{t}_2 \cdot \nabla\vec{U} \cdot \vec{t}_2\right] \tag{3.41}$$

Due to the continuous tangential speed on the interface, the following equation is satisfied:

$$\left[\vec{t} \cdot \nabla\vec{U} \cdot \vec{t}\right] = \vec{t} \cdot \left(\nabla\vec{U}\right)_G \cdot \vec{t} - \vec{t} \cdot \left(\nabla\vec{U}\right)_L \cdot \vec{t} = \left[\vec{t} \cdot \nabla\vec{U}\right] \cdot \vec{t} = 0 \tag{3.42}$$

By substituting Eq. (3.42) into Eq. (3.41), we can obtain the following equation:

$$\left[\vec{n} \cdot \nabla \vec{U} \cdot \vec{n}\right] = 0 \tag{3.43}$$

Furtherly, based on Eq. (3.43), it can be deduced:

$$\vec{n} \cdot \left(\nabla \vec{U}\right)_G \cdot \vec{n} = \vec{n} \cdot \left(\nabla \vec{U}\right)_L \cdot \vec{n} \tag{3.44}$$

Finally, the discontinuity item $\left[\mu\vec{n} \cdot \nabla \vec{U} \cdot \vec{n}\right]$ in Eqs. (3.44) and (3.35) can be directly simplified into:

$$\left[\mu\vec{n} \cdot \nabla \vec{U} \cdot \vec{n}\right] = [\mu]\vec{n} \cdot \nabla \vec{U} \cdot \vec{n} \tag{3.45}$$

Based on the principle that the tangential speed is continuous on the interface, combining with Eq. (3.35), the following series of equations can be easily deduced:

$$\left[\mu\vec{t}_1 \cdot \nabla \vec{U} \cdot \vec{n}\right] = [\mu]\vec{t}_1 \cdot \nabla \vec{U} \cdot \vec{n} \tag{3.46}$$

$$\left[\mu\vec{t}_1 \cdot \nabla \vec{U} \cdot \vec{t}_1\right] = [\mu]\vec{t}_1 \cdot \nabla \vec{U} \cdot \vec{t}_1 \tag{3.47}$$

$$\left[\mu\vec{t}_1 \cdot \nabla \vec{U} \cdot \vec{t}_2\right] = [\mu]\vec{t}_1 \cdot \nabla \vec{U} \cdot \vec{t}_2 \tag{3.48}$$

$$\left[\mu\vec{t}_2 \cdot \nabla \vec{U} \cdot \vec{n}\right] = [\mu]\vec{t}_2 \cdot \nabla \vec{U} \cdot \vec{n} \tag{3.49}$$

$$\left[\mu\vec{t}_2 \cdot \nabla \vec{U} \cdot \vec{t}_1\right] = [\mu]\vec{t}_2 \cdot \nabla \vec{U} \cdot \vec{t}_1 \tag{3.50}$$

$$\left[\mu\vec{t}_2 \cdot \nabla \vec{U} \cdot \vec{t}_2\right] = [\mu]\vec{t}_2 \cdot \nabla \vec{U} \cdot \vec{t}_2 \tag{3.51}$$

$$\left[\mu\vec{n} \cdot \nabla \vec{U} \cdot \vec{t}_1\right] = -[\mu]\vec{t}_1 \cdot \nabla \vec{U} \cdot \vec{n} - \nabla_s\sigma \cdot \vec{t}_1 \tag{3.52}$$

$$\left[\mu\vec{n} \cdot \nabla \vec{U} \cdot \vec{t}_2\right] = -[\mu]\vec{t}_2 \cdot \nabla \vec{U} \cdot \vec{n} - \nabla_s\sigma \cdot \vec{t}_2 \tag{3.53}$$

Next, by arranging the equations from (3.45) to (3.53) into the matrix, we can obtain the following equation:

$$\begin{aligned}
&\left(\vec{n}\ \vec{t}_1\ \vec{t}_2\right)^T\left(\left[\mu\nabla\vec{U}\right]\right)\left(\vec{n}\ \vec{t}_1\ \vec{t}_2\right)\\
&= \left(\begin{matrix}
\left[\mu\vec{n} \cdot \nabla \vec{U} \cdot \vec{n}\right] & \left[\mu\vec{n} \cdot \nabla \vec{U} \cdot \vec{t}_1\right] & \left[\mu\vec{n} \cdot \nabla \vec{U} \cdot \vec{t}_2\right]\\
\left[\mu\vec{t}_1 \cdot \nabla \vec{U} \cdot \vec{n}\right] & \left[\mu\vec{t}_1 \cdot \nabla \vec{U} \cdot \vec{t}_1\right] & \left[\mu\vec{t}_1 \cdot \nabla \vec{U} \cdot \vec{t}_2\right]\\
\left[\mu\vec{t}_2 \cdot \nabla \vec{U} \cdot \vec{n}\right] & \left[\mu\vec{t}_2 \cdot \nabla \vec{U} \cdot \vec{t}_1\right] & \left[\mu\vec{t}_2 \cdot \nabla \vec{U} \cdot \vec{t}_2\right]
\end{matrix}\right)
\end{aligned}$$

$$= \begin{pmatrix} \left[\mu\vec{n}\cdot\nabla\vec{U}\cdot\vec{n}\right] & 0\,0 \\ 0 & 0\,0 \\ 0 & 0\,0 \end{pmatrix} + \begin{pmatrix} 0 & 0 & 0 \\ \left[\mu\vec{t}_1\cdot\nabla\vec{U}\cdot\vec{n}\right] & \left[\mu\vec{t}_1\cdot\nabla\vec{U}\cdot\vec{t}_1\right] & \left[\mu\vec{t}_1\cdot\nabla\vec{U}\cdot\vec{t}_2\right] \\ \left[\mu\vec{t}_2\cdot\nabla\vec{U}\cdot\vec{n}\right] & \left[\mu\vec{t}_2\cdot\nabla\vec{U}\cdot\vec{t}_1\right] & \left[\mu\vec{t}_2\cdot\nabla\vec{U}\cdot\vec{t}_2\right] \end{pmatrix}$$

$$+ \begin{pmatrix} 0 & \left[\mu\vec{n}\cdot\nabla\vec{U}\cdot\vec{t}_1\right] & \left[\mu\vec{n}\cdot\nabla\vec{U}\cdot\vec{t}_2\right] \\ 0 & 0 & 0 \\ 0 & 0 & 0 \end{pmatrix}$$

$$- \begin{pmatrix} 0 & \left[\mu\vec{t}_1\cdot\nabla\vec{U}\cdot\vec{n}\right] & \left[\mu\vec{t}_2\cdot\nabla\vec{U}\cdot\vec{n}\right] \\ 0 & 0 & 0 \\ 0 & 0 & 0 \end{pmatrix} - \begin{pmatrix} 0 & \nabla_s\sigma\vec{t}_1 & \nabla_s\sigma\vec{t}_2 \\ 0 & 0 & 0 \\ 0 & 0 & 0 \end{pmatrix}$$

$$= [\mu](\vec{n}\,0\,0)^T\left(\nabla\vec{U}\right)(\vec{n}\,0\,0) + [\mu](\vec{n}\,0\,0)^T\left(\nabla\vec{U}\right)(\vec{n}\,\vec{t}_1\,\vec{t}_2)$$

$$- [\mu](\vec{n}\,0\,0)^T\left(\nabla\vec{U}\right)^T(0\,\vec{t}_1\,\vec{t}_2) - \begin{pmatrix} 0 & \nabla_s\sigma\vec{t}_1 & \nabla_s\sigma\vec{t}_2 \\ 0 & 0 & 0 \\ 0 & 0 & 0 \end{pmatrix} \tag{3.54}$$

Based on Eq. (3.54), it can be finally deduced that the viscous stress tensor discontinuity caused by thermal capillary force and metal vapor/plasma friction force is:

$$\begin{aligned} \left(\left[\mu\nabla\vec{U}\right]\right) &= [\mu](\vec{n}\,\vec{t}_1\,\vec{t}_2)(\vec{n}\,\vec{0}\,\vec{0})^T\left(\nabla\vec{U}\right)(\vec{n}\,\vec{0}\,\vec{0})(\vec{n}\,\vec{t}_1\,\vec{t}_2)^T \\ &+ [\mu]\left(\vec{n}\,\vec{t}_1\,\vec{t}_2\right)(\vec{0}\,\vec{t}_1\,\vec{t}_2)^T\left(\nabla\vec{U}\right) \\ &- [\mu](\vec{n}\,\vec{t}_1\,\vec{t}_2)(\vec{n}\,\vec{0}\,\vec{0})^T\left(\nabla\vec{U}\right)^T(\vec{0}\,\vec{t}_1\,\vec{t}_2)(\vec{n}\,\vec{t}_1\,\vec{t}_2)^T \\ &- (\vec{n}\,\vec{t}_1\,\vec{t}_2)\begin{pmatrix} 0 & \nabla_s\sigma\vec{t}_1 & \nabla_s\sigma\vec{t}_2 \\ 0 & 0 & 0 \\ 0 & 0 & 0 \end{pmatrix}(\vec{n}\,\vec{t}_1\,\vec{t}_2)^T \end{aligned} \tag{3.55}$$

If the influence of metal vapor/plasma is not considered, the boundary conditions for viscous stress on the keyhole interface caused by the thermal capillary force is:

$$\begin{aligned} \left(\left[\mu\nabla\vec{U}\right]\right)_f &= [\mu](\vec{n}\,\vec{t}_1\,\vec{t}_2)(\vec{n}\,\vec{0}\,\vec{0})^T\left(\nabla\vec{U}\right)(\vec{n}\,\vec{0}\,\vec{0})(\vec{n}\,\vec{t}_1\,\vec{t}_2)^T \\ &+ [\mu](\vec{n}\,\vec{t}_1\,\vec{t}_2)(\vec{0}\,\vec{t}_1\,\vec{t}_2)^T\left(\nabla\vec{U}\right) \\ &- [\mu](\vec{n}\,\vec{t}_1\,\vec{t}_2)(\vec{n}\,\vec{0}\,\vec{0})^T\left(\nabla\vec{U}\right)^T(\vec{0}\,\vec{t}_1\,\vec{t}_2)(\vec{n}\,\vec{t}_1\,\vec{t}_2)^T \\ &+ (\vec{n}\,\vec{t}_1\,\vec{t}_2)\begin{pmatrix} 0 & \nabla_s\sigma\vec{t}_1 & \nabla_s\sigma\vec{t}_2 \\ 0 & 0 & 0 \\ 0 & 0 & 0 \end{pmatrix}(\vec{n}\,\vec{t}_1\,\vec{t}_2)^T \end{aligned} \tag{3.56}$$

where:

Subscript f free interface;
μ density of metal liquid in weld pool.

In the following, the boundary conditions for recoil pressure and surface tension on the free surface of keyhole are modeled. Firstly, according to Eqs. (3.35) and (3.45), we can obtain:

$$\sigma\kappa = p_L - p_G + 2[\mu]\vec{n} \cdot \nabla\vec{U} \cdot \vec{n} \tag{3.57}$$

In order to simplify, without considering the influence of metal vapor/plasma viscosity, the equation above can be further expressed as:

$$\sigma\kappa = p_L - p_G - 2\mu\vec{n} \cdot \nabla\vec{U} \cdot \vec{n} \tag{3.58}$$

In the process of laser welding, it is generally believed that the pressure of metal vapor/plasma inside the keyhole is approximately equal to the recoil pressure. Then the pressure boundary conditions caused by the existence of surface tension and recoil pressure on the free interface of the keyhole are:

$$p_f = p_r + \sigma\kappa + 2\mu\vec{n} \cdot \nabla\vec{U} \cdot \vec{n} \tag{3.59}$$

where:

Subscript f free interface;
μ density of metal liquid in weld pool;
p_r recoil pressure.

In this study, the recoil pressure model proposed by Semak et al. is adopted, and the p_r is expressed as:

$$p_r = 0.54AB_0(T)^{-1/2}\exp\left(-\frac{U}{\kappa T}\right) \tag{3.60}$$

where:

A, B_0 constant related to the material;
U latent heat of evaporation of each atom;
T surface temperature of the keyhole;
k Boltzmanns constant.

The following temperature boundary conditions exist on the free interface of the keyhole due to Fresnel absorption, thermal convection, radiation and evaporation

$$k\frac{\partial T}{\partial \vec{n}} = q - h(T - T_m) - \varepsilon_r \sigma \left(T^4 - T_\infty^4\right) - \rho V_{evp} T_v \tag{3.61}$$

where:

q density of laser energy absorbed due to the Fresnel effect, which is determined through calculation based on the Eq. (3.11);
T_v evaporating temperature;
V_{evp} receding speed of the keyhole interface caused by evaporation, and this study is determined based on the formula proposed by Ki et al.

On other boundaries of the calculated region, the following temperature boundary conditions exist

$$k\frac{\partial T}{\partial \vec{n}} = -h(T - T_m) - \varepsilon_r \sigma \left(T^4 - T_\infty^4\right) \tag{3.62}$$

where:

h convection coefficient;
ε_r black body radiation coefficient;
σ Boltzmanns constant.

In this study, the flow of the metal vapor/plasma inside the keyhole is not considered. Therefore, the velocity distribution exists on one side of the free interface (i.e. the area in weld pool and workpiece). When the Level Set method is used to describe the motion on the keyhole interface, the velocity distribution exists on both sides of the keyhole interface, otherwise the grids on the keyhole interface cannot satisfy the continuity equation and the accurate transient keyhole morphology cannot be obtained either. Therefore, in each calculation time step, it is necessary to construct a virtual velocity value with certain numerical accuracy for grids inside the keyhole by using the calculated weld pool velocity value to ensure that the grids on the keyhole interface satisfy the continuity equation. When Level Set technique is used to track the free interface, the velocity extrapolation is equivalent to solving the following Hamilton–Jacobi equation.

$$\nabla \overrightarrow{U_{ext}} \cdot \nabla \phi = 0 \tag{3.63}$$

where: $\overrightarrow{U_{ext}}$—the velocity vector obtained after extrapolation.

3.5 Numerical Method

3.5.1 *Fast Solution of Laser Beam Energy Absorbed by Keyhole Wall*

3.5.1.1 Solution for Energy Absorption Based on Line Heat Source

If only the radial component of heat transfer in the weld pool is considered, according to Fourier heat transfer law, the heat flow on the hole wall will be:

$$q_v(r,\phi) = -\lambda_{th}\nabla T \approx -\lambda_{th}\frac{\partial T}{\partial r} \tag{3.64}$$

Through the derivative of r on both sides of Eq. (3.6), we can obtain

$$\frac{\partial T}{\partial r} = \frac{P'(r,\phi)}{2\pi\lambda_{th}}P_e'\left[-K_0(P_e'r)\cos\varphi + K_0(P_e'r)\right]e^{-(P_e'r)\cos\varphi} \tag{3.65}$$

In which

$$K_0'(x) = -K_1(x) \tag{3.66}$$

In the equation: K_1 (x)—modified second class first order Bessel function.

If only the radial component of heat transfer is considered, the computational formula for the heat flow on keyhole wall can be expressed as:

$$q_v(r,\phi) = -\lambda_{th}\frac{\partial T}{\partial r} = \frac{P'(r,\phi)}{2\pi}P_e'\left[K_0(P_e'r)\cos\varphi + K_1(P_e'r)\right]e^{-(P_e'r)\cos\varphi} \tag{3.67}$$

By substituting the heat source intensity $P'(r,\varphi)$ for each point on the inner wall of the keyhole into the formula above, we can obtain:

$$q_v(r,\phi) = (T_v - T_a)\lambda_{th}P_e'\left(\cos\varphi + \frac{K_1(P_e'r)}{K_0(P_e'r)}\right) \tag{3.68}$$

where:

K_1 modified second class first order Bessel function;
q_v the (radial) heat flow on keyhole wall, W m^{-2} s^{-1};
T_v evaporating temperature of metal, K.

From Eq. (3.68), it can be seen that with the change of φ, that is, at the points which have different angles with the welding direction, heat flow is very different: the heat flow on the front wall of the keyhole ($\varphi = 0°$) is much larger than that on the rear wall ($\varphi = 180°$). Therefore, the center of the laser beam is not at the center of keyhole, but near the front wall of keyhole.

3.5.1.2 Solution for Energy Absorption Based on Fresnel Absorption

According to the established physical model for the energy interaction on laser-keyhole wall, it is believed that the energy interaction mechanism between laser and keyhole is mainly the primary and multiple reflection Fresnel absorption. At present, the scholars at home and abroad in the field of laser/electron beam welding mainly use the ray tracing method to calculate and determine the Fresnel absorption coefficients of the keyhole wall.

Although some scholars at home and abroad have studied the Fresnel multiple reflection absorption mechanism of the keyhole wall surface by using ray tracing method, in most of these studies, it was assumed that the keyhole is of cone or other simple shapes. In the actual laser welding process, the shape of the keyhole is very complex. When the keyhole closes to produce bubbles or spatters, the topological shape of the keyhole will change dramatically. Therefore, in order to simulate the complicated transient evolution process of the keyhole, the established ray tracing method must have high robustness. In addition, in order to ensure the accuracy of numerical solution, the laser needs to be subdivided into a sufficient number of small rays during ray tracing calculation, but the calculation speed of ray tracing calculation will be reduced without doubt. For this reason, certain algorithm skills must be used to guarantee the ray tracing computational efficiency in the computational process of pore morphology evolution when mobile laser welding is carried out. Based on the above two considerations, a fast and robust ray tracing method is proposed and implemented in this chapter. At the same time, several examples are designed to logically verify the rationality of the method.

(1) Laser beam subdivision

To calculate the energy density of the keyhole wall by using ray tracing method, the laser beam needs to be subdivided first, and then the Fresnel absorption between each beam and the keyhole wall is calculated to obtain the aggregate value.

Considering the feasibility of calculation, the following beam subdivision method is put forward: assuming that the laser beam is incident on the surface of the material along a certain direction in the coordinate system (z-axis direction by default), since most of the energy is concentrated within 1–2 times the radius of the beam, each difference grid within the beam radius can be divided into several equal parts, each of which can be regarded as a small beam, and the energy of each beam is the average area of the laser energy on the equally divided difference grid, i.e. the energy density I' of each beam after subdivision is:

$$I' = \iint_{\Omega} I_0(r, z)ds \Big/ \iint_{\Omega} ds \tag{3.69}$$

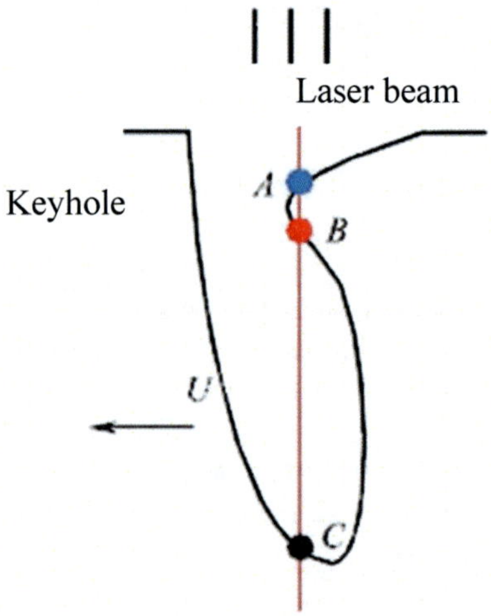

Fig. 3.1 Possible failure example of calculating the intersection point between light and complex keyhole interface by dichotomy method

In the equation: Ω—the difference grid area after equal division;
$I_0(r, z)$—laser energy density distribution function.

(2) Robust intersection calculation for intersection point of the beam and the keyhole

In the study, the intersection point of the small beam and the keyhole is generally determined by dichotomy. However, when the dichotomy is adopted to solve the intersection point of the straight line where the beam is located and the keyhole with complex shape, the special attention needs to be paid to the situation that multiple intersection points exist. An example of a failure to directly apply dichotomy to solve the intersection point of the light ray and the keyhole interface is given below. In Fig. 3.1, the bump swells on the rear wall due to the instability of the keyhole. At this time, if the intersection point of the vertical straight line where the small beam is located and the keyhole is directly calculated by using dichotomy method, three different intersection points of A, B and C may be obtained according to different situations. However, the fact is that only the first intersection point is the exact one required for the calculation of the transient evolution of the laser welding keyhole.

Considering the above condition of multiple intersection points, the good robustness must be possessed when the dichotomy is used for calculation. We shall first determine which intersection point is needed, and then determine its exact position. The required intersection point can be determined along the incidence direction of the ray, and the required intersection point can be gradually approached by the focal point or reflection point of the laser beam with a smaller step size δ. In the transient evolution process of laser welding keyhole, through many attempts, it is found that $\delta = 0.5\Delta x$ (Δx is the step size of difference grid) can better meet the requirements. In the actual numerical implementation, the following algorithm is adopted in this study, as shown in Fig. 3.2. We might as well assume that the starting point is $\overrightarrow{p_0}$, and the normal vector of the small beam is $\overrightarrow{n_0}$, the calculation process can be described

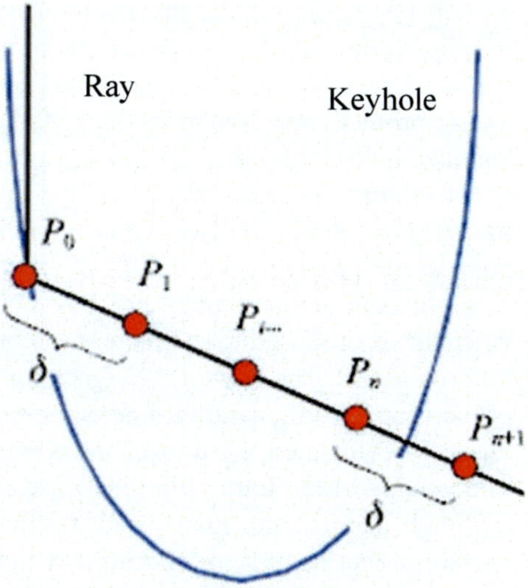

Fig. 3.2 Robust dichotomy solution algorithm

as follows: ① $\overrightarrow{p_0}$ is shifted by the distance δ along the normal direction, and this position is marked as $\overrightarrow{p_1}$. In addition, the distance sign of this position is judged according to the Level Set function; ② Starting from $\overrightarrow{p_1}$, it is continuously shifted by the distance δ until the sign of the distance field at the two positions changes; ③ It might be as well to mark the two positions when the sign of distance field just changes as $\overrightarrow{p_n}$ and $\overrightarrow{p_{n+1}}$ respectively; since the intersection point must be between these two points, then the exact position of intersection point can be obtained by using the dichotomy iteration method.

Since the ray tracing is all performed on the difference grid, considering that it is difficult for the difference grid to accurately show curves and surfaces, when the distance field of the given position is determined in the process of intersection calculation, the local distance field will be reconstructed by using the trilinear interpolation method to improve the numerical accuracy of intersection calculation.

In addition, in the transient keyhole evolution simulation, the curvature value of a few positions on the keyhole wall may be very large, that is to say, the normal vector of these positions may not be defined precisely. For these positions, although Level Set method can give a normal vector estimation value, the use of this value to calculate the energy distribution will lead to the large difference in the energy density of some adjacent positions on the keyhole wall, which is not conducive to simulating the transient keyhole evolution process. Therefore, in order to improve the robustness of the algorithm, the multiple reflection absorption at these positions is not calculated, and only the primary Fresnel absorption is calculated.

(3) Fast calculation of beam transmission path length

Calculating the beam transmission path between the current intersection point and the next intersection point of the laser beam on the keyhole wall is a prerequisite for accurately calculating the inverse bremsstrahlung absorption of the laser beam by the plasma during CO_2 laser welding, or the energy absorption of the laser beam by the metal vapor during Nd: YAG and fiber laser welding. Obviously, when the ray tracing problem is solved by using the finite difference method, the calculation of the length of intersecting line of the beam and any finite difference grid becomes the key to calculating the beam transmission path value after all the finite difference grids traversed by the beam between the current intersection point and the next intersection point are judged.

In this study, a redundant point elimination algorithm with high computational efficiency is proposed to calculate the length of intersecting line of the beam and the difference grid. The flow of the algorithm is as follows: ① The intersection point set of the beam and the straight line where the 6 sides of the finite difference grid are located is calculated. Obviously, there are 6 intersection points at most in this set. ② The volume of the finite difference grid is slightly expanded outward, such as one millionth, and then the points inside the intersection point set obtained in the first step are judged sequentially to determine whether they are located in the enlarged grid. The two points located in the grid are the intersection points of the beam and the difference grid; ③ The distance between the two intersection points is calculated, i.e. the length of the intersecting line. Although the computation method above seems complicated, the algorithm avoids too much conditional judgment when the intersection points are directly calculated, which is conducive to computer calculation with high efficiency.

We might as well assume that $I_r(r, z)$ is the power density of the laser beam at the current intersection point, if the absorption coefficient of the metal vapor/plasma to the laser beam is expressed by k_{pl}, then the energy density $I_m(r, z)$ of this beam when it is about to reach the next intersection point can be calculated by using the following formula:

$$I_m(r, z) = I_r(r, z)\exp(-\int_0^{l_m} k_{pl}dl) \tag{3.70}$$

In the equation: l_m—the beam transmission path calculated by the above method.

In this study, the influence of diffuse reflection is ignored, and it is assumed that the reflection on the keyhole wall is all specular reflection. According to the principle of specular reflection, the reflection direction of the beam can be calculated by the following formula:

$$\overrightarrow{I_r} = \overrightarrow{I_o} + 2(-\overrightarrow{I_o} \cdot \vec{n})\vec{n} \tag{3.71}$$

In the formula: $\overrightarrow{I_o}$, $\overrightarrow{I_r}$—directions of incident beam and reflected beam; $\vec{n}$—normal vector of keyhole wall.

(4) Accelerated calculation of ray tracing method when the workpiece or laser moves

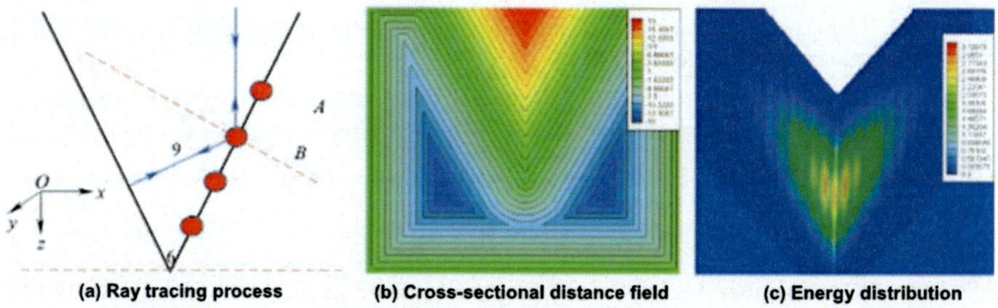

Fig. 3.3 Calculation example of V groove

In order to improve the calculation efficiency, when the workpiece or the laser moves, the area directly below the laser beam is always specified (usually a small area within twice the radius of the spot) for ray tracing calculation. For example, if the number of grids in the whole calculation area is [0–100] × [0–100] × [0–100], the area where ray tracing is located may be only concentrated in the small cuboid of [50–60] × [50–60] × 100. In the actual programming implementation, generally the small cuboid calculation area needs to be extracted separately for ray tracing calculation, which can greatly improve the calculation efficiency and realize the real-time ray tracing calculation inside keyholes with more complex shapes.

(5) Numerical example

In order to verify the effectiveness of this method, two numerical examples of V-shaped groove and tapered keyhole are designed for testing. Before the energy density is calculated, firstly 4 geometries are modeled by using the UG CAD software, and then they are divided into the finite difference grids, which are converted into the signed distance fields, and then the ray tracing calculation is carried out. In the numerical testing, it is assumed that the laser shows Gaussian distribution.

As shown in Fig. 3.3a, the V groove angle is set to 60°. When the laser hits the right surface of the groove vertically downward, according to Eq. (3.70), it can be calculated that the reflection direction must be perpendicular to the left surface of the V groove. Therefore, according to the specular reflection principle, the light will return along the original way. To verify the correctness of the method, two points A and B that do not coincide, as shown in Fig. 3.3a, are taken on the right side of the groove cross section for ray tracing calculation. The calculated values of incidence and reflection directions at each point are compared with the theoretical values.

Table 3.1 compares the calculated data with the theoretical data. It can be seen that the absolute error between the incidence direction and the reflection direction of the points A and B and the theoretical value is very small. The first incidence direction is opposite to the third reflection direction, and the second reflection direction is also opposite to the third, which proves that the proposed method indeed can be used to perform reasonable ray tracing calculation on the discrete difference grid. Figure 3.3b shows the isoline nephogram of Level Set distance field of the cross-section. It can be seen that the normal vector of the groove bottom points to (0, 0, −1), indicating

Table 3.1 Comparison between the calculated values of incidence/reflection directions and the theoretical values of 60° V groove

Location/theoretical value	Incidence direction	First reflection direction	Second reflection direction	Third reflection direction
A	(0, 0, 1)	(−0.866, 0, 0.4999)	(0.866, 0, −0.5)	(−2e−8, 0, −0.9999)
B	(0, 0, 1)	(−0.866, 0, 0.4999)	(0.865, 0, −0.501)	(6e−8, 0, −0.9999)
Theoretical value	(0, 0, 1)	(−0.866, 0, 0.5)	(0.866, 0, −0.5)	(0, 0, −1)

that there is only one reflection at the groove bottom. Figure 3.3c shows the 3D nephogram of energy distribution on the V groove surface by calculation. It can be seen from the figure, since there is only one reflection at the groove bottom, the maximum energy appears on both sides near the V groove bottom. Due to the small number of reflections, the energy distribution is still basically a Gaussian distribution. The above facts further verify the rationality of this method.

Figure 3.4a is the energy distribution diagram of the keyhole cross-section obtained by irradiation of a laser on the wall of the conical keyhole. Figure 3.4b shows the overall energy distribution on the upper surface. Figure 3.4c shows a cross-section energy distribution diagram of the keyhole by assuming that the keyhole is filled with metal vapor/plasma and the absorption coefficient of the metal vapor/plasma is constant. It can be seen from Fig. 3.4a that due to multiple reflections, the energy density at the bottom of the keyhole is relatively high, which can also be verified from Fig. 3.4b. Since the metal vapor/plasma absorption coefficient is assumed to be constant, considering the effect of multiple reflections, the energy absorbed by the metal vapor/plasma is higher as it gets closer to the bottom of the keyhole. This is well verified in Fig. 3.4c.

According to the results of calculation examples, it is shown that the ray tracing calculation method proposed in this Chapter is reasonable. It can be used for Fresnel

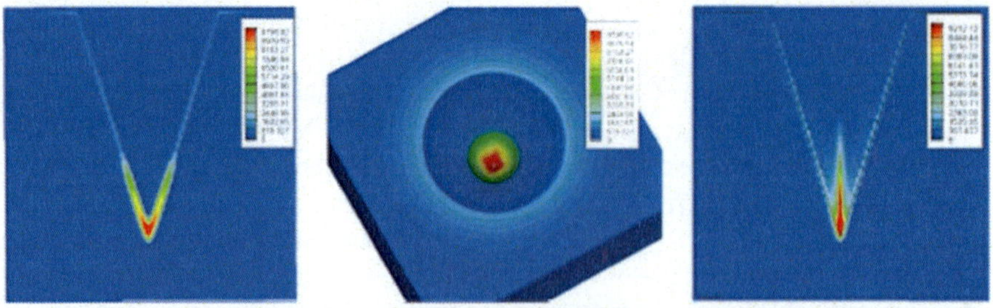

(a) Cross-section energy density distribution

(b) Overall energy density distribution

(c) Cross-section energy density distribution

Fig. 3.4 Calculation example of conical keyhole

absorption calculation of the initial and multiple laser reflections on the wall of the keyhole. When the metal vapor/plasma absorption coefficient and its distribution are given, the laser energy density absorbed by metal vapor/plasma can also be calculated reasonably.

It should be pointed out that the energy density distribution on the keyhole wall in the actual laser welding process is closely related to the shape and size of the keyhole, and therefore the actual energy density distribution on the keyhole wall may be significantly different from the results of the above two calculation examples. The above two calculation examples only discuss whether the ray tracing method proposed can be combined with the method described in the Level Set interface for calculation purely from the algorithmic perspective, and whether the calculation result is reasonable, which lays a foundation for further evolution calculation of the laser welding transient keyhole and moving weld pool.

3.5.2 *Keyhole Free Surface Tracking Method*

In the laser deep welding process, the morphology of the transient keyhole may be extremely complex, and its topological shape may also change dramatically due to the formation of bubbles and spattering. Meanwhile, using the ray tracing method to calculate the energy density on the keyhole wall and to accurately calculate the influence of surface tension on the behavior of the keyhole requires high precision requirements for the calculation of the normal vector and curvature of the transient keyhole wall. In this section, the realization process of describing the tracking keyhole interface with the Level Set method is introduced in detail. Additionally, to overcome the numerical dissipation of Level Set, the high-precision Level Set/Particle Level Set interface tracking technology has important applications. This paper expounds the technology and discusses the effectiveness of the high-precision numerical interface tracking technique by using numerical examples.

3.5.2.1 High-Precision Level Set Interface Tracking and Solving Technology

The Level Set Eq. (3.14) is a classical Hamilton–Jacobi equation. To reduce the dissipation of the Level Set method itself, usually, a high-precision discrete scheme is needed to perform a discrete solution for it. Generally, high precision means that the truncation error of the difference scheme is at least second order to the time and spatial steps. In recent years, high-precision schemes based on the idea of Total Variation Diminishing (TVD), such as Essentially Non-Oscillatory (ENO) and Weighted Essentially Non-Oscillatory (WENO), have been widely used in the Level Set method. The idea of the TVD scheme is self-adaptively adjusting the numerical dissipation of the calculation scheme. When the discrete approximate solution of an unknown function has a large gradient or amplitude trend, using the TVD scheme can appropriately enhance the numerical dissipation of approximate solution; conversely, when the discrete approximate solution is relatively flat, the TVD scheme can reduce the numerical dissipation and obtain an accurate discontinuous solution.

At present, the Runge–Kutta WENO scheme, a TVD scheme that has third-order accuracy in time and fifth-order accuracy in space, is an effective high-precision calculation scheme to solve the Level Set equation. This is also the main scheme for the discrete convection equations used in this section. The following is a detailed introduction of the discrete process of the Level Set formula using the high-precision TVD Runge–Kutta WENO scheme.

A fifth-order WENO scheme discrete $\frac{\partial \phi}{\partial x}$ is taken as an example to illustrate the discrete process of the convective term in the Level Set equation. For the node of (*i*, *j*, *k*), the direction of the velocity value at the current node position should be determined first in the discrete process. With the idea of the upwind scheme, the left-hand derivative ϕ_x^- and the right-hand derivative ϕ_x^+ of space are respectively taken according to the different directions. For the node of (*i*, *j*, *k*), its left-hand derivative ϕ_x^- is made of the node template interpolation composed of the following 6 nodes $\left\{\phi_{i-3,j,k}, \phi_{i-2,j,k}, \phi_{i-1,j,k}, \phi_{i,j,k}, \phi_{i+1,j,k}, \phi_{i+2,j,k}\right\}$, while the right-hand derivative ϕ_x^- is calculated by the node template $\left\{\phi_{i-2,j,k}, \phi_{i-1,j,k}, \phi_{i,j,k}, \phi_{i+1,j,k}, \phi_{i+2,j,k}, \phi_{i+3,j,k}\right\}$. Assuming that the computing grid is a uniform finite difference grid, when calculating the left-hand derivative ϕ_x^-, the following are set respectively

$$a_1 = \frac{\phi_{i-2,j,k} - \phi_{i-3,j,k}}{\Delta x} \tag{3.72}$$

$$a_2 = \frac{\phi_{i-1,j,k} - \phi_{i-2,j,k}}{\Delta x} \tag{3.73}$$

$$a_3 = \frac{\phi_{i,j,k} - \phi_{i-1,j,k}}{\Delta x} \tag{3.74}$$

$$a_4 = \frac{\phi_{i+1,j,k} - \phi_{i,j,k}}{\Delta x} \tag{3.75}$$

$$a_5 = \frac{\phi_{i+2,j,k} - \phi_{i+1,j,k}}{\Delta x} \tag{3.76}$$

When calculating the right-hand derivative ϕ_x^+, the following are set

$$a_1 = \frac{\phi_{i+3,j,k} - \phi_{i+2,j,k}}{\Delta x} \tag{3.77}$$

$$a_2 = \frac{\phi_{i+2,j,k} - \phi_{i+1,j,k}}{\Delta x} \tag{3.78}$$

$$a_3 = \frac{\phi_{i+1,j,k} - \phi_{i,j,k}}{\Delta x} \tag{3.79}$$

$$a_4 = \frac{\phi_{i,j,k} - \phi_{i-1,j,k}}{\Delta x} \tag{3.80}$$

$$a_5 = \frac{\phi_{i-1,j,k} - \phi_{i-2,j,k}}{\Delta x} \tag{3.81}$$

Three variables are then defined to self-adaptively adjust the numerical dissipation of the scheme.

$$S_1 = \frac{13}{12}(a_1 - 2a_2 + a_3)^2 + \frac{1}{4}(a_1 - 4a_2 + 3a_3)^2 \tag{3.82}$$

$$S_2 = \frac{13}{12}(a_2 - 2a_3 + a_4)^2 + \frac{1}{4}(a_2 - a_4)^2 \tag{3.83}$$

$$S_3 = \frac{13}{12}(a_3 - 2a_4 + a_5)^2 + \frac{1}{4}(3a_4 - 4a_4 + a_5)^2 \tag{3.84}$$

Let β be a relatively small constant; in this study, its value is $10^{-3}\Delta x$. Finally, three weight factors are defined respectively as

$$\lambda_1 = \frac{1}{10}\frac{1}{(S_1 + \beta)^2} \tag{3.85}$$

$$\lambda_2 = \frac{6}{10}\frac{1}{(S_2 + \beta)^2} \tag{3.86}$$

$$\lambda_3 = \frac{3}{10}\frac{1}{(S_3 + \beta)^2} \tag{3.87}$$

Let the weight variable be

$$w_i = \frac{\lambda_i}{\lambda_1 + \lambda_2 + \lambda_3} \tag{3.88}$$

Then according to different upwind directions, the convective item $\frac{\partial \phi}{\partial x}$ can be calculated using the following equation

$$\frac{\partial \phi}{\partial x} = w_1\left(\frac{1}{3}a_1 - \frac{7}{6}a_2 + \frac{11}{6}a_3\right) + w_1\left(-\frac{1}{6}a_2 + \frac{5}{6}a_3 + \frac{1}{3}a_4\right) + w_3\left(\frac{1}{3}a_3 + \frac{5}{6}a_4 - \frac{1}{6}a_5\right) \tag{3.89}$$

Similarly, $\frac{\partial \phi}{\partial y}$ and $\frac{\partial \phi}{\partial z}$ can also be calculated according to the above mentioned high-precision WENO scheme.

After discretizing the space item, it is also required to discretize the time item $\frac{\partial \phi}{\partial t}$ of the Level Set equation. It can be conducted by calculating $L(\phi) = -\overrightarrow{U} \cdot \nabla \phi$. If the Level Set function at the moment of t is denoted as $\phi^{(n))}$, and as $\phi^{(n+1)}$ at the moment of $t + \Delta t$, then the third-order TVD Runge–Kutta scheme can be used. The process of discretizing the time item of the Level Set can be described as

$$\varphi^{(1)} = \varphi^{(0)} + \Delta t L(\varphi^{(0)}) \tag{3.90}$$

$$\varphi^{(2)} = \frac{3}{4}\varphi^{(0)} + \frac{1}{4}\varphi^{(1)} + \frac{1}{4}\Delta t L(\varphi^{(1)}) \tag{3.91}$$

$$\varphi^{(3)} = \frac{1}{3}\varphi^{(0)} + \frac{2}{3}\varphi^{(2)} + \frac{2}{3}\Delta t L(\varphi^{(2)}) \tag{3.92}$$

The study shows that even if the TVD Runge–Kutta WENO high order scheme is used to discretize the Level Set equation, under some special conditions, after several time steps, the Level Set function can no longer be maintained as a signed distance field, and therefore errors occur when tracking the numerical interface. To overcome this problem, after solving the Level Set equation, it is generally necessary to re-initialize the value of the Level Set function to ensure that the function is always a signed distance field. There are generally two ways of reinitialization, one of which is by solving the following time-dependent partial differential equation

$$\phi_\tau = sign(\phi_0)(1 - |\nabla \phi|) \tag{3.93}$$

and shall meet the initial conditions: $\phi(x, 0) = \phi_0$. Or reinitialization can be achieved by solving the time-independent Eikonal equation

$$1 - |\nabla \phi| = 0 \tag{3.94}$$

In the study, it is found that even if high order schemes, the third TVD Runge–Kutta and the fifth-order WENO are combined to solve Eq. (3.93), when the keyhole interface is very complex or there is a serious topological deformation, some small errors tend to happen in solving for the signed distance field, while with Eq. (3.94), a better interface tracking effect can be obtained. Therefore, Eq. (3.94) is used for reinitialization.

The classical method for solving Eq. (3.94), is the fast marching method which features NlogN complexity. In this study, the more efficient fast sweeping method is adopted to solve Eq. (3.94). The complexity of this method is only O (N), which is one of the fastest methods to solve the Eikonal equation, and this method is very suitable for parallel solution. In the process of using the fast sweeping method to solve Eikonal Eq. (3.94), in this study a first-order implicit scheme is firstly used to solve the equation discretely. After reinitialization of the Level Set, then a third-order implicit scheme is used to solve Eq. (3.94) for higher precision numerical calculation results.

In the process of solving implicit equations, Gauss–Seidel iterative method is used in this section for iterative solution. Meanwhile, in order to speed up the convergence process, different grid sequences are adopted for iterative solution. It is proved in the study that for the Eikonal equation, with the first-order scheme discrete Eq. (3.94), the convergence solution can be generally obtained after 8 steps through Gauss–Seidel iteration in different grid sequences. For third-order schemes, it is found in this study that generally only a few steps are needed to obtain the convergent solution satisfying the accuracy. The Gauss–Seidel iterative method needs a small calculation amount for each iteration process, and it is very suitable for parallel calculation, so the reinitialization efficiency is high when using the fast sweeping method.

3.5.2.2 Particle Level Set Interface Tracking and Solving Technology

Numerical dissipation is the biggest inherent defect of the Level Set method. In recent years, the academia has proposed the CLSVOF method in which the Level Set is combined with the VOF, and the Particle Level Set Method in which the front tracking method is combined with the Level Set to correct the defect of the Level Set Method. Both of the above methods have been proved to be effective in suppressing the mass non-conservation problem of Level Set. However, the CLSVOF method is less effective than the Particle Level Set method in the interface tracking effect. In this study, the Particle Level Set method is adopted to overcome the quality loss problem of the Level Set method.

The Particle Level Set method artificially arranges some massless particles with different radii on both sides of the interface described by the Level Set, and makes the particles move passively with the flow field, where the particle motion equation is determined by Eq. (3.80). In the process of solving the particle motion equation, the third-order TVD Runge–Kutta scheme is also adopted to integrate Eq. (3.80). Since the particles can always reflect the characteristic information of the flow field, the numerical dissipation error of the Level Set can be corrected by the position information of the particles. In this study, particles are only arranged in two time steps on both sides of the interface, and the initial radius of these particles x_p is set as

$$r_p = \begin{cases} r_{\max} & if\ s_p\phi(x_p) > r_{\max} \\ s_p\phi(x_p) & if\ r_{\min} \le s_p\phi(x_p) \le r_{\max} \\ r_{\min} & if\ s_p\phi(x_p) < r_{\min} \end{cases} \tag{3.95}$$

where

$r_{\min} = 0.1\Delta x,\ r_{\max} = 0.5\Delta x$	minimum and maximum radius of the particle;
$\phi(x_p)$	Level Set function value at the particle position;
s_p	function value of the corresponding sign of $\phi(x_p)$.

After updating or re-initializing the position of each step of Level Set function, the value of the Level Set function needs to be corrected with the information of particles. The basis for the correction is that if the particle that started on the positive side of the interface goes to the negative side of the interface, then the Level Set function shall be corrected; conversely, if the particle with a negative distance value goes to the side with a positive distance value, a correction shall be conducted.

The most difficult part of adopting the Particle Level Set method is the increase, deletion and re-seeding of particles. In the process of numerical calculation, due to the influence of numerical error at the interface, the previously seeded particles need to be deleted and re-seeded at a specific time. Obviously, there is no prescribed rule about when to delete a particle and re-seed one. This is only related to specific problems. In this study, it is stipulated that all particles shall be deleted and re-seeded every certain computational time step. Besides, when the area of the tracked interface changes greatly, it may be needed to add some particles to correct the dissipation of the Level Set values. When the Level Set is used to track the interface, the surface area of the interface can be accurately determined according to the following formula

$$S = \int \delta(\varphi)|\nabla\varphi|dV \tag{3.96}$$

where $\delta(\varphi)$—Dirac function, which only has a value on the interface, and the value is 1; and the value is 0 at any other position.

In this study, it is stipulated that when the area of the current interface is compared with the area in the last particle seeding, once the ratio is larger than a certain critical value, a certain number of particles will be appropriately added on both sides of the interface on the original basis.

3.5.2.3 Numerical Examples

To verify the rationality of the Level Set and Particle Level Set interface tracking technologies developed by the institute, two classical numerical examples, namely, the shear flow field and notched Zalesak disc, are used to test the interface tracking technologies developed by the institute respectively.

The example of shear flow field is mainly used to test the interface tracking method, to check the mass conservation of the interface when the interface has shear deformation in the flow field. The disc radius may as well be set to be 15, and the center of the circle is at (5,75). The flow function of the flow field is

$$\psi = \frac{1}{\pi}\sin^2(\pi x)\sin^2(\pi y) \tag{3.97}$$

Figure 3.5 shows the disc and shear flow field. The black part indicates the streamline of the flow function formula (3.97), while the blue part is the disc. 100 × 100 grid is adopted, and the time step $\Delta t = 0.0004$ and the space step $\Delta x = \Delta y = 0.01$

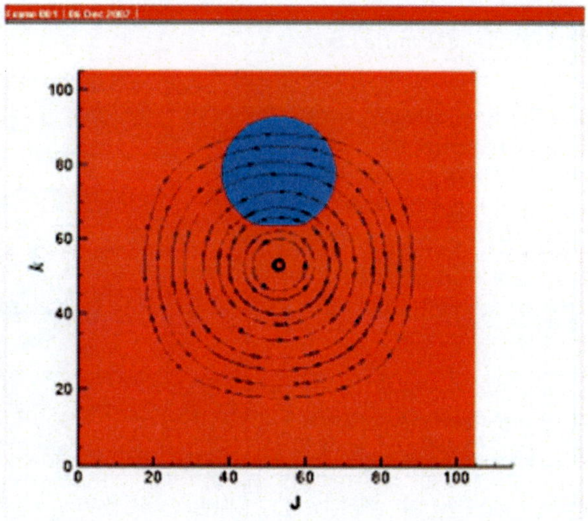

Fig. 3.5 Disc and shear flow field

are taken. 9950 steps are figured out using the Level Set method and Particle Level Set method respectively. Figure 3.6 shows the shape of the interface calculated by the Level Set method. Figure 3.7 and Fig. 3.8 respectively show the interface shape and the particle shape calculated by the Particle Level Set method. As shown in Fig. 3.6, only a very small piece of the original blue disk is left, and most of the disk area is dissipated. The actual shape after the 9950 steps should be the same as the interface between the blue and red particles shown in Fig. 3.8, because the particles can always truly reflect the real information of flow field and motion interface. Figure 3.6 indicates that there is a relatively serious non-conservation phenomenon when the

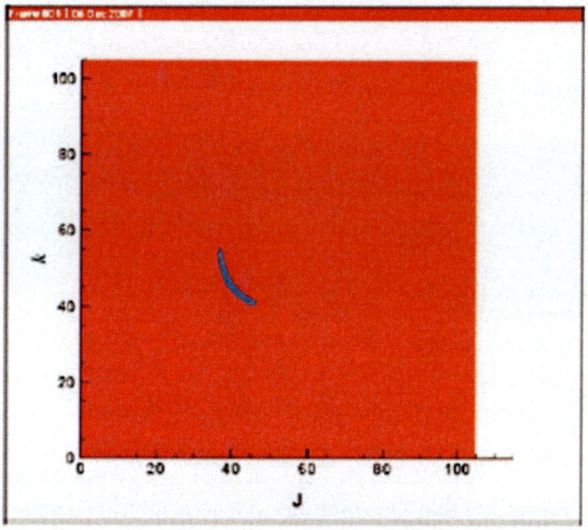

Fig. 3.6 After 9950 steps are calculated (Level Set)

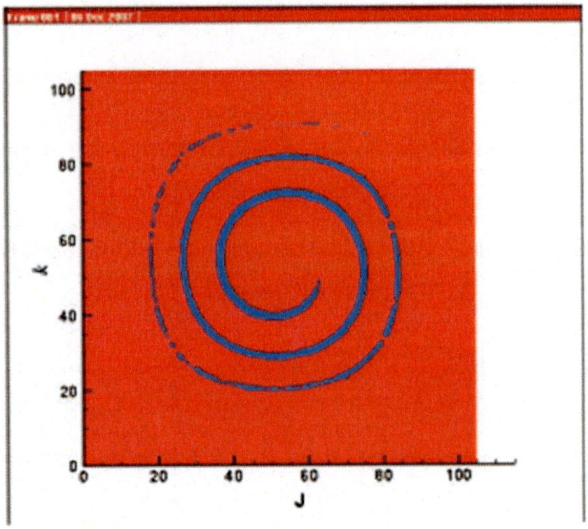

Fig. 3.7 After 9950 steps are calculated (Particle Level Set)

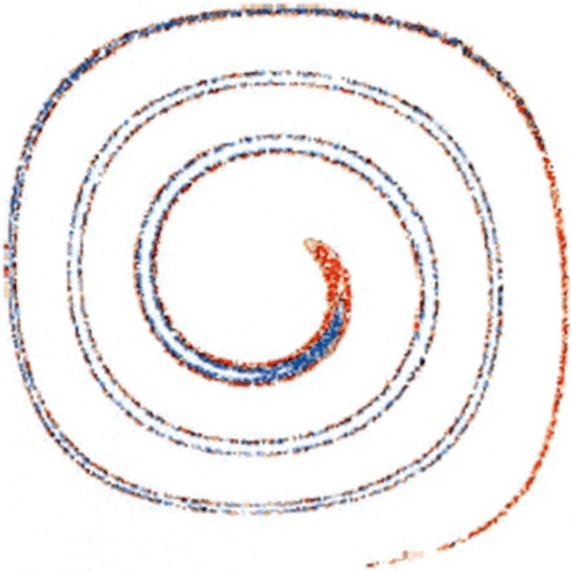

Fig. 3.8 Particle shapes corresponding to Fig. 3.7

interface is tracked by the Level Set. Figure 3.7 shows that the Particle Level Set method can be used to better simulate the severe tear deformation of the moving interface.

In order to further test the developed interface tracking technology, the numerical test is carried out using the classic calculation example of Zalesak sphere rigid motion. The so-called Zalesak disc is actually a three-dimensional sphere with a square notch. Assuming that the sphere is in a computational space [0, 100] × [0, 100] × [0, 100]

and the radius of the sphere is defined to be 30 unit lengths, the width and depth of the notch are respectively 5 unit lengths and 55 unit lengths. Assuming that there is a three-dimensional flow field in the calculation space as follows:

$$\begin{aligned} u(x, y, z) &= \pi(50 - y)/314 \\ v(x, y, z) &= \pi(x - 50)/314 \\ w(x, y, z) &= 0 \end{aligned} \tag{3.98}$$

Due to the effect of convection speed, the sphere will rotate in the flow field. For the sphere is rigid, the shape of it ideally will not change, but as a matter of fact, the sharp corner of the notched sphere may be severely rounded due to the influence of numerical dissipation. In the numerical test, with the calculation grid equal to 1 unit length, the interface of the Zalesak sphere is tracked using the Level Set method and the Particle Level Set method respectively. Figure 3.9 and Fig. 3.10 show the numerical results calculated by the Level Set method and the Particle Level Set method respectively. In the two figures, the left-to-right graphs sequentially represent the initial interface of Zalesak sphere, and the interface after one rotation, after two rotations, and after three rotations. It can be seen from the two figures that both the Level Set and Particle Level Set method can get good results in tracking the Zalesak interface, however it is obvious that the result of the Level Set method leads to large

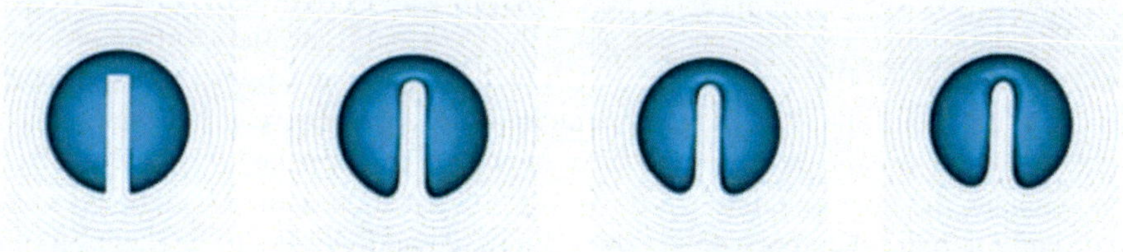

(a) Initial interface (b) Interface after one rotation (c) Interface after two rotations (d) Interface after three rotations

Fig. 3.9 Computation result of Level Set method for Zalesak calculation example

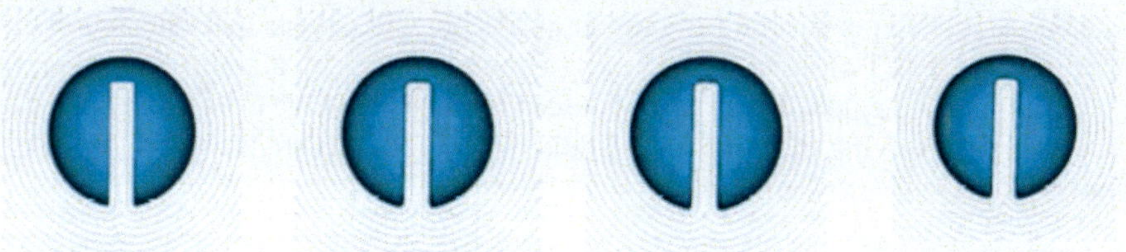

(a) Initial interface (b) Interface after one rotation (c) Interface after two rotations (d) Interface after three rotations

Fig. 3.10 Computation result of Particle Level Set method for Zalesak calculation example numerical dissipation, while the result of the Particle Level Set method is almost the same as the original shape.

The results of the above two calculation examples show that both Level Set and Particle Level Set methods can achieve relatively good interface tracking effect, but the Level Set method has larger numerical dissipation, while the Particle Level Set method can better overcome the numerical dissipation indicated in the Level Set method and achieve a more reasonable numerical interface tracking effect. The realization of the above-mentioned high-precision Level Set and Particle Level Set solutions provides an essential technology foundation for tracking the evolution of transient pores in the laser welding process.

3.5.3 Solutions for Coupling Free Surface Flow and Heat Transfer

The temperature field and the flow field in the motion weld pool affect each other. The factors linking the flow field with the temperature field are the recoil pressure at the free interface, the surface tension and the thermal capillary force, and the buoyancy on the boundary and in the weld pool. It is difficult to solve the nonlinear, strongly coupled free-interface flow and heat transfer model, such as deep penetration laser welding.

The traditional solutions for coupling the flow field and the temperature field mainly include SIMPLE method or SOLA method, and the explicit discrete method is often adopted for solving the temperature field. It is generally believed that it's not efficient to use the pressure–velocity coupling correction methods such as SOLA to calculate the flow field in the laser weld pool; it will be more efficient to use the currently popular incompressible flow solving method—Projection. Therefore, the Projection method is adopted for solving the flow field in this paper. In addition, in the laser welding process, the physical dimensions of the pore and the weld pool are very small. If an explicit method is used to discretely solve the temperature field, the absolute time step will be 10^{-12}–10^{-14} s. In this case, at least 100 billion calculation steps are required to calculate the actual laser welding process within 1 s. If one time step is calculated per second (for current quad-core CPU, the actual calculation time per step exceeds 1 s), it will take years to complete the calculation. Therefore, the semi-implicit scheme with a relatively large time step available to be taken is adopted for discretely solving the temperature field equation in this study.

3.5.3.1 Projection Method for Solving the Flow Field of the Motion Weld Pool

As an efficient step-by-step method for solving the incompressible Navier–Stokes equation, an outstanding advantage of the Projection method is that when the Navier–Stokes equation is being solved, the velocity and the pressure are decoupled during the solution. The Projection method is based on the well-known Helmholtz-Hodege decomposition principle: that is to say, any vector field can always be decomposed into a Solenoidal part and an Irrotational part. In general, the incompressible Navier–Stokes equation being solved by the Projection method can be divided into three steps: the first step is to calculate an intermediate velocity field that does not meet the incompressible condition; the second step is to use the intermediate velocity field to calculate the pressure field of the next moment; at last, the intermediate velocity field is projected into a velocity space with zero divergence by using the pressure, to obtain an accurate velocity field. The procedure for solving the coupled Eqs. (3.1) and (3.2) using the Projection method based on the staggered difference grid technique will be described in detail below. Figure 3.11 shows a staggered grid diagram used in this study. In this grid, the velocity is located on the 6 planes of a finite difference grid, while variables such as pressure, concentration and temperature are stored in the center of the grid.

By using the time term of the discrete Eq. (3.2) of first-order forward difference scheme, irrespective of the influence of the pressure term, the semi-discrete form of Eq. (3.2) can be expressed as:

$$\rho^n \frac{\vec{U}^* - \vec{U}^n}{\Delta t} + \widetilde{C}(\vec{U}^n) = \widetilde{D}(\mu^n \vec{U}^n) + \widetilde{K}(\vec{U}^n) + \widetilde{F}^n \tag{3.99}$$

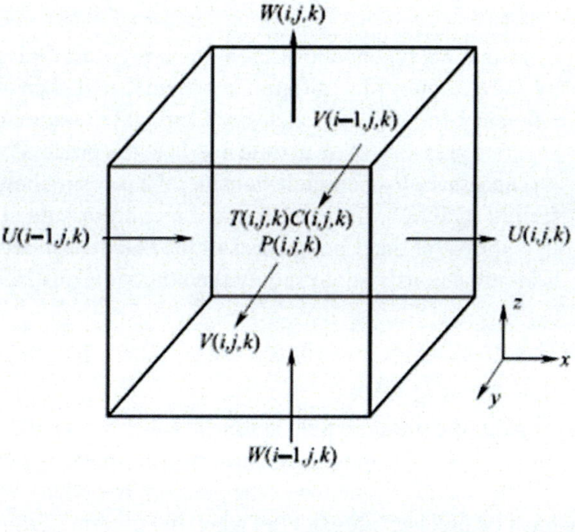

Fig. 3.11 Staggered grid diagram

where:

$\overrightarrow{U}^*$	Intermediate velocity field;
$\widetilde{C}(\cdot)$, $\widetilde{D}(\cdot)$, $\widetilde{K}(\cdot)$ and $\widetilde{F}^n$	Mathematical operators for convection terms, viscous terms, Darcy terms, and buoyancy terms in discrete momentum conservation equation.

In order to improve the accuracy of the numerical calculation, the fifth-order WENO scheme is used to discretize the operator in this study $\widetilde{C}\left(\overrightarrow{U}^n\right)$, and in order to facilitate the loading of the viscous stress boundary conditions caused by thermal capillary force (see Eq. (3.56)), the second-order central difference format is used to discretize the viscosity term $\widetilde{D}\left(\overrightarrow{U}^n\right)$. At the same time, in order to speed up the calculation, both Darcy and buoyancy terms are discretized using the explicit scheme. In particular, the loading of viscous stress boundary conditions on a staggered difference grid is very complicated and will be discussed in detail in the next section. Besides, there will be certain requirements on the time step if both Darcy and buoyancy terms are explicitly processed, which will be discussed in more details in the subsequent chapters.

Considering the effect of the pressure term, the equation is written to be a complete discrete form, so

$$\rho^n \frac{\overrightarrow{U}^{n+1} - \overrightarrow{U}^*}{\Delta t} = -\nabla p^{n+1} \tag{3.100}$$

Considering that the velocity field at time n + 1 must satisfy the continuity Eq. (3.1), the divergence is taken from both sides of the Eq. (3.100), so:

$$\nabla^2 p^{n+1} = \frac{\rho^n}{\Delta t} \nabla \cdot \overrightarrow{U}^* \tag{3.101}$$

In this study, the above-mentioned pressure Poisson's Eq. (3.101) is discretized by the second-order difference scheme, and then iteratively solved by the high-efficiency Incomplete Cholesky Conjugate Gradient (ICCG) method to obtain the accurate pressure value at time n + 1. In the process of solving the pressure equation, it is necessary to accurately consider the pressure boundary conditions caused by the factors such as surface tension and recoil pressure at the free interface of the pores. In addition, at other boundaries of the calculation area, the boundary of the pressure is simultaneously set to satisfy the homogeneous Neumann condition.

$$\frac{\partial p}{\partial n} = 0 \tag{3.102}$$

After the pressure field at time n + 1 is calculated, the pressure field is substituted into Eq. (3.100), and the velocity field at time n + 1 can be calculated. It is important to note that the influence of the pressure boundary condition Eqs. (3.58) and (3.102) must be precisely considered when the pressure gradient in Eq. (3.100) is calculated during the back substitution process, in order to take into account of the effect of physical conditions, such as recoil pressure and surface tension, on the motion of the weld pool.

3.5.3.2 Solution for Temperature Field and Latent Heat Treatment Method in the Calculation Area

Based on the above discussion, the semi-implicit method is adopted for solving the temperature field equation (i.e., Eq. (3.3)) in this study. The semi-implicit scheme means that the explicit scheme is adopted for discretization of all the terms except the diffusion term. The purpose is to theoretically ensure that the time step of the temperature field is not limited and that the system of linear equations obtained after the discretization is symmetric positive definite. Like the discrete flow field equation (i.e., Eq. (3.2)), the convection term in the discrete temperature field equation of the fifth-order WENO scheme (i.e., Eq. (3.3)) and the discrete diffusion term of the implicit central difference scheme are used in this paper, so the semi-discrete form of Eq. (3.3) can be expressed as:

$$\rho^n C_p^n \frac{T^{n+1} - T^n}{\Delta t} + C'(T^n) = D'(k^{n+1} T^{n+1})$$
$$\nabla \vec{U}_{ext} \cdot \nabla \phi = 0 \tag{3.103}$$

where: $C'(\cdot)$ and $D'(\cdot)$—Convection and diffusion operators in the discrete energy equation (i.e., Eq. (3.3)).

When the Eq. (3.103) is discretized, the free boundary condition Eq. (3.60) and the boundary condition equation on other boundary surfaces of the calculation area (3.61) must be considered. After this operation, a typical symmetric, positively definite large sparse matrix can be obtained. Here, the ICCG iterative method is again used to solve the obtained temperature field matrix, and the new temperature field distribution on the workpiece at time n + 1 is obtained.

In the laser welding process, the absorption of latent heat of fusion and the release of latent heat of solidification occur all the time. In practical research, it is found that both release and absorption of the latent heat affect the temperature at the interface of the pores and the size of the weld pool all the time. If the latent heat is not considered, the dynamic behavior of transient pores and weld pool during welding cannot be accurately simulated.

In general, for latent heat phenomena such as fusion and solidification, studies have shown that simulation can be well realized by using the temperature recovery/fall method, the equivalent specific heat method, and the heat enthalpy method. In this study, the temperature recovery method is used to simulate how to treat the physical phenomenon of fusion and solidification.

3.5.3.3 Solution for the Velocity Boundary Conditions of the Free Surface of Pores

In general, to deal with the velocity boundary conditions of a free interface is the most difficult step in solving the problem of free interface flow. The purpose of dealing with the velocity boundary conditions of a free interface is to make the free interface grid close to the pore meet the continuity equation. For the problem of the free interface flow in tracking the interface motion using Level Set and Particle Level Set methods, processing of this velocity boundary conditions can be equivalent to solving the partial differential equation.

$$\nabla \vec{U}_{ext} \cdot \nabla \phi = 0 \tag{3.104}$$

Equation (3.104) is a typical time-independent Hamilton–Jacobi equation that can be quickly solved by using the Fast Sweeping Method. It is found that the first-order upwind scheme is generally used to implicitly discretize and solve Eq. (3.104), to achieve the condition that the velocity divergence of the free boundary is zero, i.e., the continuity equation condition.

3.5.4 Numerical Calculation Process

In the laser welding process, laser energy distribution can be represented by the schematic diagram Fig. 3.12. Only Fresnel absorption is initially considered during the calculation process of the numerical solution, and the preliminary depth and shape of the pore can be obtained. Then, based on the first step, the multi-reflection Fresnel absorption and the inverse bremsstrahlung absorption of the plasma are comprehensively considered to calculate the cross-sectional shape of the pore and the weld pool.

Based on the mathematical model and numerical method of this paper, the numerical simulation system for the corresponding laser welding transient pore and motion weld pool is programmed in C++ language. In order to speed up the calculation, the corresponding code is parallelized in OpenMP language. The calculation process of the software system is shown in Fig. 3.13.

3.6 Summary

In this chapter, the energy interaction mechanism between material and laser during deep penetration laser welding and the related factors affecting the dynamics of pores and weld pool are systematically analyzed. With consideration of physical factors such as Fresnel absorption, evaporation, solidification, fusion and coupled convection

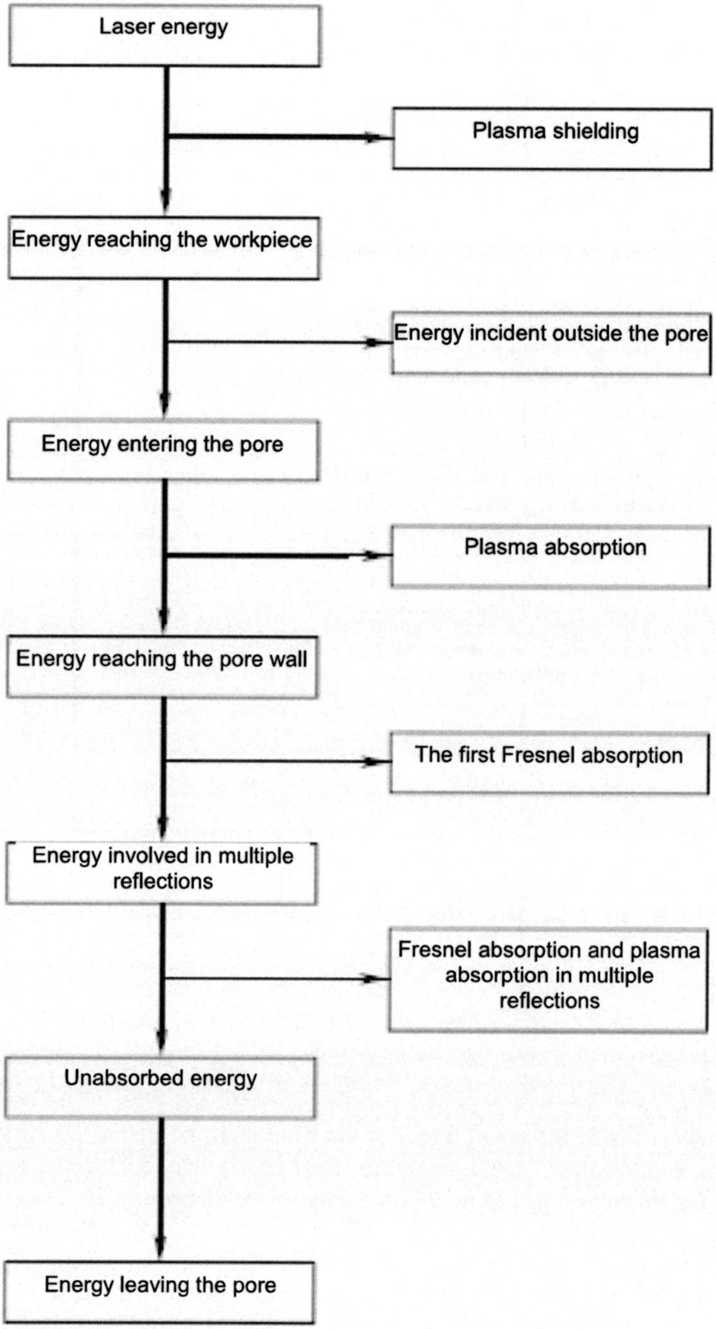

Fig. 3.12 Schematic diagram of laser welding energy absorption mechanism

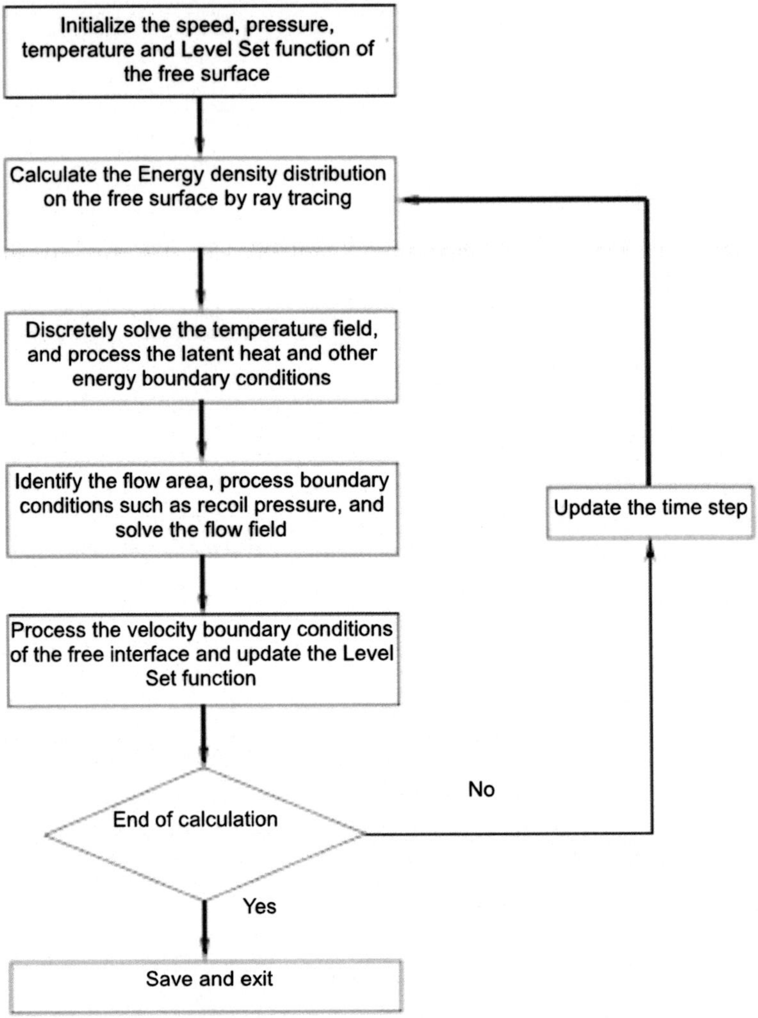

Fig. 3.13 Flow chart for numerical solution procedure

heat transfer inside the metal liquid of the weld pool, as well as important factors such as recoil pressure, surface tension, thermal capillary force, etc., the physical model for the deep penetration laser welding is established.

According to the physical model, the Level Set/Particle Level Set method is used to track the free interface of the pore, and the discontinuous mathematical model describing the coupling behavior of the transient pore and the motion weld pool of deep penetration laser welding is innovatively established. Based on the principle of viscous fluid mechanics, the discontinuous boundary conditions describing factors such as recoil pressure, surface tension, and thermal capillary force, as well as boundary conditions, such as energy and free interface, are derived in detail. The model can simultaneously couple and simulate transient phenomena such as three-dimensional heat transfer, flow, and evolution of free interface of the pore in deep penetration laser welding.

The high-precision Level Set/Particle Level Set method is used to track the transient pore moving interface. Based on this, a fast and robust computation method is proposed to solve the energy density distribution on the wall of the pores with any complex shapes, and the high-precision implicit scheme is used to discretely solve the corresponding energy equation and momentum equation. Furthermore, the corresponding discontinuous boundary conditions, such as temperature, pressure and viscous stress, are accurately processed.

Chapter 4
Simulation of Transient Keyhole and Weld Pool

Abstract This chapter introduces the three-dimensional transient keyhole and weld pool dynamics simulated by using the self-consistent laser keyhole welding model. Transient keyhole dynamics and its physical characteristics as a function of welding parameters are presented. Transient weld pool dynamics under a stable keyhole as well as an unstable keyhole condition are also characterized by high resolution numerical simulations. The influence of physical factors such as multiple Fresnel reflections, thermophysical parameters and welding parameters on the coupled keyhole and weld pool dynamics are presented.

4.1 Introduction

In deep penetration laser welding, the shape and temperature of the keyhole change in real time with obvious instantaneity, even in the quasi-steady phase. At the same time, the keyhole and its surrounding metal liquid extend forward with the movement of the laser heat source, while the weld pool sees drastic and complex heat transfer and flow and phase shift. At present, it is difficult to quantitatively observe and understand the dynamics of the transient keyhole and moving weld pool by applying the test method. Therefore, based on the effective mathematical model of laser welding, the quantitative visual simulation analysis in this process can promote the understanding of the physical process of laser welding and provides guidance on the welding process.

In this chapter, we take the fiber laser welding of aluminum alloy as the research object, and conduct numerical simulation research on the dynamics of the transient keyhole and moving weld pool, as well as its mechanism based on the sharp model of laser welding described in Chap. 3. By making use of the generally accepted literature result and high-speed camera CCD test result, we at first conduct verification calculation on the sharp interface model of laser welding established in the laser welding process, and analyze and discuss the dynamics features of transient coupling of keyhole and weld pool. Then, we systematically discuss the impact of different physical factors (recoil pressure, surface tension, thermal capillary force and multiple reflections and absorptions), different thermophysical parameters (heat conductivity

S. Gong et al., *Weld Pool Dynamics in Deep Penetration Laser Welding*,
https://doi.org/10.1007/978-981-16-0788-2_4

and kinematic viscosity) and different welding technologies (spot radius, welding speed and laser power) on the keyhole and moving weld pool coupling, and introduce the stability mechanism of laser welding keyhole and weld pool, providing the theoretical guidance for the optimization of laser welding technology in the actual industrial application.

4.2 Transient Coupling Dynamics of Keyholes and Weld Pool

4.2.1 Dynamic Evolution Process and Characteristics of Keyholes

In order to study the process, a numerical simulation study was carried out on the laser welding process of aluminum alloy. Figure 4.1 shows the keyhole depth variation curves with the welding time when the simulated laser power is 2.5 kW and the welding speed is 2 m/min. Table 4.1 shows the physical parameters used in the simulation. Figure 4.2 shows the keyhole depth variation curves between 40 and 45 ms under the same process conditions. As the welding process proceeds, the depth variation of the keyhole under the current process conditions can be divided into three characteristic stages: (I) rapid linear depth growth; (II) fluctuating depth growth. The growth rate in this stage is slower than that in the first stage, and decreases gradually as the welding time increases; (III) tendency to average depth, which is, however, subject to high-frequency oscillation. Stage I lasts only a short time, less than 1 ms under current process conditions. The duration of stage II is approximately 12 ms. In addition, as can be seen from Fig. 4.2, under the current process conditions, the oscillation frequency of the keyhole depth ranges from 2 to 5 kHz. The above

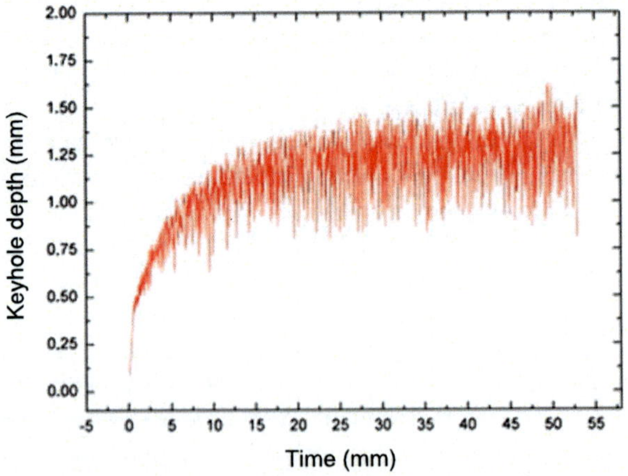

Fig. 4.1 Keyhole depth variation curves with the welding time when the simulated laser power is 2.5 kW and the welding speed is 2 m/min

Table 4.1 Default parameters used in the numerical simulation

Parameter	Symbol	Value
Density (kg/m^3)	ρ	2600
Specific heat (J/(kg K))	C_p	1000
Heat conductivity (W/(m K))	k	90
Solidus temperature (K)	T_s	831
Liquidus temperature (K)	T_l	889
Evaporation temperature (K)	T_b	2790
Melting latent (J/kg)	T_m	5.03×10^5
Evaporation latent (J/kg)	T_v	1.07×10^7
Surface tension (N/m)	σ	1.0
Dynamic viscosity (N m^{-2} s)	μ_l	1.6×10^{-3}
Coefficient of thermal expansion (1/K)	β	2.36×10^{-5}
Surface tension temperature coefficient (N/(m K))	$\frac{d\sigma}{dT}$	-3×10^{-4}
Black-body radiation coefficient	ε_r	0.3
Convective coefficient (W/(m K))	h	60
Fresnel absorption correlation constant	ε	0.20
Ambient temperature (K)	T_∞	300

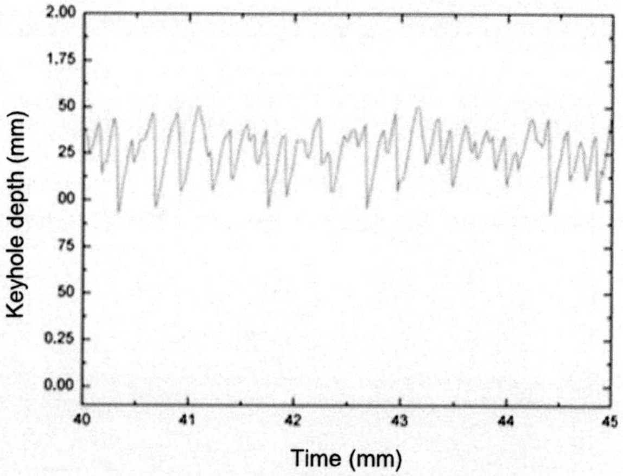

Fig. 4.2 Keyhole depth variation curves between 40 and 45 ms under the same process conditions

simulation results regarding the formation time of the keyhole agree well with the results of the X-ray test on the order of magnitude. In addition, the results of the keyhole depth oscillation frequency are also in good agreement with the theoretical results of Kroos and Klein et al. and Lee et al. regarding the keyhole oscillation frequency.

4.2.2 Flow Characteristics of Weld Pool with Stable Keyhole

According to the research on the mechanism of keyhole stability, the scale of the change in keyhole depth decreases when the welding speed increases or the linear energy decreases at constant power. Therefore, deep penetration laser welding with a stable keyhole can be implemented at a lower line energy. The simulation results show that the welding performed at the laser power of 1.3 kW and the welding speed of 3 m/min is deep penetration laser welding that produces stable keyholes. The keyhole morphology and weld pool flow under this process condition are analyzed and discussed below (Table 4.2).

Figure 4.3 illustrates the evolution of the free surface morphology of the keyhole and its temperature field under current process conditions. During welding, the keyhole is clearly visible and always open. Overall, its temperature field has a similar distribution pattern to Gaussian heat source. Moreover, according to Fig. 4.3, the temperature at the keyhole center and walls is high, approaching the boiling temperature of the material. These areas experience intense evaporation, producing a recoil pressure great enough to prevent surface tension and fluid impact force from making

Table 4.2 Laser welding process parameters

Laser type	Protective gas	Laser power	Focus radius	Welding speed	Air velocity	Shooting speed
Fiber	Ar_2	2500 W	0.2 mm	2 m/min	15 L/min	4000 f/s

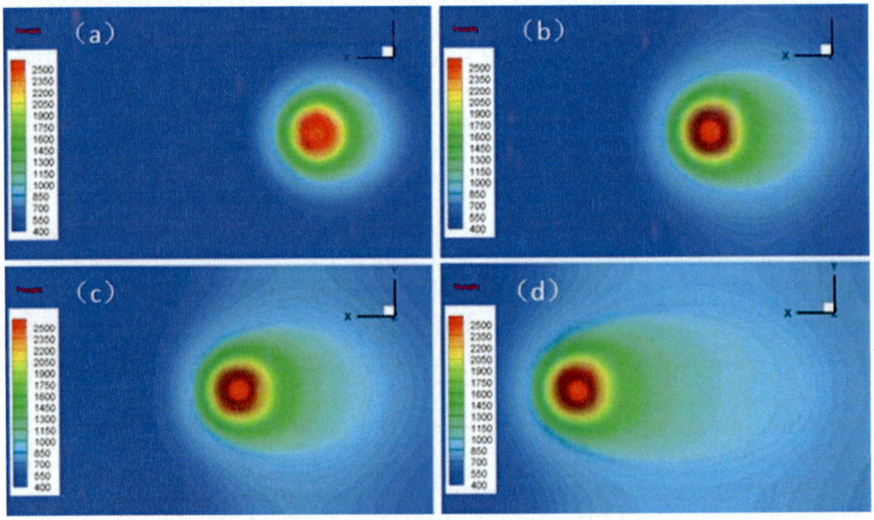

Fig. 4.3 Evolution of the transient free surface morphology of stable keyhole and its temperature field **a** 4.99 ms; **b** 11.55 ms; **c** 16.40 ms; **d** 30.75 ms

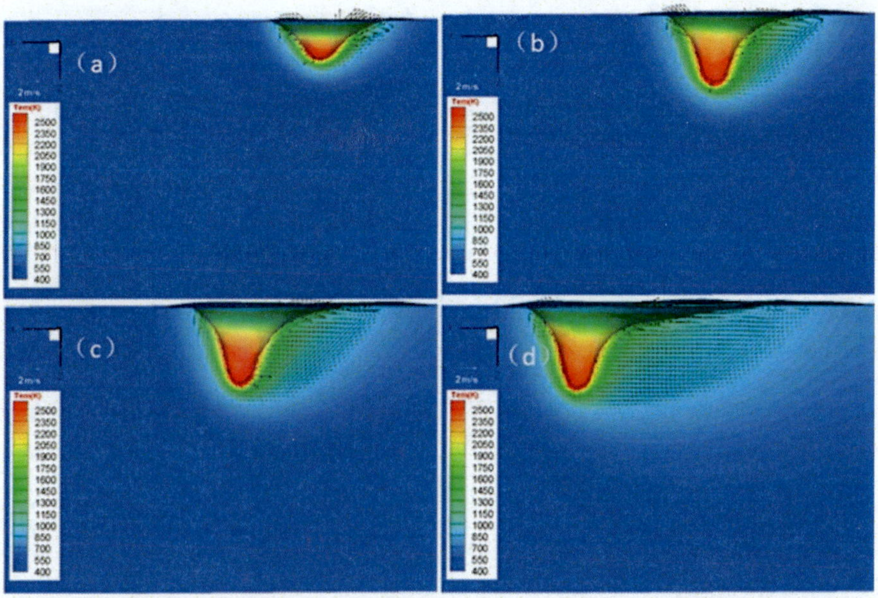

Fig. 4.4 Velocity distribution pattern on the longitudinal cross section of stable keyhole **a** 4.99 ms; **b** 11.55 ms; **c** 16.40 ms; **d** 30.75 ms

keyhole walls close. As a result, a strong flow momentum, driven by the recoil pressure, could force the fluid flow in the weld pool to gush out along the keyhole walls. Hence, the fluid flow directions near the keyhole walls are upwards and approximately parallel to the keyhole walls. Meanwhile, because of the conservation of mass, the outgoing fluid is forced to return when meeting the solid surface of the weld pool. This flow pattern is shown in Fig. 4.4.

In recent years, Japanese scholars have studied welding processes with CO_2, Nd:YAG, LD-pumped solid-state laser and fiber laser using X-ray imaging equipment. They found that when the keyhole is stable and remains open, the weld pool velocity field near keyhole walls goes upwards and nearly parallel to the walls. The present simulation results of weld pool dynamics under the stable keyhole condition agree with these x-ray imaging results, as shown in Fig. 4.5. This agreement suggests that when the welding wire has low energy in deep penetration laser welding, the keyhole keeps stable and open. It also provides evidence that the discontinuous mathematical model and keyhole stability mechanism put forward by the author for deep penetration laser welding are reasonable.

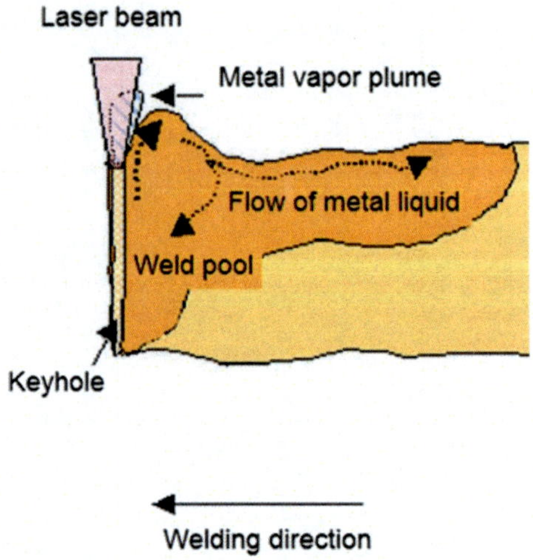

Fig. 4.5 Internal flow trend of weld pool with keyhole stability obtained by X-ray penetration test

4.2.3 Flow Characteristics of Weld Pool with Unstable Keyhole

In the welding process, it is impossible to maintain a balance between the recoil pressure on some keyhole walls and the surface tension and the impact force of metal liquid, so the keyhole begins to oscillate. The oscillation is manifested in the appearance of multiple convex platforms on the keyhole walls, as shown in Figs. 4.6 and 4.7. Once a convex platform appears, its upper surface absorbs more energy because of direct exposure to laser, while its lower surface is blocked by the convex platform and cannot receive direct laser exposure. As a result, there is a significant temperature difference between the upper and lower parts of the convex platform, as shown in Fig. 4.7. In the upper part of the convex platform, the temperature is very high and may approach or exceed the boiling point of aluminum alloy so that this upper part experiences intense evaporation, which produces recoil pressure. The recoil pressure will drive the metal liquid in the weld pool near the platform to move downwards at a high speed in an attempt to close the keyhole, as shown in Fig. 4.6b–h. At the same time, because the curvature of the keyhole walls of the convex platform is usually large, the surface tension on these walls is also large. In addition, because of the small keyhole radius of the platform, the above-mentioned keyhole walls are more sensitive to metal liquid impact of the same degree. As a result of the combined action of the above reasons, the keyhole eventually collapses and becomes two pieces. Figure 4.7d, f show two keyholes that are about to collapse, while Fig. 4.7g shows a keyhole that has been broken into two parts. If air enters the keyhole before fracture, the closed area of the keyhole at the bottom will contain a bubble, and if the bubble is captured by the solidification front, a pore defect will occur. When the keyhole is closed, the keyhole becomes obviously shallow. However, due to the effect of laser irradiation, the recoil pressure forces the keyhole open and drive the free interface

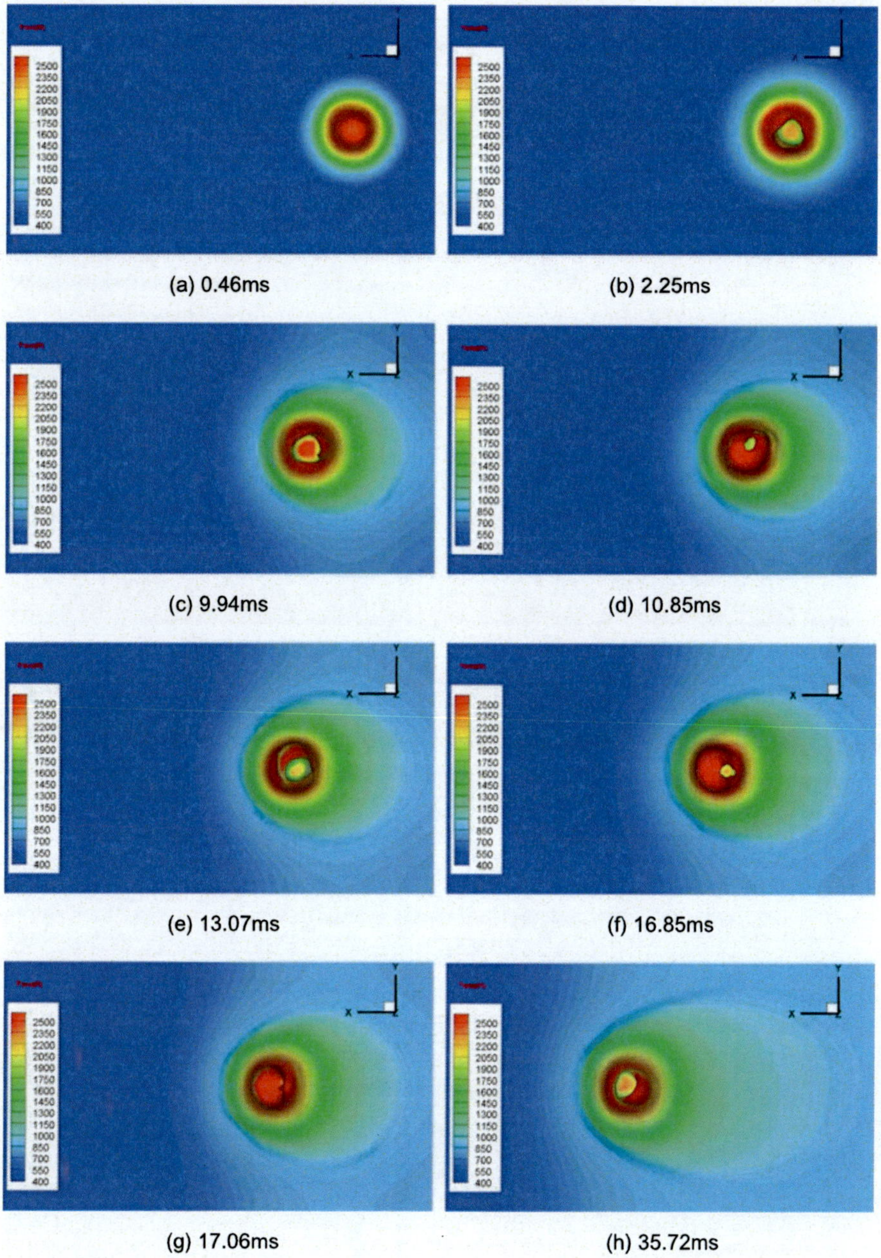

Fig. 4.6 Evolution of the temperature field on the surface at laser power of 2.5 kW and welding speed of 2 m/min

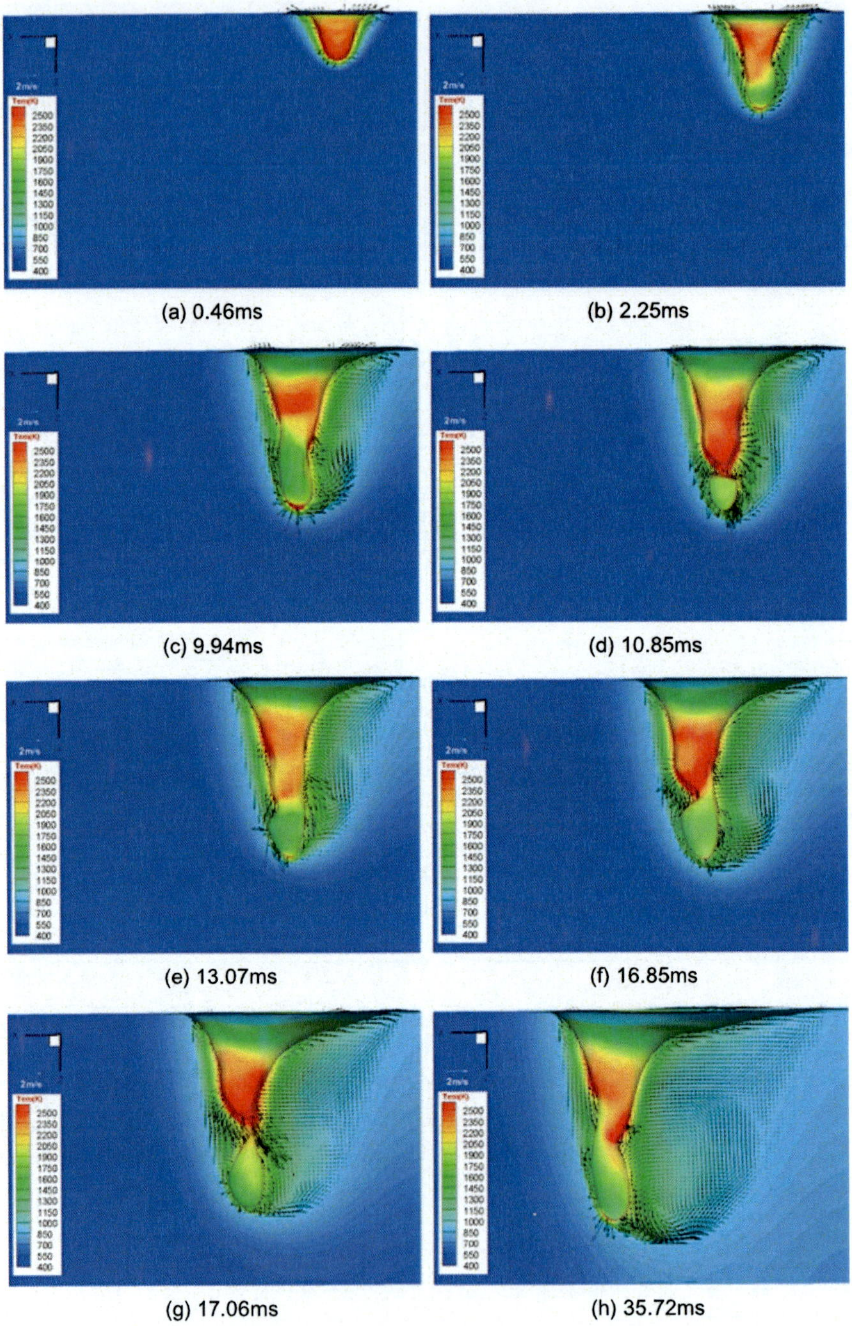

Fig. 4.7 Evolution of free interface and the distribution of weld pool velocity field

to move in the direction of the depth of penetration, thus increasing the depth of the keyhole again. On the other hand, the convex platform will appear again on the keyhole walls, which facilitates the next keyhole oscillation. From Figs. 4.1 and 4.2, it can be seen that in deep penetration laser welding, the above oscillation process of the keyhole is repeated periodically, and because of keyhole oscillation, bubbles are easy to form at the bottom or the middle of the keyhole.

In the past numerical simulation of the coupling behavior between the transient keyhole and moving weld pool, foreign scholars have done some excellent research work. Semak et al. are the first to confirm the phenomenon of convex platforms on a keyhole wall through numerical simulation. However, their numerical model is one-dimensional and considers only the phenomenon of convex platform on the front keyhole wall. Lee et al. have theoretically proved that the interaction between laser and convex platform may lead to keyhole closure and the generation of bubbles through numerical simulation. However, this model is two-dimensional and only suitable for stationary laser welding. Ki et al. carried out a theoretical study on the coupling behavior between three-dimensional transient keyholes and moving weld pools by numerical simulation for the first time. In this study, however, they failed to visualize the process of pore generation. By comparing the present simulation results and findings of previous researches, it is proved that the discontinuous mathematical model proposed in this paper can be used to simulate keyhole instability and bubble formation in deep penetration laser welding.

Figure 4.7 also shows the flow field distribution on the longitudinal section of the weld pool during welding. The flow field in the weld pool is very complex when the keyhole oscillates. The following analyzes and discusses several typical flow mechanisms in the moving weld pool. First, there is a periodic high-speed downward flow field driven by recoil pressure in the weld pool near the keyhole. Under the current technological conditions, the maximum velocity of the characteristic fluid dynamics can reach above 10 m/s. That characteristic flow pattern exists not only on the front wall and both sides of the keyhole, but also on its rear wall. From Fig. 4.7c, f, g, h, it can be seen that the high velocity hydrodynamics results in a strong vortex flow state in the weld pool near the back edge at the bottom of the keyhole. The mechanism of the vortex flow is caused by the interaction between the high-speed downward flow and the boundary of the weld pool. In addition, generally speaking, the vortex flow is not conducive to bubble floating, so controlling the characteristic flow pattern may help to reduce the pore defects in the weld.

Secondly, the recoil pressure on the walls of the keyhole will cause some liquid to flow laterally around the keyhole. This characteristic flow is distributed on both sides of the weld pool, as shown in Fig. 4.9. Due to the existence of temperature gradient, some liquid flows under the effect of thermal capillary force near the keyhole, as shown in Fig. 4.8. In addition, in the vicinity of the keyhole, some liquid flows upwards along the back wall of the keyhole due to the driving effect of the recoil pressure, as shown in Figs. 4.8 and 4.9.

There is a strong vortex flow on both sides of the weld pool due to the interaction of the boundary of the weld pool with transverse flow, thermal capillary flow and upward flow driven by recoil pressure, as shown in Fig. 4.8. Finally, because the

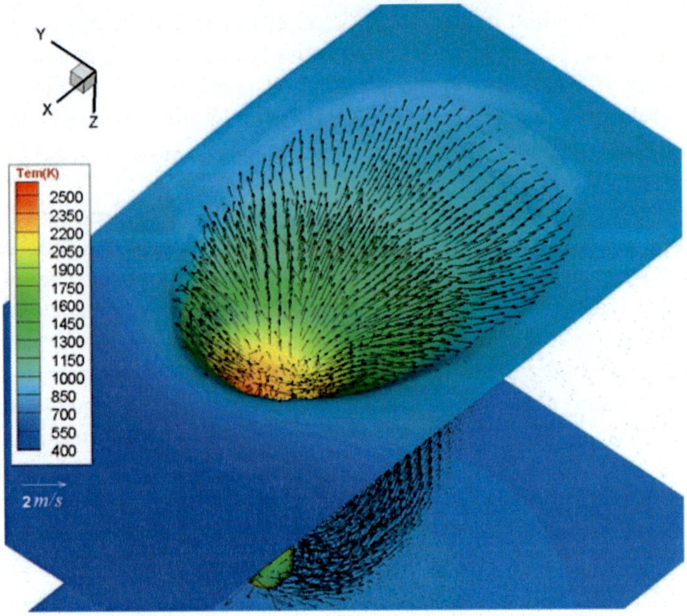

Fig. 4.8 A top view of velocity fields of weld pool surface at 35.72 ms in laser welding

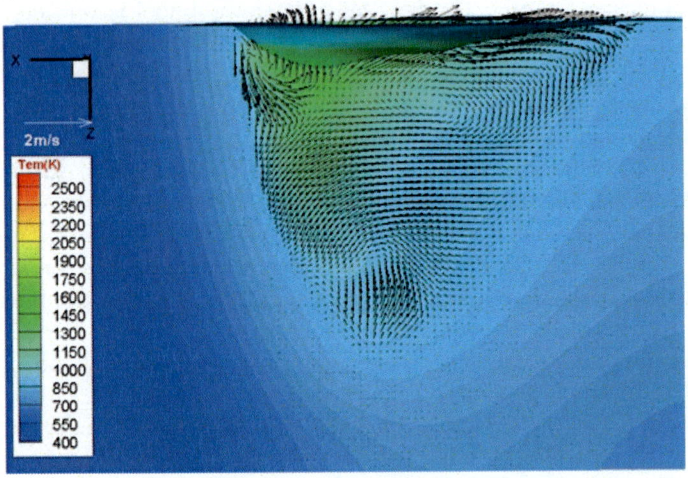

Fig. 4.9 A longitudinal section view of velocity fields at 35.72 ms in laser welding

keyhole is always in the continuous oscillatory closure, the interaction between the recoil pressure and the periodic convex platform on the walls of the keyhole will produce a special periodic, characteristic flow moving upward along the back wall of the keyhole. This characteristic motion will result in a wave-like flow state on the

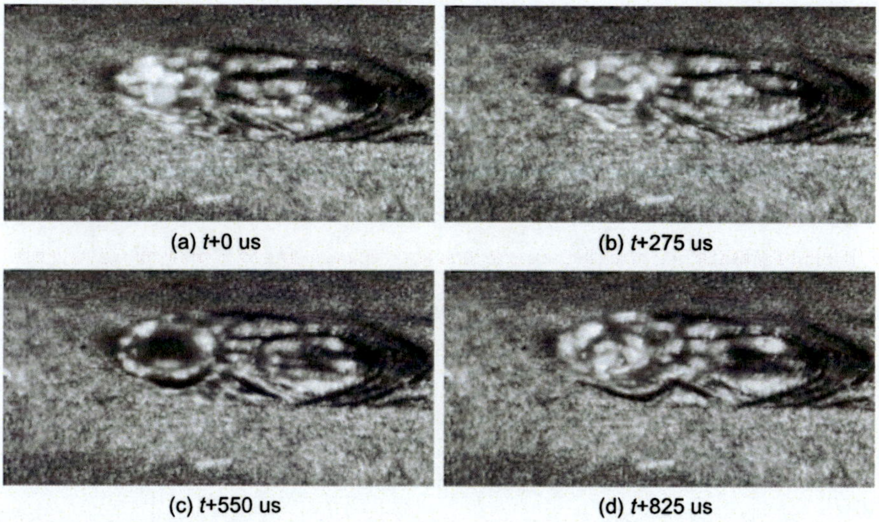

Fig. 4.10 High-speed CCD imaging results of velocity field of weld pool surface

weld pool surface. This characteristic flow state can be observed directly from the corrugated features on the free surface near the rear side of the weld pool in Fig. 4.7h, or from the hierarchical velocity field distribution on the weld pool surface in Fig. 4.9. In order to verify this special wave-like flow state, the surface images of the weld pool in a cycle from keyhole closure to re-opening are observed by using high-speed CCD photography. The wave-like flow pattern caused by the above mechanism can be seen, as shown in Fig. 4.10c.

Recent X-ray experiments by Japanese scholars have shown that despite the complexity of deep penetration laser welding, when the keyhole oscillates, there is always a strong vortex flow inside the moving weld pool near the back edge at the bottom of the keyhole, and there is also a strong vortex on both sides of the weld pool, as shown in Fig. 4.11. In the simulation study of weld pool dynamics at home and abroad, the rapid downward flow near the front and both side walls of the keyhole was confirmed theoretically by Semak et al., Lee et al. and Ki et al. However, none of these research efforts carry out reasonable numerical simulation and predictions of the internal vortex flow near the back edge of the keyhole. In this paper, the numerical simulation results consistent with the characteristic flow of the weld pool observed by the X-ray test are obtained for the first time by using the discontinuous numerical method.

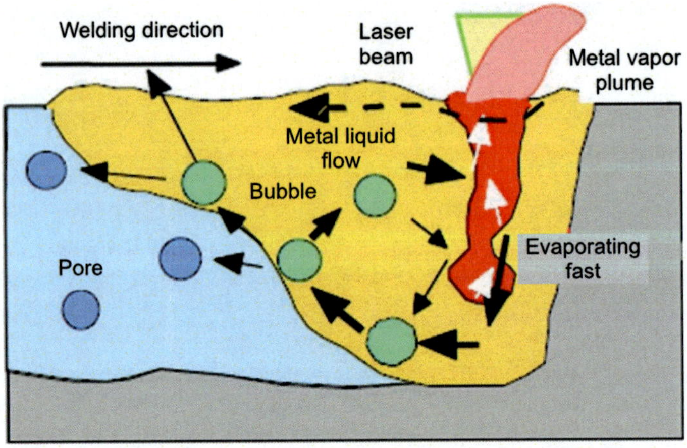

Fig. 4.11 Internal flow trend of weld pool under keyhole instability

4.3 Effects of Physical Factors on the Coupling Behavior

4.3.1 Interfacial Force and Multiple Reflections and Absorptions

4.3.1.1 Influence of Recoil Pressure on Coupling Between Keyhole and Moving Weld Pool

The discussion in this section is based on two conditions, one with recoil pressure considered and the other without recoil pressure considered. In addition, the materials and technical parameters applied are the same as those in Sect. 4.2.2. We mainly analyze and compare the transient shape and depth of keyholes and the flow tendency of weld pools as well as the transient internal temperature evolution of workpieces. In the discussion on the above recoil pressure conditions, we fully consider the effect of other physical factors, such as surface tension, thermal capillary force, free surface evolution, and multiple reflections on the welding process.

Figure 4.12 shows the transient free surface (can be deemed as the keyhole) shape at 20 ms of the welding process and the simulation result of weld pool flow field on the workpiece longitudinal section. In the figures, the color codes indicate the temperature; the arrows show the speed and direction, while their lengths reflect the speed value. Figure 4.13 shows the distribution of weld pool temperature field at the same time and is the top view of the free surface shape.

It can be seen from Fig. 4.12a that without considering recoil pressure, the free surface is in the shape of a shallow pit, and this process can be considered similar to the conduction welding technology. There is one small vortex driven by thermal capillary force near both the front and rear walls of the free surface, with the maximum

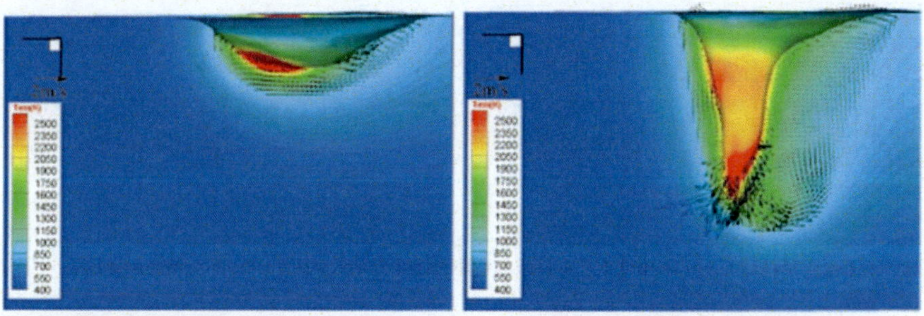

(a) With recoil pressure not considered (b) With recoil pressure considered

Fig. 4.12 Free surface shape and speed distribution of weld pool (side view) under different recoil pressure conditions at 20 ms

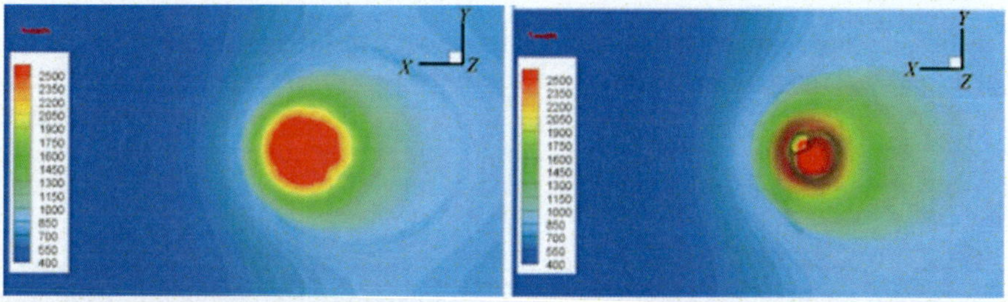

(a) With recoil pressure not considered (b) With recoil pressure considered

Fig. 4.13 Free surface shape and temperature field distribution of weld pool (top view) under different recoil pressure conditions at 20 ms

speed of 1–2 m/s. With recoil pressure, the free surface (namely, the keyhole) is thin and long, normally vibrating and much wider (refer to Fig. 4.14) than that without considering recoil pressure. The great difference between the two conditions shows that the recoil pressure is the root cause of deep penetration welding. Compared with the condition without considering recoil pressure, the condition with recoil pressure considered sees very complex fluid dynamics of weld pools during welding, and the fluid near the keyhole free surface flows very fast, as shown in Fig. 4.12b. From the comparative analysis, it is clear that the recoil pressure significantly influences the dynamics of a weld pool. It is difficult to predict the high-speed dynamics of weld pool fluid only based on thermal capillary force but without considering recoil pressure.

According to Figs. 4.12b and 4.13b, the temperature distribution on the keyhole wall is complex and uneven; the wall with direct laser irradiation has the highest temperature. In consideration of recoil pressure, the keyhole vibration makes it impossible for the laser to irradiate all the keyhole walls. In contrast, from Figs. 4.12a and 4.13a, it can be seen that all the free surfaces are exposed to direct laser irradiation if recoil pressure is not considered. Nevertheless, the size of a weld pool under

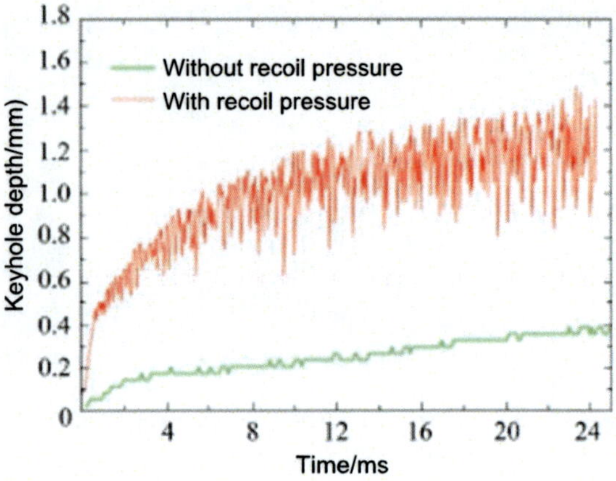

Fig. 4.14 Variation curve of the keyhole depth with welding time change under different recoil pressure conditions

recoil pressure is much larger than that without considering recoil pressure. This indicates that recoil pressure promotes the absorption of laser energy by workpieces, as shown in Fig. 4.15. From the above discussion, it is evident that recoil pressure changes the absorption of laser by materials, and deeply influences the heat-transfer mechanism during laser welding.

Figure 4.16 compares transient welding shape of a weld joint at 20 ms under two above conditions. It can be seen that the weld joint without considering recoil pressure is shallow and wide and similar to the traditional conduction weld joint; the weld joint in consideration of recoil pressure is deep and narrow, and its shape is a typical weld joint of deep penetration laser welding. It can be inferred that the distinction between conduction welding and deep penetration welding lies mainly in whether the intensive and constant evaporation occurs. In addition, the speed comparison indicates that two symmetrical vortexes exist on the weld joint root with recoil pressure considered; the root vortex does not exist without considering recoil

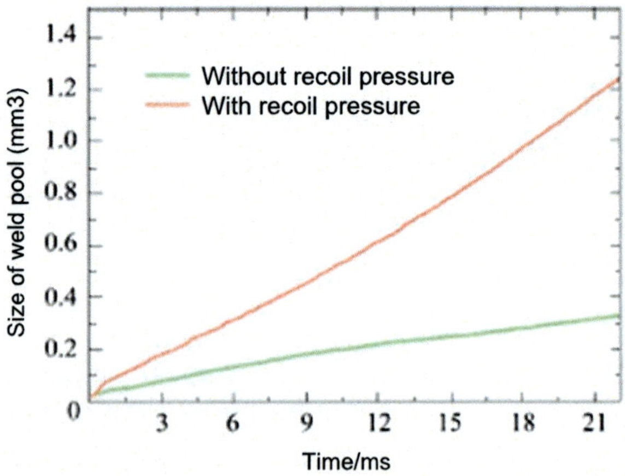

Fig. 4.15 Variation curve of weld pool size with welding time change under different recoil pressure conditions

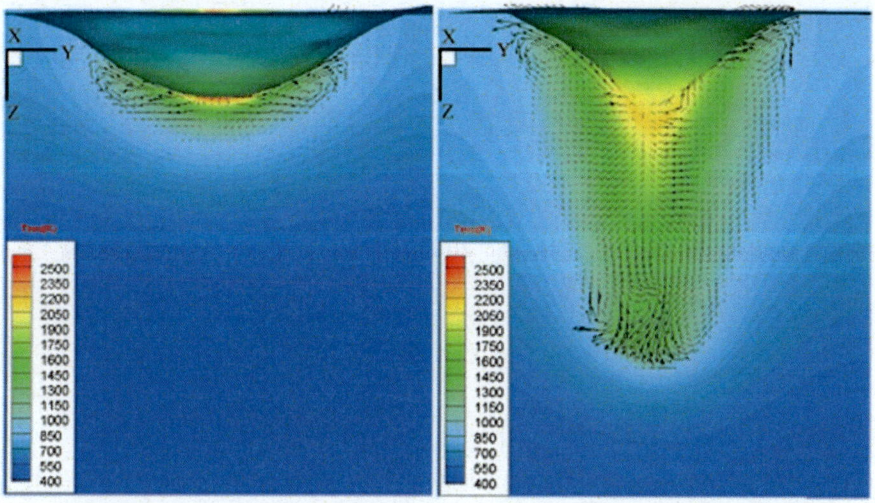

(a) With recoil pressure not considered (b) With recoil pressure considered

Fig. 4.16 Weld joint shape comparison at 20 ms of transient welding process

pressure, further proving a significant influence of recoil pressure on the flow of a weld pool.

Based on a comparison of the above simulation results, it can be concluded as follows: ① recoil pressure is the key factor for the formation of a keyhole and the implementation of deep penetration laser welding; ② intensive and constant evaporation is the main basis for distinguishing between laser conduction welding and deep penetration laser welding; ③ it is difficult to predict the internal high-speed dynamics of a weld pool by considering all key physical factors related to laser welding except for recoil pressure; ④ the recoil pressure changes the absorption of laser energy by materials, influencing the heat-transfer mechanism during laser welding.

4.3.1.2 Influence of Surface Tension on Coupling Between Keyhole and Moving Weld Pool

In this section, we mainly compare and discuss the behavior of keyhole and moving weld pool with the surface tensions of 0.2, 0.6 and 1.0 N/m. In addition to the surface tension coefficient, other parameters applied are the same as the parameters in Sect. 4.2.2.

Figure 4.17 indicates the shape of keyhole free interface at 5 ms during the welding with the surface tension of 0.2 N/m and 0.6 N/m, respectively. It can be seen from the figure that as the surface tension increases, the level of the fluid flowing from the edge of the keyhole decreases. The main reason is that a larger surface tension indicates a smaller recoil pressure and, therefore, a smaller force driving the metal

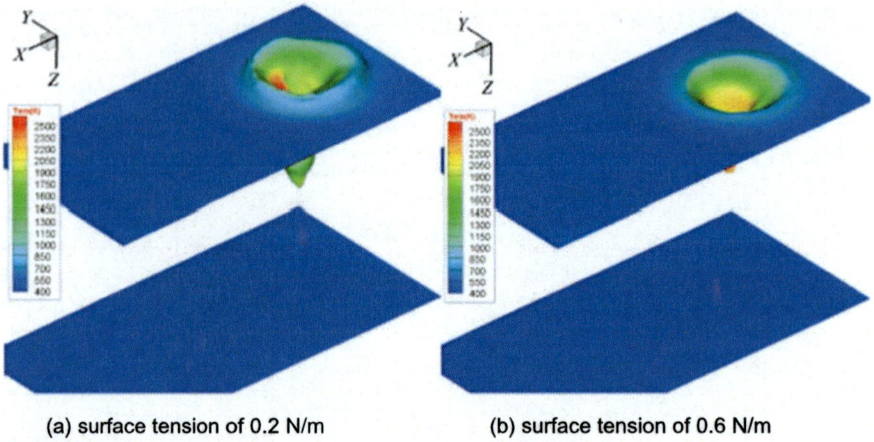

Fig. 4.17 Transient free surface shape of keyhole at 5 ms during welding under different surface tension conditions

fluid out of the workpiece surface at the same time. This theoretically proves that surface tension is the factor hindering the formation of a keyhole.

To discuss the influence of surface tension on the keyhole vibration, we studied the depth variation of keyhole within 17 ms during the welding with the surface tension of 0.6 and 1.0 N/m (refer to Figs. 4.18 and 4.19). It can be seen from the figures that, the depth evolution of keyhole can be divided into three phases: the linear increase phase, increase phase with small amplitude vibration and high-frequency vibration phase surrounding a relatively fixed depth. At around 12 ms, the mean depth of keyhole under two conditions reaches a fixed value, which match with the actual welding process.

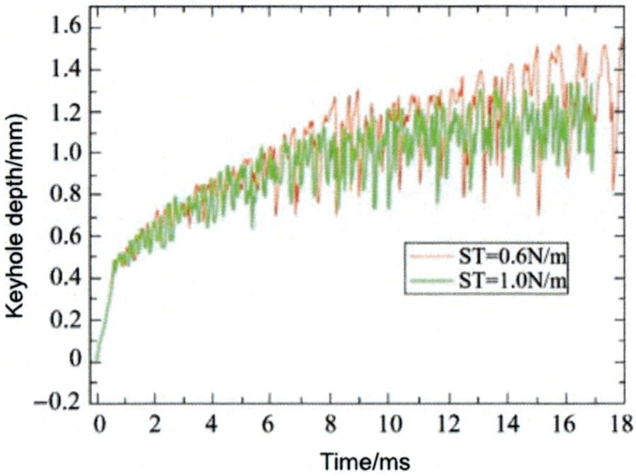

Fig. 4.18 Variation curve of keyhole depth with welding time under different surface tension conditions

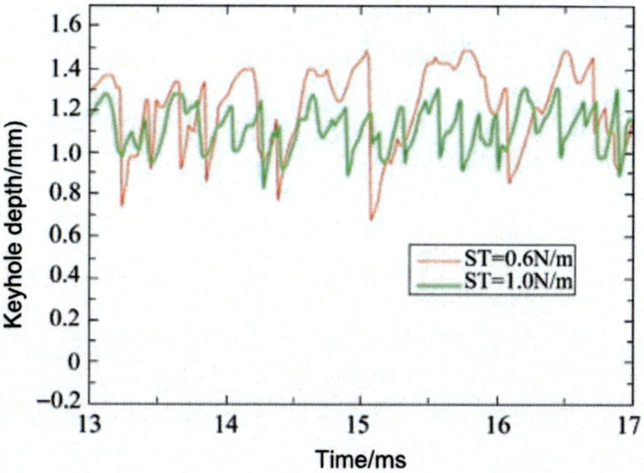

Fig. 4.19 Variation curve of depth with welding time after the formation of keyhole

During the actual laser welding, the keyhole depth can be relatively fixed within milliseconds. Therefore, the depth variation of keyhole at (12–17 ms) after the keyhole depth is relatively stable is applied to analyze the influence of surface tension on the transient keyhole action. It can be seen from Fig. 4.19 that, the largest depth and smallest depth of the keyhole are the maximum with the smaller surface tension, resulting from the fact that smaller surface tension means larger recoil pressure, larger weld penetration depth and depth vibration of keyhole with other conditions certain. This phenomenon can partly explain the fact that the aluminum alloy welding with low surface tension is less stable than the steel and titanium alloy with high surface tension. In addition, it can be seen from the figure that, the oscillation period of keyhole depth is gradually increasing with the smaller surface tension. This phenomenon can be explained as: the smaller the surface tension of keyhole is, the harder it is for the keyhole to close; therefore, the longer the closure of the keyhole takes, and the larger the oscillation period of keyhole depth is.

Figure 4.20 and Fig. 4.21 indicate the evolution curve of depth and size of moving weld pool with welding time with the surface tension of 0.6 N/m and 1.0 N/m, respectively. It can be seen from the figures that, from 8 ms of welding, the same linear energy input realizes the relatively larger recoil pressure with smaller surface tension, making the increase of penetration depth with the surface tension of 0.6 N/m faster than that with the surface tension of 1.0 N/m; however, the sizes of weld pool under two surface tension conditions during the initial welding phase are nearly the same. Therefore, it can be considered that, at the initial phase of deep penetration laser welding, certain change of surface tension with other parameters the same have a greater influence on the geometric dimension of keyhole than on weld pool dimension.

The analysis on the above simulation result indicates that: ① the surface tension can hinder the formation of keyhole during deep penetration laser welding; ② under certain conditions, larger keyhole depth, weld penetration depth and lower oscillation frequency of keyhole depth be obtained with smaller surface tension; however, the oscillation amplitude of the keyhole depth is larger; ③ the surface tension has serious

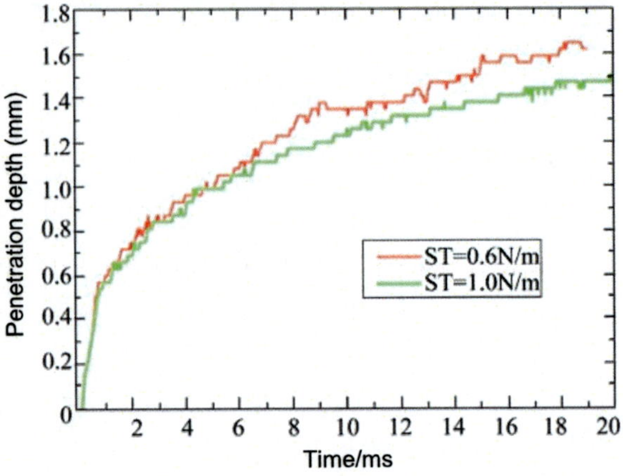

Fig. 4.20 Variation curve of the penetration depth with welding time

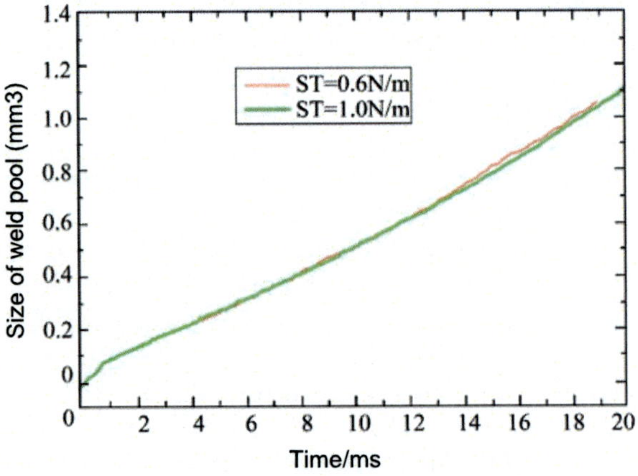

Fig. 4.21 Variation curve of the weld pool size with time

influence on the variation of the keyhole depth, and due to the fact that the variation of keyhole depth indicates the keyhole instability, the surface tension has important influence on the keyhole stability; ④ compared with the keyhole dimension, the variation of the weld pool dimension during the initial welding phase under the studied technical condition is not sensitive to the variation of surface tension size.

4.3.1.3 Influence of Thermal Capillary Force on Coupling Between Keyhole and Moving Weld Pool

In this section, we compare and study the transient coupling of keyhole and weld pool with the surface tension temperature coefficient of −0.5 and −0.1 N/(m K). In addition to the surface tension temperature coefficient, other parameters are the same as the parameters in Sect. 4.2.2.

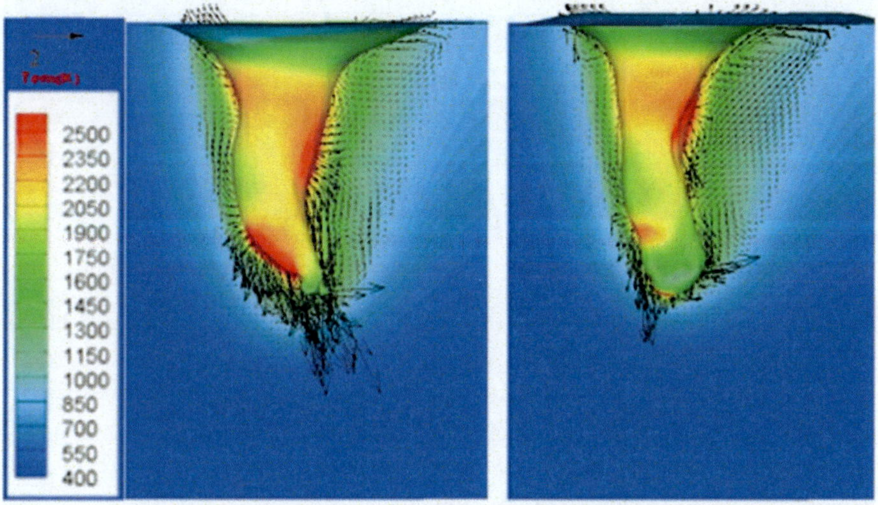

(a) Coefficient of -0.1 N/(m•K) (b) Coefficient of -0.5 N/(m•K)

Fig. 4.22 The keyhole shape and speed field distribution of weld pool at 10 ms of welding with different surface tension temperature coefficients

Figure 4.22 is the comparison figure of keyhole shape and speed field distribution of weld pool at 10 ms of welding under two surface tension temperature coefficient conditions. It can be seen from the figure that, the larger the surface tension temperature coefficient (the thermal capillary force) is, the more intensively the metal fluid vortex in weld pool flows at the rear keyhole edge near the upper surface of workpieces, indicating that the thermal capillary force can promote the flow of weld pool in the deep penetration laser welding. This can be verified by the size of keyhole open throat. The larger the thermal capillary force is, the stronger the impact force of vortex flow in weld pool at the rear keyhole edge near the upper surface of workpieces is, and the relatively smaller the keyhole open aperture is. In addition, the more intensively the metal fluid vortex flows at the rear keyhole edge near the upper surface of workpieces, the stronger the convective heat dissipation of the metal fluid there to the keyhole surrounding is, making it difficult for the laser energy to be transferred to the deep workpieces, thus causing the relatively shallower penetration depth with larger thermal capillary force.

Figure 4.23 is the comparison figure of surface speed distribution tendency of weld pool at 15 ms of welding under two thermal capillary force conditions. It can be seen from the figure that the surface speed of the weld pool is obviously larger with the surface tension temperature coefficient of −0.5 N/(m K) than that with the surface tension temperature coefficient of −0.1 N/(m K). This again certifies that the thermal capillary force has an important influence on the moving weld pool dynamics of deep penetration laser welding. However, even though the thermal capillary forces under two conditions differ by a factor of 5, the sizes of weld pools are not much different;

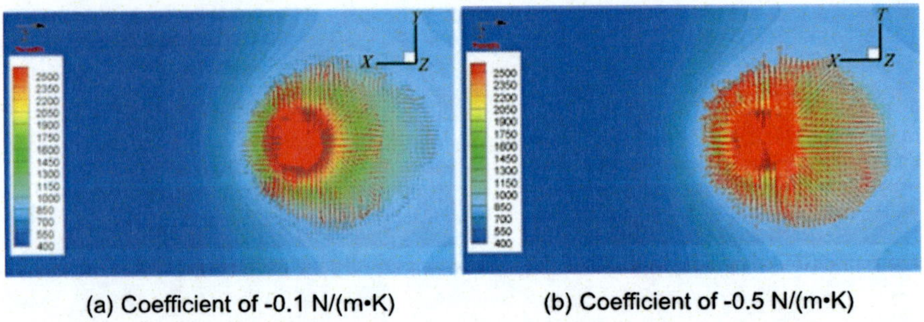

Fig. 4.23 Transient surface speed field comparison of weld pool at 15 ms of welding with different surface tension temperature coefficients

therefore, the thermal capillary force is not the decisive factor for the formation and growth of the weld pool, mainly due to the important effect of recoil pressure and other factors on the formation of the weld pool.

Figure 4.24 and Fig. 4.25 show the variation curves of the keyhole depth and the penetration depth with welding time under two kinds of thermal capillary force conditions respectively. The smaller the absolute value of the surface tension temperature coefficient, the larger the amplitude of the keyhole depth oscillation, as shown in Fig. 4.24. The analysis suggests that it is mainly because the maximum depth of the keyhole is relatively large at this moment. In spite of little difference in the growth rate of the penetration depth under two different conditions, a smaller absolute value of the surface tension temperature coefficient leads to weaker convection on the surface of the weld pool, further resulting in greater penetration depth, as shown in the Fig. 4.25.

It can be concluded based on the foregoing results that: ① The thermal capillary force exerts an important influence on the surface velocity of the weld pool. The

Fig. 4.24 Variation curves of the keyhole depth with time under different thermal capillary force conditions

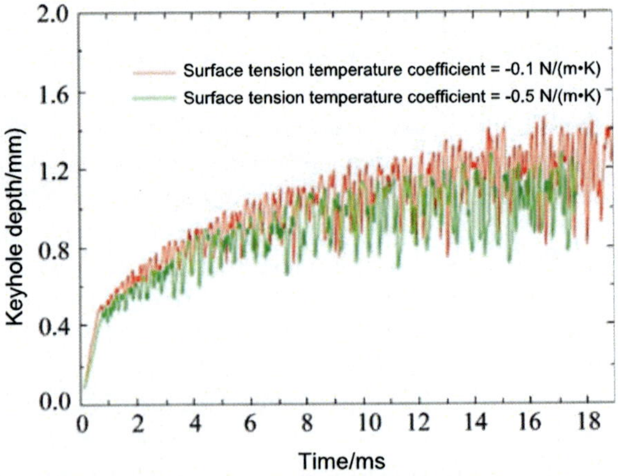

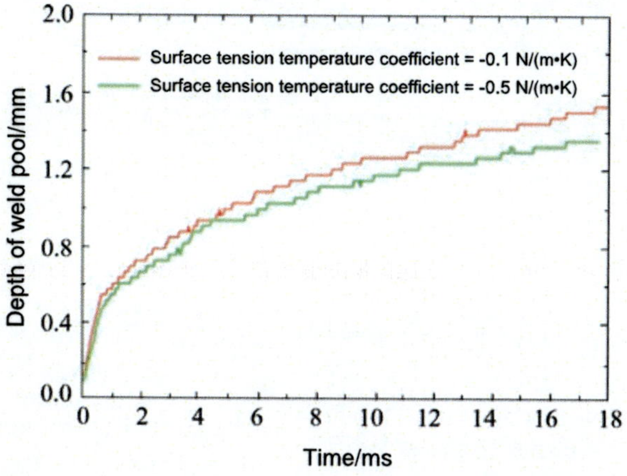

Fig. 4.25 Variation curves of the penetration depth with time under different thermal capillary force conditions

greater the thermal capillary force, the more intense the surface motion of the weld pool, but the smaller the penetration depth; ② The thermal capillary force affects the shape and size of the keyhole to some extent. The larger the thermal capillary force, the smaller the radius of the keyhole opening, the keyhole depth and the change of the keyhole depth; ③ The thermal capillary force is not the decisive factor to form the weld pool, despite its important influence on the coupling between the keyhole and the weld pool.

4.3.1.4 Influence of Multiple Reflections on Coupling Between Keyhole and Moving Weld Pool

With the deep penetration laser welding process studied in Sect. 4.2.2 as the object, the coupling between the keyhole and the moving weld pool is studied under one-time multiple reflections and absorptions and complete multiple reflections and absorptions, respectively. In this section, the term "complete multiple reflections" refer to that when reflected repeatedly inside the keyhole, any of the beams has its energy attenuated to 1% of its initial value or leaves away from the inside of the keyhole. In either condition, the coefficient ε in Fresnel absorption formula is 0.25.

Figure 4.26 shows the variation curves of the keyhole depth with time under two multiple reflection conditions. As shown in the figure, the keyhole depth presents the basically same variation trends with time under both the one-time reflection condition and the multiple reflection conditions; the same is true for the maximum and frequency of the keyhole depth oscillation. This means that under the current process conditions, the energy absorbed by the initial Fresnel and the first reflection Fresnel is significantly higher than that absorbed by the subsequent multiple reflections.

Figure 4.27 and Fig. 4.28 show the variation curves of the weld penetration depth and the weld pool volume with welding time under two multiple reflection absorption conditions respectively. According to the figures, the variation curves of the weld

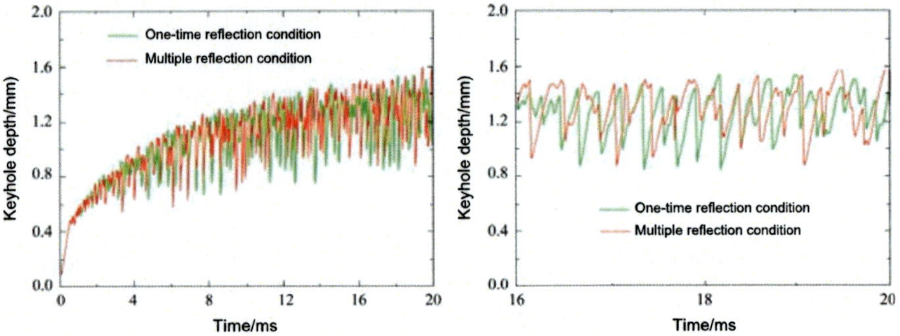

(a) (b)

Fig. 4.26 Variation curves of the keyhole depth with welding time under different multiple reflection absorption conditions

Fig. 4.27 Variation curves of the penetration depth with time under different multiple reflection conditions

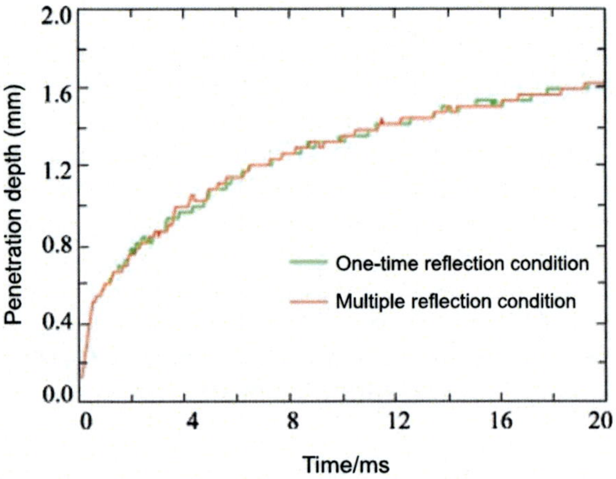

Fig. 4.28 Variation curves of the weld pool volume with time under different multiple reflection conditions

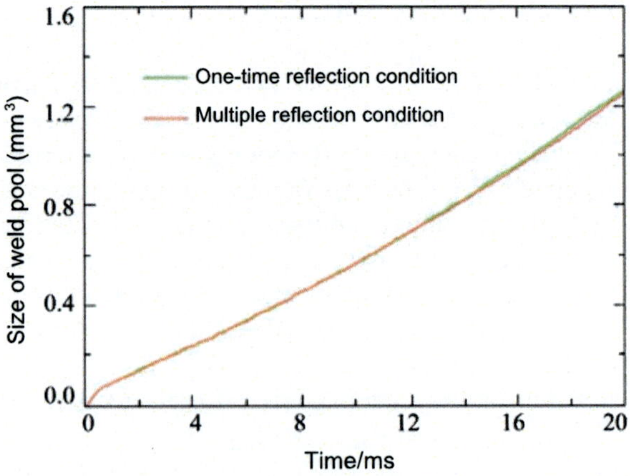

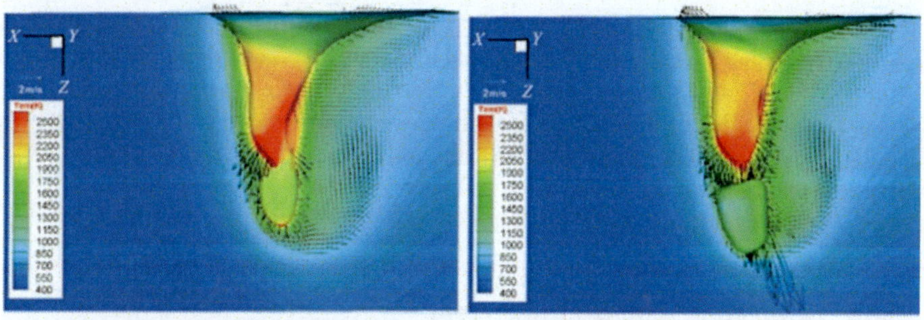

Fig. 4.29 Shape of the keyhole and velocity field distribution of the weld pool at 20 ms of the welding under different multiple reflection absorption conditions

penetration depth and the weld pool volume with time when only one-time multiple reflection absorption is considered substantially coincide with those when complete multiple reflection absorption is considered. The two results demonstrate again that the energy absorbed by the initial Fresnel and the first reflective Fresnel accounts for a dominant position in the total laser energy absorbed under current process conditions.

Figure 4.29 shows the comparison between the longitudinal section velocity distribution of the weld pool and the shape of the keyhole at the welding moment of 20 ms under two multiple reflection absorption conditions. It can be seen from the figure that the size of the weld pool on the cross section, as well as the speed of the weld pool, is basically the same under the two conditions; meanwhile, the partial shape of the keyhole is also basically the same.

The foregoing simulation results reveal that the energy absorbed by the wall surface of keyhole during deep penetration laser welding under certain conditions may depend heavily on the initial Fresnel absorption and the first and second multiple reflection Fresnel absorptions. In recent years, in-depth research has been conducted by Jin Xiangzhong et al. on the Fresnel absorption mechanism of deep penetration laser welding of CG17 glass based on the keyhole shape observed through tests. Their results also prove that the keyhole depth has an insignificant impact on the Fresnel absorption during deep penetration laser welding, while the bending degree (namely, angle of incidence) of the keyhole greatly affects the Fresnel absorption; besides, the repeated multiple reflections contribute little to the total energy absorbed by the wall surface of the keyhole. The experimental study launched by Fabbm et al. recently indicates that the front wall surface of the keyhole achieves a laser absorption rate of 60–80%. Therefore, they also believe that it is enough to consider the one-time multiple reflection absorption under general process conditions. The study analysis suggests that the aforesaid results are attributable to the following two reasons: ① In deep penetration laser welding, the keyhole will always be in a state of oscillation, and the periodic bump on the wall surface of the keyhole will block the incidence of some laser beams in the direction of depth, thus keeping such beams above the bump, regardless of multiple reflection or direct incidence; ② Fresnel absorption is closely associated with the angle of incidence, and the beams incident along the wall surface of the keyhole tend to reflect repeatedly inside the keyhole. Such beams are usually far from the center of the light spot, and thus carry less energy. Consequently, the multiple reflection absorption of such beams contributes little to the energy absorption of the wall surface of the keyhole.

4.3.2 Thermophysical Parameters

4.3.2.1 Influence of Thermal Diffusion Coefficient on Coupling Between Keyhole and Moving Weld Pool

This section mainly studies the influence of thermal diffusion coefficient α ($\alpha = k/(\rho c_p)$) on the transient coupling between a keyhole and moving weld pool during deep penetration laser welding. The following two cases of $\alpha = 1.769 \times 10^{-5}$ m^2/s and $\alpha = 6.923 \times 10^{-5}$ m^2/s are mainly considered. Other parameters involved in the simulation process are consistent with those in Sect. 4.2.2.

Figure 4.30 shows the shape of the keyhole and the longitudinal section velocity distribution of the weld pool at the welding moment of 15 ms under two different thermal diffusion coefficient values. The smaller the thermal diffusion coefficient, the greater the keyhole depth and the penetration depth, and in the event of a large thermal diffusion coefficient, the weld pool significantly narrows close to the front wall surface and the back edge of the keyhole, as shown in the figure. Besides, the lower the thermal diffusion coefficient, the faster the flow rate of the weld pool near the keyhole. Figure 4.31 shows the temperature field and velocity field distribution on the surface of the weld pool at the same moment shown in Fig. 4.30. The smaller the thermal diffusion coefficient, the smaller the width and length of the weld pool,

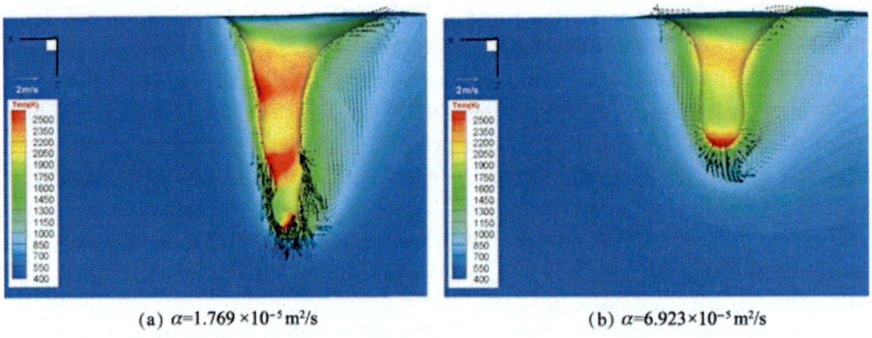

(a) α=1.769 ×10^{-5} m^2/s (b) α=6.923×10^{-5} m^2/s

Fig. 4.30 Shape of the keyhole and the longitudinal section velocity field distribution of the weld pool at the welding moment of 15 ms under different thermal conductivity values

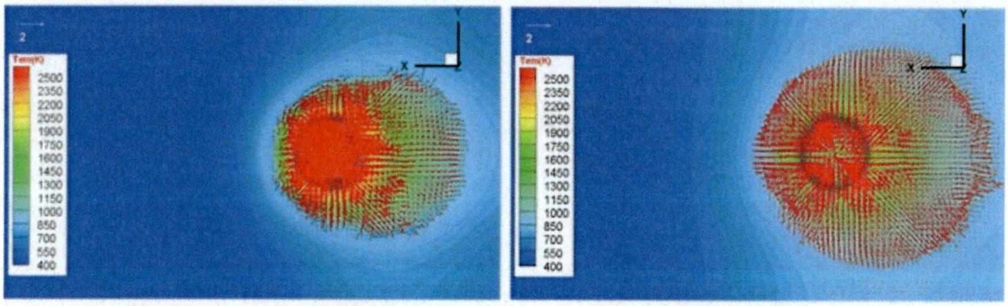

(a) α=1.769×10^{-5} m^2/s (b) α=6.923×10^{-5} m^2/s

Fig. 4.31 Temperature field and velocity field distribution on the surface of the weld pool at the welding moment of 15 ms under different thermal conductivity values

but the greater the temperature gradient on the surface of the weld pool, as shown in Fig. 4.31.

The process curves of keyhole depth and weld width with time are indicated in Fig. 4.32 and Fig. 4.33 respectively. A smaller thermal diffusion coefficient leads to a larger keyhole depth, which may further result in more violent oscillation of depth, as shown in Fig. 4.32. However, it can be seen from the figure that a smaller thermal diffusion coefficient leads to a smaller change in the keyhole depth, which means that the welding process is steadier at this time. Further analysis suggests that a smaller thermal diffusion coefficient results in a smaller range of flow of the weld pool in the process of welding weld pool formation, and the metal liquid inside the weld pool is greatly affected by the solid–liquid boundary layer effect of the weld pool, which mitigates the positive impact of the weld pool metal liquid on the wall surface of the keyhole, thereby improving the stability of the keyhole to a certain extent. As shown in Fig. 4.33, the larger the thermal diffusion coefficient, the faster the increase rate

Fig. 4.32 Variation curves of the keyhole depth with time under different thermal conductivity conditions

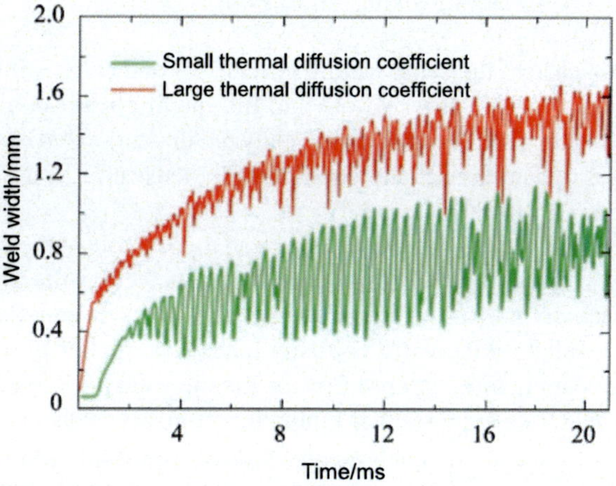

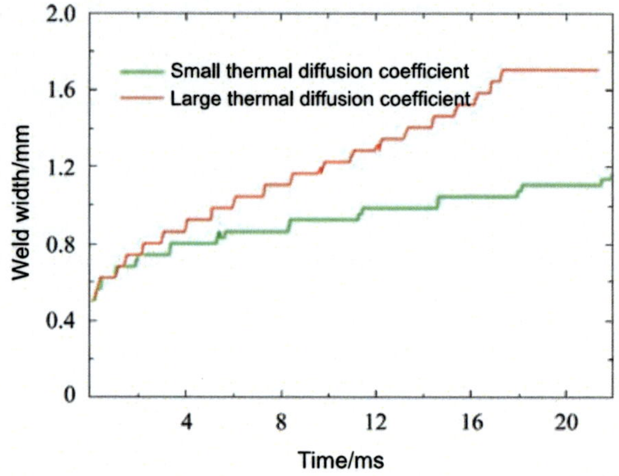

Fig. 4.33 Variation curves of the weld width with time under different thermal conductivity conditions

of the weld pool width (that is, the weld width), meaning the relative stable width value can be achieved at an earlier time.

The above results indicate that: ① The thermal diffusion coefficient exerts a great influence on the transient behavior of the keyhole and the moving weld pool during deep penetration laser welding; ② In case that other parameters are same, the alloy with lower thermal diffusion coefficient (such as stainless steel and titanium alloy) is easier to form deep and narrow weld profile (namely, penetration fusion welding) than the alloy with higher thermal diffusion coefficient (such as aluminum alloy); ③ A reduction in the thermal diffusion coefficient can improve the stability of the keyhole depth oscillation under certain conditions.

4.3.2.2 Influence of Kinematic Viscosity on Coupling Between Keyhole and Moving Weld Pool

In case of the same welding processes and calculation parameters as Sect. 4.2.2, the behaviors of the keyhole and the moving weld pool during deep penetration laser welding are compared and analyzed under the two cases of kinematic viscosity (ratio of dynamic viscosity coefficient to density), $\nu = 1.846 \times 10^{-6}$ m^2/s and $\nu = 15.38 \times 10^{-6}$ m^2/s.

Figure 4.34 shows the shape of the keyhole at the welding moment of 15 ms under two kinematic viscosity conditions, as well as the longitudinal section velocity and temperature distribution of the weld pool. It can also be seen from the figure that a smaller kinematic viscosity leads to a greater flow velocity inside the weld pool, meaning more intense flow inside the weld pool. Because of the effect of convective heat transfer, a smaller kinematic viscosity leads to a faster heat dissipation near the keyhole area, further resulting in a shallower depth of the keyhole.

Figure 4.35 shows the temperature field distribution on the surface of the weld pool at the welding moment of 15 ms under different kinematic viscosity values. It can

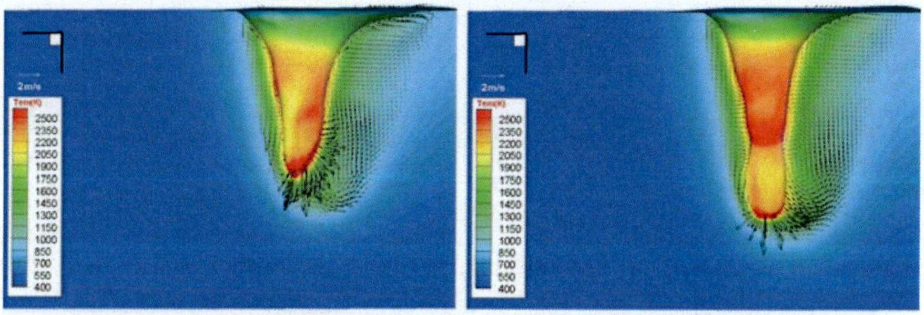

(a) $\nu=1.846\times10^{-6}\,m^2/s$ (b) $\nu=15.38\times10^{-6}\,m^2/s$

Fig. 4.34 Shape of the keyhole and the longitudinal section velocity distribution of the weld pool at the welding moment of 15 ms under different kinematic viscosity values

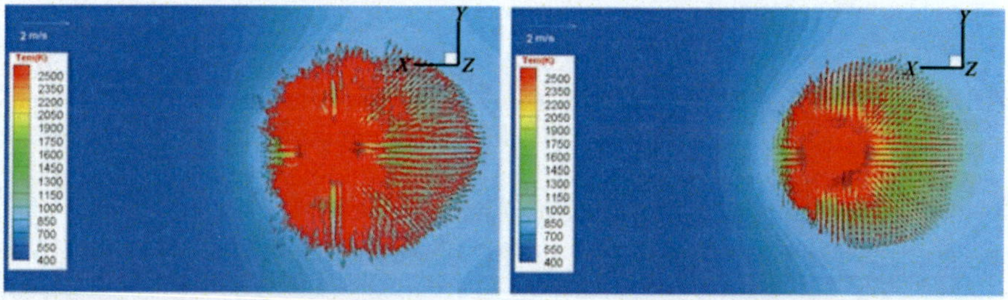

(a) $\nu=1.846\times10^{-6}\,m^2/s$ (b) $\nu=15.38\times10^{-6}\,m^2/s$

Fig. 4.35 Temperature field and velocity field distribution on the surface of the weld pool at the welding moment of 15 ms under different kinematic viscosity values

also be seen from the figure that, in case of small kinematic viscosity, the velocity on the surface of the weld pool is significantly high, which means the flow on the surface of the weld pool is more intense. Because of the effect of convective heat transfer, a smaller kinematic viscosity results in a larger surface area of the weld pool at the exact same moment. Figure 4.36 shows the variation curves of the keyhole depth with welding time under different kinematic viscosity values. In the stage of 4–10 ms of keyhole formation, the larger the kinematic viscosity, the larger the amplitude of the keyhole depth oscillation, as shown in Fig. 4.36a. After the average depth of the keyhole becomes relatively stable, the amplitude of the keyhole depth oscillation gets smaller in the event of large kinematic viscosity under the current study conditions, as shown in Fig. 4.36b. It is because that, in the early stage of keyhole formation, the flow with small kinematic viscosity inside the weld pool exerts a small impact force on the wall surface of the keyhole, thanks to the effect of friction between solid and liquid phases; after formation of the keyhole, the flow range inside the weld pool increases. At this point, the metal flow with small kinematic viscosity inside the weld pool exerts a more obvious impact force on the wall surface of the keyhole, thus causing an increase in the amplitude of the keyhole depth oscillation.

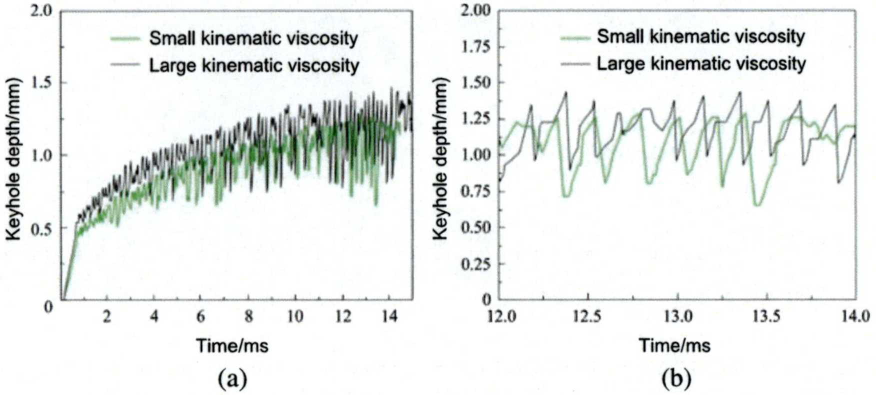

Fig. 4.36 Variation curves of the keyhole depth with welding time under different kinematic viscosity values

Therefore, the kinematic viscosity has no absolute monotonic function relationship with the keyhole depth oscillation, despite its important influence on the keyhole depth oscillation.

The foregoing results show that: ① The kinematic viscosity can greatly affect the flow behavior of the weld pool in case of deep penetration laser welding. The smaller the kinematic viscosity, the larger the weld penetration depth, but the gentler flow on the surface and interior of the weld pool; ② The kinematic viscosity also can greatly affect the behavior of the keyhole depth oscillation in case of deep penetration laser welding, but there is no monotonic function relationship between the kinematic viscosity and the keyhole depth oscillation.

4.3.3 Welding Process Parameters

4.3.3.1 Influence of Welding Speed on Coupling Between of Keyhole and Moving Weld Pool

In the event of the constant laser power of 3 kW in the simulation, the behaviors of the keyhole and the weld pool are studied under the three process conditions of welding speed of 2, 3 and 4 m/min. Other parameters used under the three process conditions are consistent with those in Sect. 4.2.2.

Figure 4.37 shows the change process of the keyhole depth with welding time under a constant power and different welding speeds. It can be seen from the figure that a lower welding speed leads to a larger average depth of keyhole. Moreover, in the stage of high-frequency oscillations around a relatively fixed depth, the keyhole depth oscillation has a small periodic variation under the current several process conditions, but the amplitude of oscillation gradually decreases with the increase of

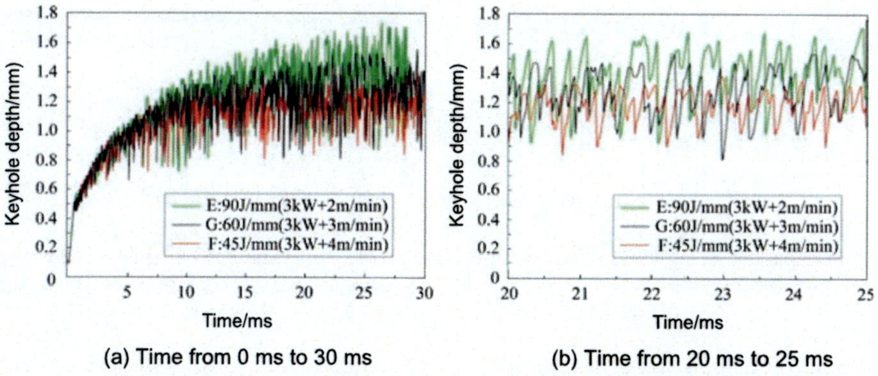

Fig. 4.37 Variation curves of the keyhole depth with welding time at different welding speeds

welding speed. It is mainly because that the increase in the welding speed lowers the average depth of the keyhole and improves the stability of the keyhole, making a small change in the amplitude. The constant surface tension coefficient under the three conditions makes a small periodic variation of the keyhole depth oscillation.

Figure 4.38 shows the temperature distribution on the surface of the workpiece at the moment of 25 ms at the welding speed of 2 m/min and 3 m/min respectively. According to the figure, with the increase of the welding speed, the weld pool gets narrower and longer at the same time. Besides, from the perspective of the temperature gradient, the temperature gradient on the surface of the weld pool becomes smaller with the increase of the welding speed. In spite of different welding speeds, the laser energy presents Gaussian distribution, which makes the surface temperature field distribution close to the moving Gaussian distribution pattern at each welding speed condition.

Figure 4.39 is the contrast diagram of keyhole shape and velocity distribution of weld pool longitudinal section at 25 ms with the welding speed of 2 m/min and 4 m/min respectively. According to the figure, as the welding speed increases, the penetration depth gradually decreases, and the flow of the weld pool is also

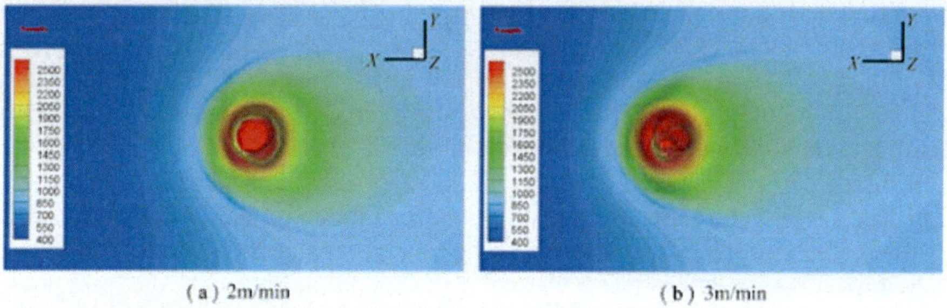

Fig. 4.38 Comparison diagram of temperature field distribution on the surface of the weld pool at the welding moment of 25 ms at different welding speeds

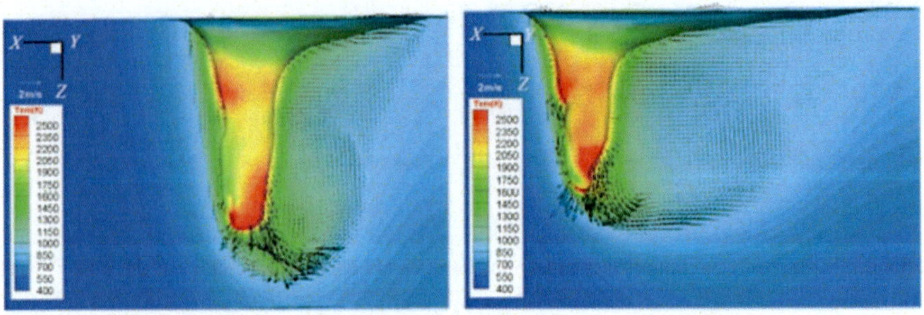

(a) 2m/min (b) 4m/min

Fig. 4.39 Comparison of keyhole shape and velocity distribution of weld pool longitudinal section at 25 ms with different welding speeds

significantly weakened. More notably, the vortex flow near the trailing edge of the bottom of the keyhole gradually decreases as the welding speed increases. Over-sized penetration depth and strong vortex flow at the bottom of the weld pool go against bubble floating, thus increasing the welding speed may contribute to an increase in the probability of bubbles overflowing from the weld pool.

The above simulation results can be concluded as follows. ① welding speed has a significant effect on the depth dimension and variation of depth dimension of the keyhole, the size of the weld pool, the velocity of the weld pool and the temperature gradient inside the workpiece; ② under certain conditions, as welding speed increases, the amplitude of the keyhole depth oscillation significantly decreases, but the influence on the oscillation period of the keyhole depth is not obvious; ③ under certain conditions, as welding speed increases, the penetration depth and the flow speed of the weld pool decrease, and the vortex flow in near the trailing edge of the bottom of the keyhole decreases.

4.3.3.2 Influence of Laser Power on Coupling Between Keyhole and Moving Weld Pool

In the numerical simulation, the laser power was 1.3, 1.5 and 2.0 kW and the welding speed was 3 m/min. Other parameters are the same as those in Sect. 4.2.2

Figure 4.40 illustrates the variation of the keyhole depth with the welding time under different laser powers. According to the figure, the higher the laser power, the earlier the keyhole forms. And the growth rate of the keyhole depth also increases with the increase of laser power in the forming stage. In addition, when the keyhole depth reaches a relatively stable average value, the oscillation of the hole depth increases with the increase of laser power, but the change of the oscillation period is not obvious. It mainly lies in that under the same condition, the larger the laser power, the deeper the average keyhole depth, thus the greater the oscillation of keyhole depth. As can be seen from Sect. 4.3.1, the oscillation period of keyhole depth is

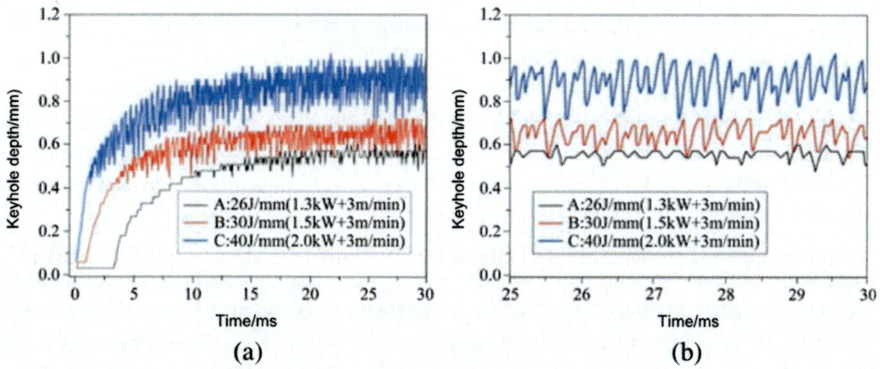

Fig. 4.40 Variation curve of keyhole depth under different laser powers

greatly influenced by the surface tension coefficient. Nevertheless, the surface tension coefficient of the aforementioned technical conditions is a certain value, resulting in less obvious periodic change. Figure 4.41 compares the keyhole shape and velocity field of the weld pool longitudinal section at 20 ms with the laser power of 1.5 kW and 2.0 kW respectively. According to the figure, as the laser power increases, the weld penetration depth increases; meanwhile, the vortex flow in the weld pool near the trailing edge of the bottom of the keyhole significantly increases.

According to the comparison and discussion of the simulation results above, it can be concluded as follows. ① with the increase of the laser power, the growth rate of the keyhole depth increases in the keyhole formation stage; ② under certain conditions, with the increase of laser power, the amplitude of keyhole depth oscillation gradually increases, but the change of the oscillation period is not obvious; ③ under certain conditions, the vortex flow in the weld pool near the trailing edge of the bottom of the keyhole significantly increases with the increase of laser power.

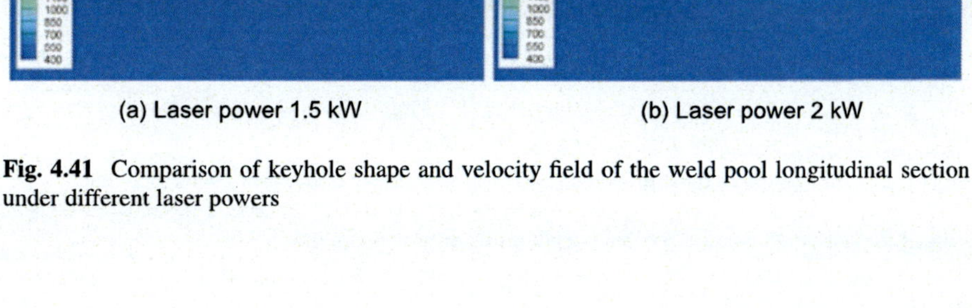

Fig. 4.41 Comparison of keyhole shape and velocity field of the weld pool longitudinal section under different laser powers

4.3.3.3 Influence of Spot Radius on Coupling Between Keyhole and Moving Weld Pool

Under the condition that other welding technical parameters and calculation parameters are the same with that in Sect. 4.2.2, the following study mainly focuses on the transient behavior of the keyhole and the moving weld pool under the condition that the spot radius is 0.3 mm and 0.4 mm, respectively.

Figure 4.42 illustrates the variation of the keyhole depth with time under two spot radii. As can be seen from the figure, with smaller spot radius, the laser power density is higher, thus the speed of keyhole forming is faster and the average depth and maximum depth of the keyhole are deeper. In addition, as the light spot is small, the depth of the keyhole is deeper, and the amplitude of the keyhole depth oscillation increases compared with the condition of the larger spot, as shown in Fig. 4.42a. However, under the current technical condition, the variation trend of the frequency of the keyhole depth oscillation is not obvious with the change of the spot radius, as shown in Fig. 4.42b. According to Sect. 4.3.1, the frequency of keyhole depth oscillation is greatly influenced by the surface tension coefficient. Thus, under the current technical condition with the same surface tension coefficient, there is less obvious depth oscillation frequency with two spot radii.

Figures 4.43 and 4.44 show the variation of weld penetration depth and weld pool volume with welding time under two spot radii. According to the figures, as the spot radius increases, the distribution of laser power is more scattered and the power density is lower, leading to lower weld penetration depth and its growth speed. In addition, under the current technical condition, when the spot radius increases, although the penetration depth at this time decreases, the width of the weld pool will be larger, making a slight difference of weld pool volume under two spot radii in the keyhole formation stage (welding time <12 ms). However, as the welding process goes on, the contribution of penetration depth to volume takes advantage. Therefore, at 12–20 ms after the average keyhole depth stabilizes, with the smaller spot, the volume of weld pool is larger and the volume value grows faster. At the same time,

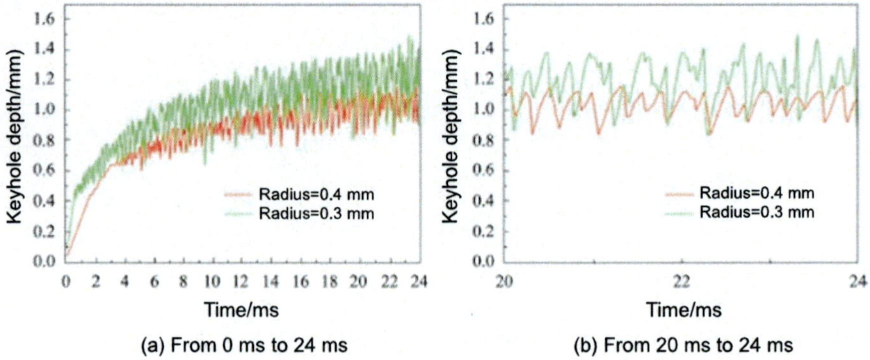

Fig. 4.42 Variation curve of keyhole depth under different spot radii

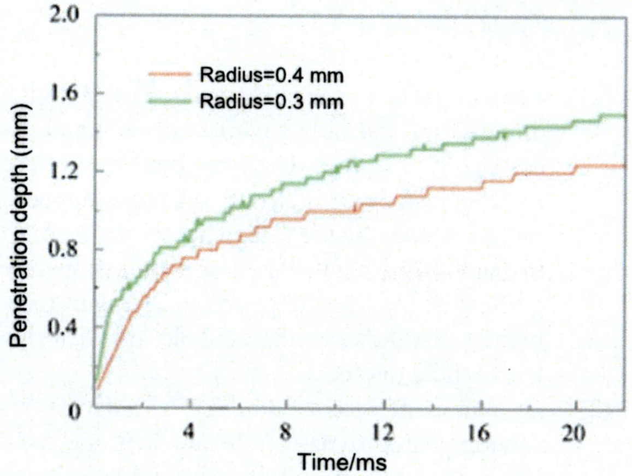

Fig. 4.43 Variation curve of penetration depth with time under different spot radii

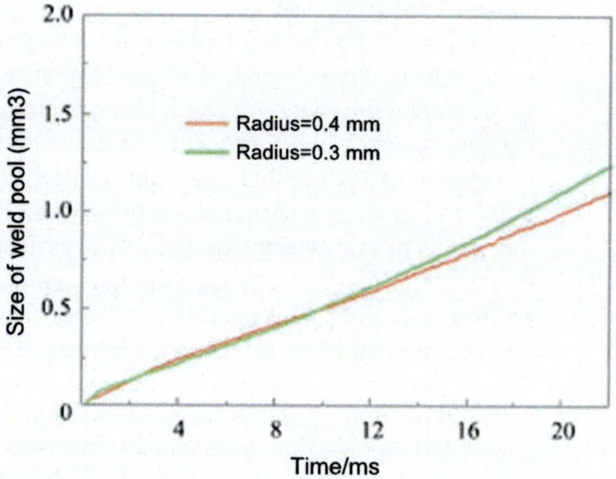

Fig. 4.44 Variation curve of weld pool volume with time under different spot radii

the change of spot also influences the penetration depth and width. And according to Fig. 4.44, it is clear that there is no monotonic relationship between the weld pool volume and spot radius.

According to the comparison and discussion of the simulation results above, it can be concluded as follows. ① spot radius is highly related to penetration depth; specifically, the larger the spot, the shallower the penetration depth; ② the larger the spot, the less amplitude of keyhole depth, but the change of the oscillation period is not obvious; ③ spot radius has a great influence on weld pool volume, but there is no simple monotonic relationship between them.

4.4 Summary

(1) For the first time, the numerical simulation visually reveals the unsteady three-dimensional transient keyhole behavior and the flow field inside the weld pool that are in line with the X-ray penetration test result, and it is found that the keyhole instability is closely related to the periodic boss on the pore wall. The boss is caused by the imbalance of the recoil pressure, and surface tension of the wall surface of the keyhole and the impact force of the fluid. The instability of the keyhole leads to periodic closing of the keyhole, forming bubbles at the bottom or middle of the keyhole, thus causing high-frequency oscillation of the keyhole depth.

(2) By means of visual numerical simulation, stable keyhole shape and corresponding fluid dynamics behavior in the weld pool that are in line with the X-ray are penetration test result obtained for the first time. When the keyhole is stable, the flow velocity near the keyhole wall is almost parallel to the hole wall and moves upward.

(3) Recoil pressure is the key factor to form keyhole and obtain deep penetration laser welding. Intense and persistent evaporation is the main factor to distinguish conduction welding from deep-penetration welding. With less surface tension, the oscillation frequency of keyhole depth is low, but the amplitude of keyhole depth oscillation is higher. The thermal capillary force has great influence on the speed of weld pool surface. The larger the thermal capillary force, the more intense the weld pool surface movement, but the penetration depth is smaller. Although thermal capillary force has great influence on keyhole and weld pool coupling, it is not the decisive factor for the formation of the weld pool. In general, the energy absorbed by the keyhole wall is mainly determined by the initial Fresnel absorption and the first and second multiple reflection Fresnel absorptions. With the same other parameters, the alloy welding process with low thermal conductivity is more stable, but there is no simple monotonic function relationship between the dynamic viscosity and keyhole stability. Under certain conditions, appropriate improvement of welding speed, reduction of welding power, and an increase of spot radius, i.e., reduction of line energy and power density can effectively reduce the amplitude of keyhole depth oscillation, to obtain deep penetration laser welding without keyhole oscillation. However, appropriately reducing line energy and power density has a slight influence on the oscillation period of keyhole depth.

Chapter 5
Dynamic Behaviors of Metal Vapor/Plasma Plume Inside Transient Keyhole

Abstract This chapter introduces the time dependent plasma/plume dynamics inside the keyhole during laser welding. Mathematical models of the transient high-speed plasma/plume dynamics based on compressible Euler flow assumptions are given. Characteristics of the time dependent plasma/plume behaviors such as uneven distribution, multiple flow directions, strong oscillations and local evaporations at keyhole wall are found and discussed by numerical simulations.

5.1 Introduction

At present, investigation of metal vapor dynamic behaviors in transient keyholes during deep penetration laser welding is very limited. Due to small size of the keyhole, and high temperature and drastic changing of metal vapor in the keyhole, it is quite difficult to detect the physical state of metal vapor inside the keyhole during welding process by experimental method. Numerical simulation of metal vapor in the keyhole may provide an effective method for better understanding the welding process. Based on the physical process and experimental results of laser welding and the keyhole profile and weld pool flow patterns consistent with the experimental results of X-ray, and with compressibility of the metal vapor and effect of recoil pressure and ambient pressure taken into account, a mathematical model is established to describe the dynamical behaviors of metal vapor inside a keyhole during deep penetration laser welding, and corresponding boundary conditions for compressible gas–liquid interface in line with the physical reality are derived.

During laser welding, the moving weld pool, transient keyhole and compressible metal vapor are closely related. The dynamic behaviors of metal vapor in the transient keyhole exert an effect on the variation of the weld pool and the evolution of the keyhole, thus affecting the quality of the whole welding process. The moving state of metal vapor in the keyhole is a link between the shape of the vapor outside the keyhole and the morphology of the keyhole and the welding quality, and its kinetic study is helpful to realize on-line monitoring by high-speed video recording. The simulation study on the properties of metal vapor in the keyhole and the experimental study can guide and draw lessons from each other. The dynamic state of metal vapor in the

S. Gong et al., *Weld Pool Dynamics in Deep Penetration Laser Welding*,
https://doi.org/10.1007/978-981-16-0788-2_5

keyhole is compared under different welding process conditions. At the same time, the spouting velocity and oscillating degree of the metal vapor at the opening are studied under both steady and unsteady welding conditions. The results are helpful to guide the design and optimization of laser welding process.

5.2 Dynamic Model of Metal Vapor/Plasma in Transient Keyhole

Because the process of deep penetration laser welding is very complicated, in this model, the main influencing factors in deep penetration laser welding process are considered, the secondary factors are ignored, and the model is simplified on the premise of accurately and reasonably describing the welding process. In order to improve the efficiency of simulation calculation. The following specific assumptions are made:

(1) The molten metal liquid is considered incompressible and the mixed phase model is used for the solid–liquid interface treatment.
(2) The metal vapor produced during welding is an ideal compressible gas, the density of which varies with time and position, with the effect of viscosity ignored.
(3) At the gas–liquid interface of the keyhole, the effect of the Knudsen layer is not considered for the gas phase boundary, while the refraction, reflection and absorption of metal vapor to the laser are taken into account for the laser energy distribution in the keyhole.
(4) It is assumed that the main driving force of metal vapor is the surface pressure of coupling ambient pressure and recoil pressure.
(5) The effect of the protective gas added in the actual welding process on the whole laser welding process is ignored in the study.

5.2.1 *Governing Equations*

(1) Governing equations of transient keyhole and moving weld pool behaviors

Accurate dynamic keyhole profiles and surface temperature distributions serve as the basis for modeling the metal vapor dynamics during laser welding. These physical characteristics can be determined by free surface heat transfer and fluid flow calculations of weld pool. The equations of mass conservation and momentum conservation depicting the heat transfer and fluid flow of the liquid metal in the study are as follows:

$$\nabla \cdot \vec{U}_l = 0 \tag{5.1}$$

$$\rho_l(\frac{\partial \vec{U}_l}{\partial t} + (\vec{U}_l \cdot \nabla)\vec{U}_l) = \nabla \cdot (\mu_l \nabla \vec{U}_l) - \nabla p_l - \frac{\mu_l}{K}\vec{U}_l - \frac{c_p}{\sqrt{K}}\left|\vec{U}_l\right|\vec{U}_l + \rho_l \vec{g}\beta(T_l - T_{ref}) \tag{5.2}$$

$$\rho_l c_p(\frac{\partial T_l}{\partial t} + (\vec{U}_l \cdot \nabla)T_l) = \nabla \cdot (k_l \nabla T_l) \tag{5.3}$$

where: Subscript l—Molten liquid in weld pool;

$\vec{U}_l, \rho_l, p_l, T_l$—Velocity vector, density, pressure, temperature of the liquid metal in the weld pool;

μ_l—Viscosity;

k_l—Thermal conductivity;

T_{ref}—Reference temperature;

$\vec{g}$—Acceleration of gravity;

β, c_p—Diffusion coefficient, specific heat capacity;

K—Seepage coefficient.

In laser welding, drastic topological variations of keyhole profiles could occur throughout the welding process. Here, the Level Set method is used to track the evolution of keyhole's free surface. The free interface of the transient keyhole can be depicted by the following equation:

$$\frac{\partial \phi}{\partial t} + \vec{U}_l \cdot \nabla \phi = 0 \tag{5.4}$$

where ϕ—Distance function based on Level Set method.

The normal vector $\vec{n}$ and the curvature value κ at any point on the surface of the keyhole can be obtained by the following formulas:

$$\vec{n} = \frac{\nabla \phi}{|\nabla \phi|} \tag{5.5}$$

$$\kappa = \nabla \cdot \frac{\nabla \phi}{|\nabla \phi|} \tag{5.6}$$

(2) Dynamic governing equations of compressible metal vapor in transient keyhole

Here, Euler equations are used to depict the compressible inviscid metal vapor in the keyhole, and the matrix form of the governing equations is expressed as follows:

$$\frac{\partial \vec{Q}}{\partial t} + \frac{\partial \vec{F}}{\partial x} + \frac{\partial \vec{G}}{\partial y} + \frac{\partial \vec{H}}{\partial z} = \vec{S} \tag{5.7}$$

where

$$\vec{Q} = \begin{bmatrix} \rho \\ \rho u \\ \rho v \\ \rho w \\ E \end{bmatrix}, \vec{F} = \begin{bmatrix} \rho u \\ \rho u^2 + p \\ \rho uv \\ \rho uw \\ u(E+p) \end{bmatrix}, \vec{G} = \begin{bmatrix} \rho v \\ \rho uv \\ \rho v^2 + p \\ \rho vw \\ u(E+p) \end{bmatrix}, \vec{H} = \begin{bmatrix} \rho w \\ \rho wu \\ \rho wv \\ \rho w^2 + p \\ u(E+p) \end{bmatrix},$$

$$\vec{S} = \begin{bmatrix} 0 \\ 0 \\ 0 \\ 0 \\ \nabla(k_g \nabla T) + \eta I_R \end{bmatrix};$$

ρ—Density of metal vapor;
u—Velocity component in X direction;
v—Velocity component in Y direction;
w—Velocity component in Z direction;
p—Pressure;
T—Temperature;
E—Total energy;
k_g—Thermal conductivity of metal vapor;
I_R—Laser energy density, usually assumed as a Gauss distribution;
η—Coefficient with absorption and scattering of laser energy by metal vapor taken into account.

From the state equation of ideal gas, we can get:

$$p = \rho R_a T = (\gamma - 1) \times (E - \frac{1}{2}\rho(u^2 + v^2)) \tag{5.8}$$

where $R_a = R/M_a$;
R_a—Constant of ideal gas state;
M_a—Molar mass of the metal vapor;
γ—Specific heat ratio.

5.2.2 Boundary Conditions

Figure 5.1 is a schematic illustration of the boundary conditions of the weld pool, keyhole, and metal vapor coupling in the deep penetration laser welding process. The laser welding process involves multiple physical phenomena and can be divided into multiple phases, and features complex momentum and energy coupling boundary conditions between laser beam, material and the ambient environment, including energy increased due to Fresnel absorption of the laser beam and vapor energy loss due to radiation and convection. Moreover, evaporation and condensation usually

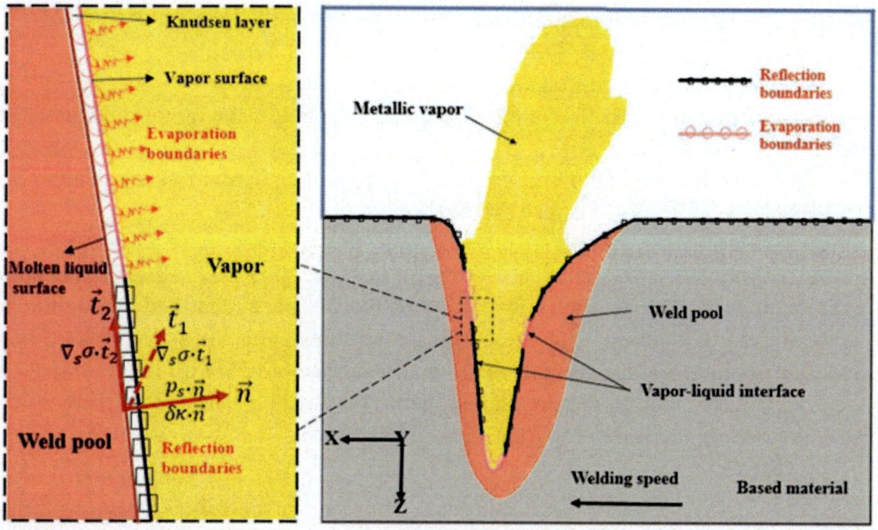

Fig. 5.1 Schematic illustration of coupling boundary conditions in deep penetration laser welding process

take place simultaneously on different locations due to the difference of temperature distribution on the wall of the keyhole. The detailed settings of coupling boundary conditions for the proposed multiphase model of laser welding are as follows.

5.2.2.1 Boundary Conditions for Transient Weld Pool and Keyhole

The latest research shows that the discontinuous boundary method can obtain better simulation results of the weld pool and keyhole. Based on the discontinuous boundary model proposed by Pang Shengyong, the effects of ambient pressure, recoil pressure, surface tension and Marangoni force on the boundary of weld pool and keyhole are coupled, so the momentum boundary conditions of the wall of the keyhole in the model can be expressed as follows:

$$p_r = p_s + \delta\kappa + 2\mu_l \vec{n} \cdot \vec{U_l} \cdot \vec{n} \tag{5.9}$$

$$\begin{aligned}(\mu_l \nabla \vec{U_l}) = {} & \mu_l(\vec{n} \quad \vec{t_1} \quad \vec{t_2})(\vec{n} \quad 0 \quad 0)(\nabla \vec{U_l})(\vec{n} \quad 0 \quad 0)(\vec{n} \quad \vec{t_1} \quad \vec{t_2})^T \\ & + \mu_l(\vec{n} \quad \vec{t_1} \quad \vec{t_2})(0 \quad \vec{t_1} \quad \vec{t_2})^T(\nabla \vec{U_l}) \\ & - \mu_l(\vec{n} \quad \vec{t_1} \quad \vec{t_2})(\vec{n} \quad 0 \quad 0)(\nabla \vec{U_l})(0 \quad \vec{t_1} \quad \vec{t_2})(\vec{n} \quad \vec{t_1} \quad \vec{t_2})^T \\ & + (\vec{n} \quad \vec{t_1} \quad \vec{t_2})\begin{pmatrix} 0 & \nabla_S\sigma \vec{t_1} & \nabla_S\sigma \vec{t_2} \\ 0 & 0 & 0 \\ 0 & 0 & 0 \end{pmatrix}(\vec{n} \quad \vec{t_1} \quad \vec{t_2})^T \end{aligned} \tag{5.10}$$

where p_s—Pressure at free interface;
p_r—Recoil pressure;
δ—Surface tension coefficient;
$\vec{n}$—Normal unit vector of area differential element of the liquid metal near the free interface of the keyhole;
$\vec{t_1}$, $\vec{t}_2$—Two tangential unit vectors that are perpendicular to each other;
$\nabla_s = (I - \vec{n}\vec{n})\nabla$—Surface gradient operator;
$\nabla_s\sigma$—Marangoni force.

Most of previous mathematical models for laser welding did not address the effect of ambient pressure. Recently, however, this effect has been identified as a significant factor for keyhole evaporation. According to studies Pang Shengyong and others, dynamic evolution of the keyhole is driven by surface pressure of the interface, i.e., the joint force of recoil pressure and ambient pressure. The surface pressure p_s can be expressed by the following formula:

$$p_s = \begin{cases} p_{atm} & 0 \le T_{gb} \le T_L \\ \frac{1+\beta_R}{2} P_{atm} \exp(\frac{\Delta H_v}{k_B T_v}(1 - \frac{T_v}{T_{gb}})) & +\infty > T_{gb} \ge T_R \\ p_c(T_{gb}) & T_L \le T_{gb} < T_R \end{cases} \tag{5.11}$$

where p_{atm}—Standard atmospheric pressure;
k_B—Boltzmann constant;
T_v—Boiling point under normal atmospheric pressure;
β_R—Fraction of re-condensed vapor;
T_{gb}—Temperature of metal vapor near the free interface of the keyhole;
ΔH_v—Entropy of evaporative phase change.
ΔH_v and $p_c(T)$ can be calculated by formulas (5.11) and (5.12).

$$\Delta H_v = mL_v \tag{5.12}$$

where m—Atomic mass;
L_v—latent heat of vaporization.

$$p_c = (T_{gb}) = ax^3 + bx^2 + cx + d \tag{5.13}$$

$p_c(T_{gb})$ is a geometric smooth curve, as indicated by the red line in Fig. 5.2. T_L and T_R represents temperature values at tangent points respectively. The values of T_L, T_R, a, b, c and d are given in Table 5.1.

Multiple reflections, convection, radiation, and evaporation and cooling on laser energy of the surface of the keyhole are considered, and the temperature boundary conditions can be expressed as follows:

$$k_l \frac{\partial T_l}{\partial \vec{n}} = q - h(T_l - T_m) - \varepsilon_r \sigma_s (T_l^4 - T_\infty^4) - \rho_l V_{\exp} L_v \tag{5.14}$$

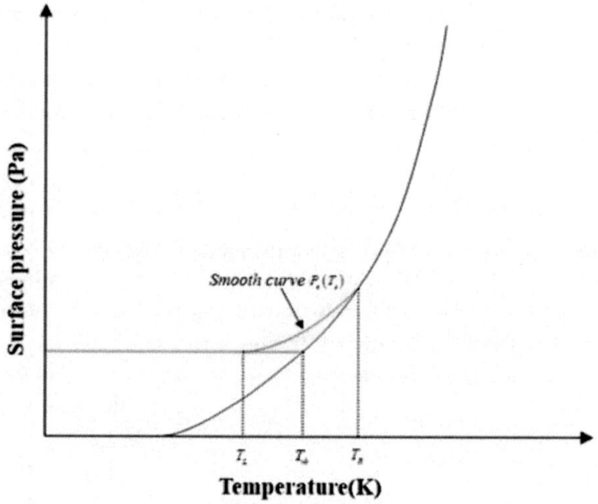

Fig. 5.2 Surface pressure versus temperature curve in laser welding process

Table. 5.1 Relevant parameters of surface pressure model in laser welding of 304 stainless steel

Ambient pressure	T_L/K	T_R/K	a	b	c	d
1 bar[a]	2950	3600	9.88E−4	−8.63	2.51E4	−2.43E7

[a]1 bar = 100 kPa

where q—Heat flux due to multiple Fresnel absorption;

h—Heat transfer coefficient for air convection;

ε_r—Black-body radiation coefficient;

σ_s—Stefan-Boltzmann constant;

T_∞—Ambient temperature;

V_{evp}—Interface recession speed due to evaporation.

Boundary temperature conditions for other boundaries can be expressed as follows:

$$k\frac{\partial T_l}{\partial \vec{n}} = -h(T_l - T_\infty) - \varepsilon_r\sigma_s(T_l^4 - T_\infty^4) \tag{5.15}$$

5.2.2.2 Dynamic Boundary Conditions on the Hole Walls of Metal in the Keyhole

The keyhole keeps oscillating in the deep penetration welding process resulting in real-time dynamic changing of temperature distributions on the wall of the keyhole. Therefore, the area of evaporation on the wall of the keyhole is changing with the time. To treat this phenomenon rationally, a temperature-dependent dynamic boundary conditions are proposed for dynamic modeling of compressible vapor in the keyhole. The principle is as follows: The vapor–liquid interface is divided into evaporation region and condensation region for each transient state; based on the temperature value on the wall of the keyhole, if the temperature of a certain local position is higher than a certain value (set around the boiling point in this paper), it is set as the evaporation boundary; otherwise, simple reflection boundaries are imposed for local positions of the keyhole with temperatures lower than the threshold.

For the evaporation boundaries, the metal vapor generated due to evaporation of the liquid metal on the wall of the keyhole enters inside the keyhole, and the metal generated is driven by the surface pressure described by formula (5.11).

Assuming the mass conservation law is satisfied during the evaporation of molten metal, the vapor density ρ_g on the evaporation boundary can be obtained approximately by the following formula:

$$\rho_g = (\frac{M_a}{N_a k_B})\frac{p_r}{T_{gb}} \tag{5.16}$$

where M_a—molar mass;

N_a—Avogadro constant.

The velocity of the metal at the evaporation boundary approximates to the velocity of the vapor at the nearest neighboring grid, the direction is assumed to be perpendicular to the wall. It can be expressed as follows:

$$\frac{\partial \overrightarrow{U_g}}{\partial \vec{n}} = \vec{0} \tag{5.17}$$

where $\overrightarrow{U_g}$—Velocity vector of metal vapor.

For the condensation area on the wall of the keyhole, the velocity of the metal vapor meets the reflection boundary conditions. Assuming the velocity perpendicular to the tangent direction of the boundary is reversed, the velocity direction along the tangential direction of the boundary line remains unchanged, let $\overrightarrow{U_g} = \left(u_g\ v_g\ w_g\right)$, u_g is the velocity component value in normal direction of $\vec{U}_g$, v_g and w_g are the velocity component values in two tangential directions of $\vec{U}_g$, i and $i+1$ represent two adjacent grid nodes at the boundary, and the velocity boundary condition on the reflection boundary can be expressed as:

$$\left(u_g\right)_{i+1} = -(u_g)_i,\ (v_g)_{i+1} = (v_g)_i,\ (w_g)_{i+1} = (w_g)_i \tag{5.18}$$

With the effect of the Knudsen layer ignored, the temperature of metal vapor on the evaporation boundary approximates to that of the wall of the keyhole.

$$T_{gb} = T_k \tag{5.19}$$

where T_k—Wall temperature of the keyhole for the point.

The outermost layer of the calculation area above the keyhole is set as the outlet, and its pressure boundary is set to the atmospheric pressure by the first type boundary condition.

5.3 Dynamics of Metal Vapor/Plasma in Transient Keyhole

5.3.1 Uneven Distribution and High Transient

5.3.1.1 Unevenness

Figure 5.3 illustrates the transient temperature distribution of the metal vapor inside the keyhole during laser welding at different times (0.350 ms, 1.841 ms, 2.990 ms, 5.308 ms, 15.277 ms and 20.844 ms) with laser power at 1.5 kW and welding speed at 3 m/min (The black line represents the outline of the keyhole, the inside of the keyhole indicates the temperature distribution of the metal vapor, and the outside indicates the temperature distribution of the weld pool and the unmelted substrate). The temperature distribution of the metal vapor is also uneven because the location in the keyhole directly heated by the laser is near the fore edge of the keyhole and the energy distribution on the wall of the keyhole is uneven due to multiple reflections and Fresnel absorption. Figure 5.3c–f shows that there is a high temperature region in the lower part of the keyhole. In particular, as shown in Fig. 5.3f, the lower temperature of the metal vapor is about 3000 °C, while the temperature in higher temperature region can reach above 4500 °C, and you can see the temperature difference in the keyhole can reach 1000 °C.

Figure 5.4 illustrates the transient pressure distribution of the metal vapor inside the keyhole during laser welding at different times (0.350 ms, 1.841 ms, 2.990 ms, 5.308 ms, 15.277 ms and 20.844 ms) with laser power at 1.5 kW and welding speed at 3 m/min (The black line represents the outline of the keyhole, the inside of the keyhole indicates the pressure distribution of the metal vapor, and the outside indicates the pressure distribution of the weld pool and the unmelted substrate). At the free interface of the keyhole under the action of a high power density laser beam, the molten metal vapor forms a huge recoil pressure, which drives the flow of the metal vapor in the keyhole. Because the gradient of the recoil pressure on the wall of the keyhole is very large (up to several atmospheric pressure) and the keyhole experiences complicated morphological changes, the vapor pressure distribution in the keyhole is very uneven. From Fig. 5.4b–f, it can be seen that there is a negative pressure (lower than the atmospheric pressure) inside the keyhole near the opening, and from Fig. 5.4b–e, it can be seen that there is a high pressure region near the wall of the keyhole where the pressure is higher than the atmospheric pressure by about 20 kPa.

Figure 5.5 illustrates the transient density distribution of the metal vapor inside the keyhole during laser welding at different times (0.350 ms, 1.841 ms, 2.990 ms,

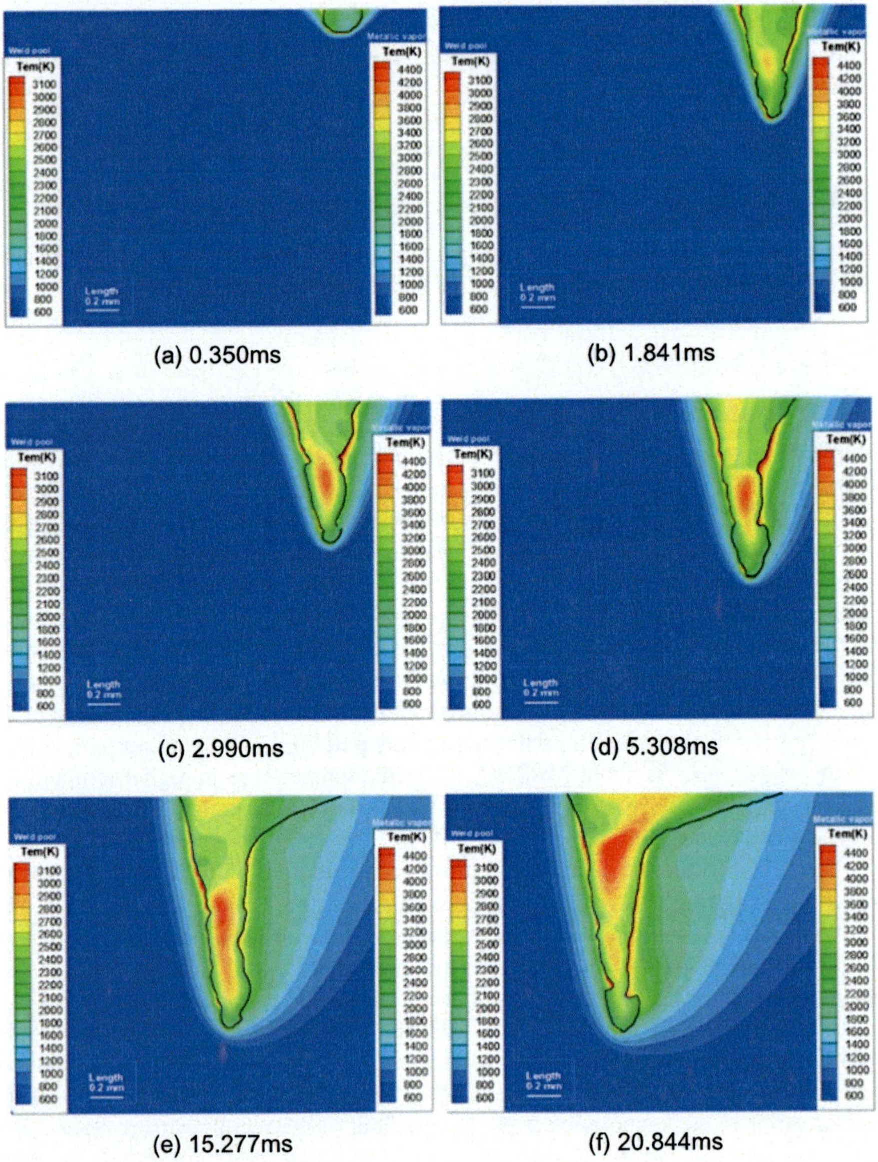

Fig. 5.3 Evolution of temperature distribution of metal vapor with time in the keyhole (P = 1.5 kW, V = 3 m/min)

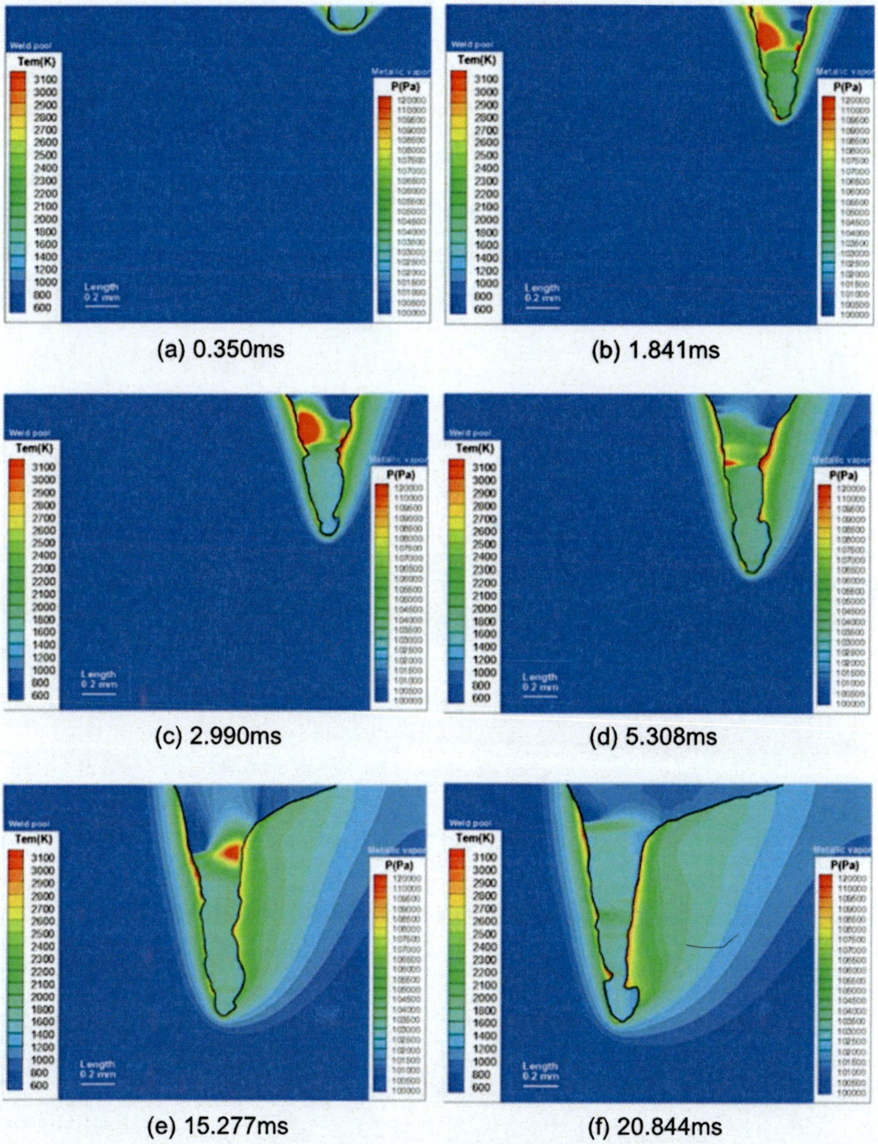

Fig. 5.4 Evolution of pressure distribution of metal vapor with time in the keyhole (P = 1.5 kW, V = 3 m/min)

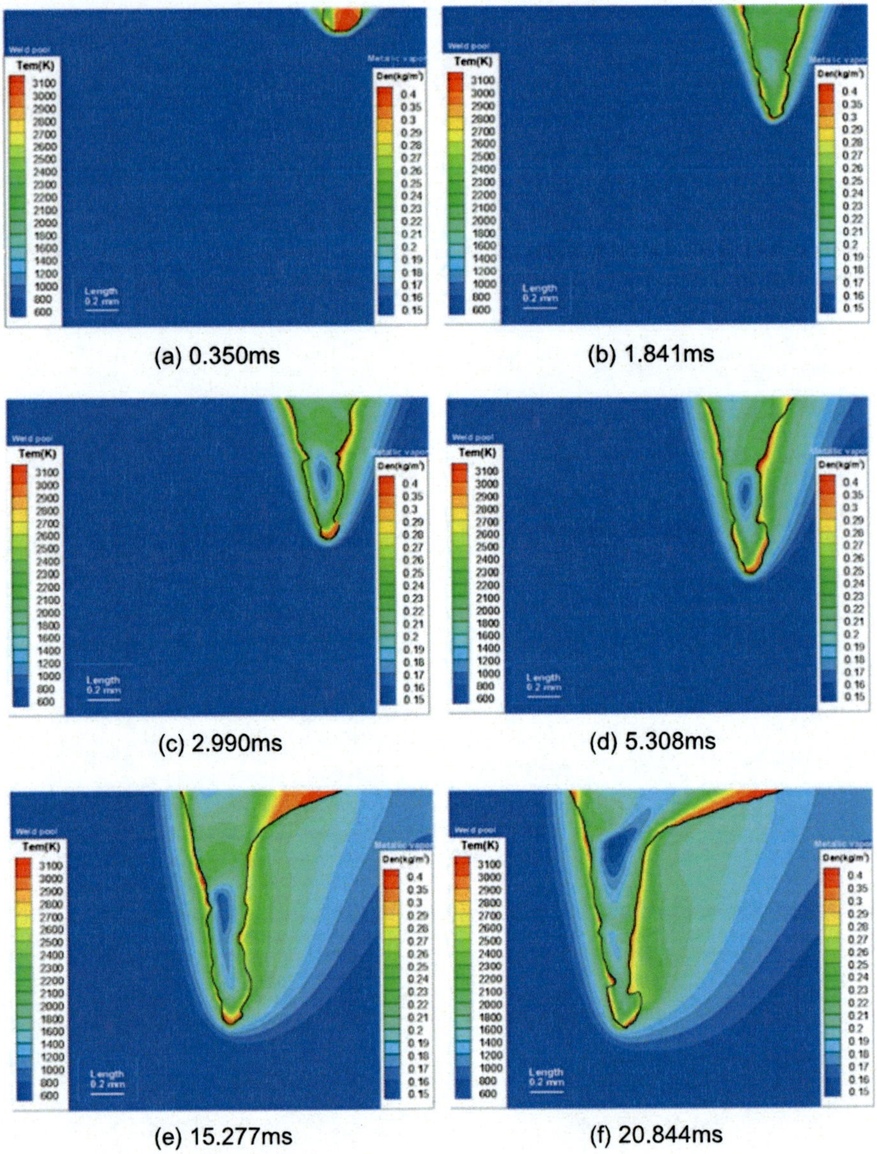

Fig. 5.5 Evolution of density distribution of metal vapor with time in the keyhole (P = 1.5 kW, V = 3 m/min)

5.308 ms, 15.277 ms and 20.844 ms) with laser power at 1.5 kW and welding speed at 3 m/min. (The black line represents the outline of the keyhole, the inside of the keyhole indicates the density distribution of the metal vapor, and the outside indicates the density distribution of the weld pool and the unmelted substrate). As we know from the state equation of ideal gas $\rho = p/R_a T$, under the conditions of uneven pressure distribution (Fig. 5.2) and temperature distribution (Fig. 5.1), the density distribution of vapor in the keyhole is also uneven. There exists a low density region in the center of the keyhole where the density is lower than 0.2 kg/m^3, far lower than the ambient gas (air and argon) density. The low density might be caused by the high-velocity flow which carries away excessive metal vapor from the region. The density at the opening of the keyhole is higher, which might be the result of the action of ambient gases (protective gas and air) and compression of the metal vapor. The density of the metal vapor is higher at the bottom of the keyhole, which might the result of compression of the metal vapor.

Figure 5.6 illustrates the transient Mach number distribution of the metal vapor inside the keyhole during laser welding at different times (0.350 ms, 1.841 ms, 2.990 ms, 5.308 ms, 15.277 ms and 20.844 ms) with laser power at 1.5 kW, and welding speed at 3 m/min. (The black line represents the outline of the keyhole, the inside of the keyhole indicates the Mach number distribution of the metal vapor, and the outside indicates the Mach number distribution of the weld pool and the unmelted substrate). As shown in Fig. 5.6, the Mach number distribution of vapor in the keyhole is also uneven. The Mach number of vapor at the bottom of the keyhole is smaller and can be as small as 0.01, but due to the presence of high temperature on the wall of the local keyhole, the molten metal vaporizes and spews out metal vapor at a very high speed, resulting in a relatively large Mach number distribution, and the largest Mach number can be larger than 0.8. Meanwhile, even the keyhole depth reaches a quasi-steady state (Fig. 5.6c, d), the Mach number still varies violently due the drastic changing of the profile of the keyhole.

The distributions of temperature, pressure, density and Mach number of the vapor in the keyhole and its evolution with time shown in Figs. 5.3, 5.4, 5.5 and 5.6 indicates that the metal vapor plume in the transient keyhole features nonuniformity during deep penetration laser welding.

5.3.1.2 High Transient Property of Metal Vapor

Figure 5.7 shows the evolution of velocity of vapor in the keyhole during the 1.256 us transient period from 11.717949 to 11.719205 ms with laser power at 1.5 kW and welding speed at 3 m/min. In Fig. 5.7a, there is a small hump on the right lower part of the wall of the keyhole, and the temperature of the hump is rather high, about 3100 °C. As shown in Fig. 5.7a–f, violent evaporation occurs at that region, and strong metal vapor are generated there. The time that the velocity experiences significant changes is only 396 ns, as shown in Fig. 5.7a, b. With the progress of the welding process, at 11.718604 ms, the velocity field inside the keyhole approaches to be a quasi-steady state except that the metal vapor at the hump continue to increase slightly.

Figure 5.8 and Fig. 5.9 respectively show the evolution of pressure and Mach Number of vapor in the keyhole during the 1.256 us transient period from 11.717949 ms to 11.719205 ms with laser power at 1.5 kW and welding speed

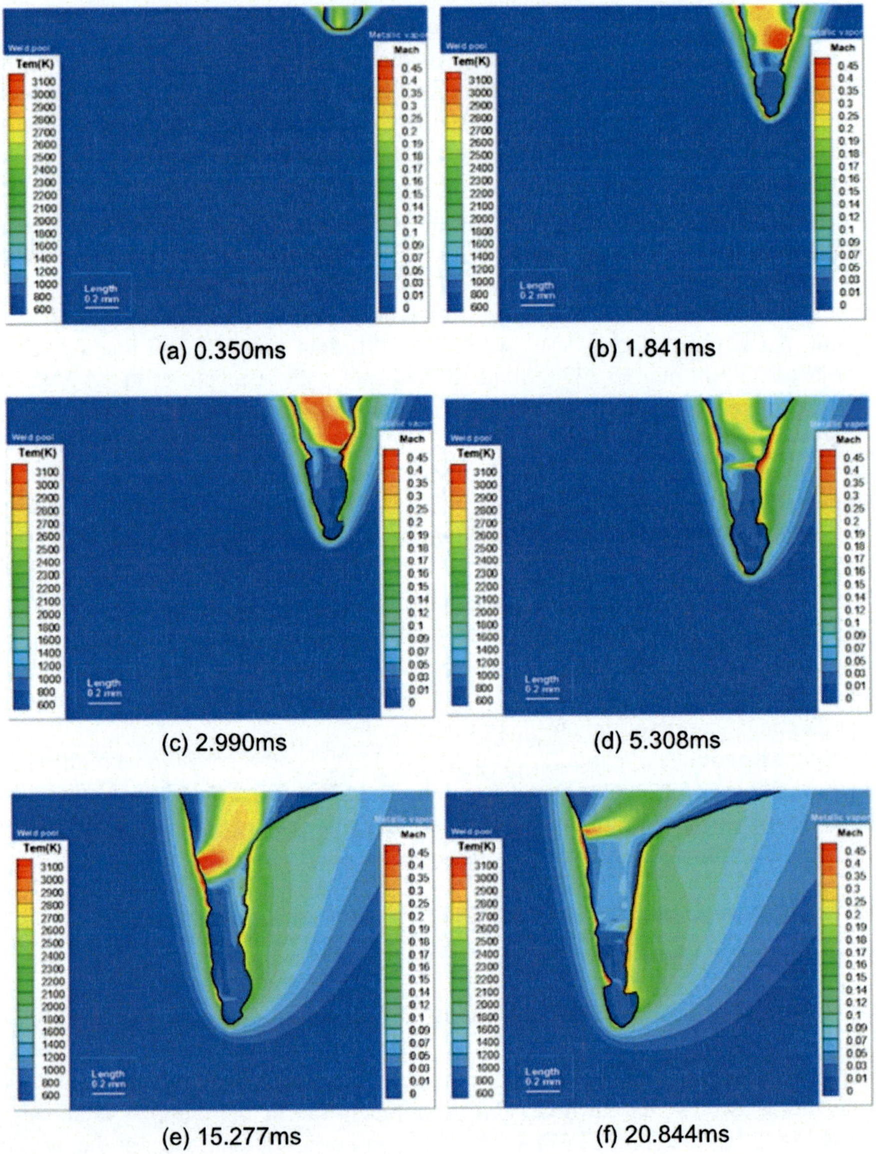

Fig. 5.6 Evolution of Mach number distribution of metal vapor with time in the keyhole (P = 1.5 kW, V = 3 m/min)

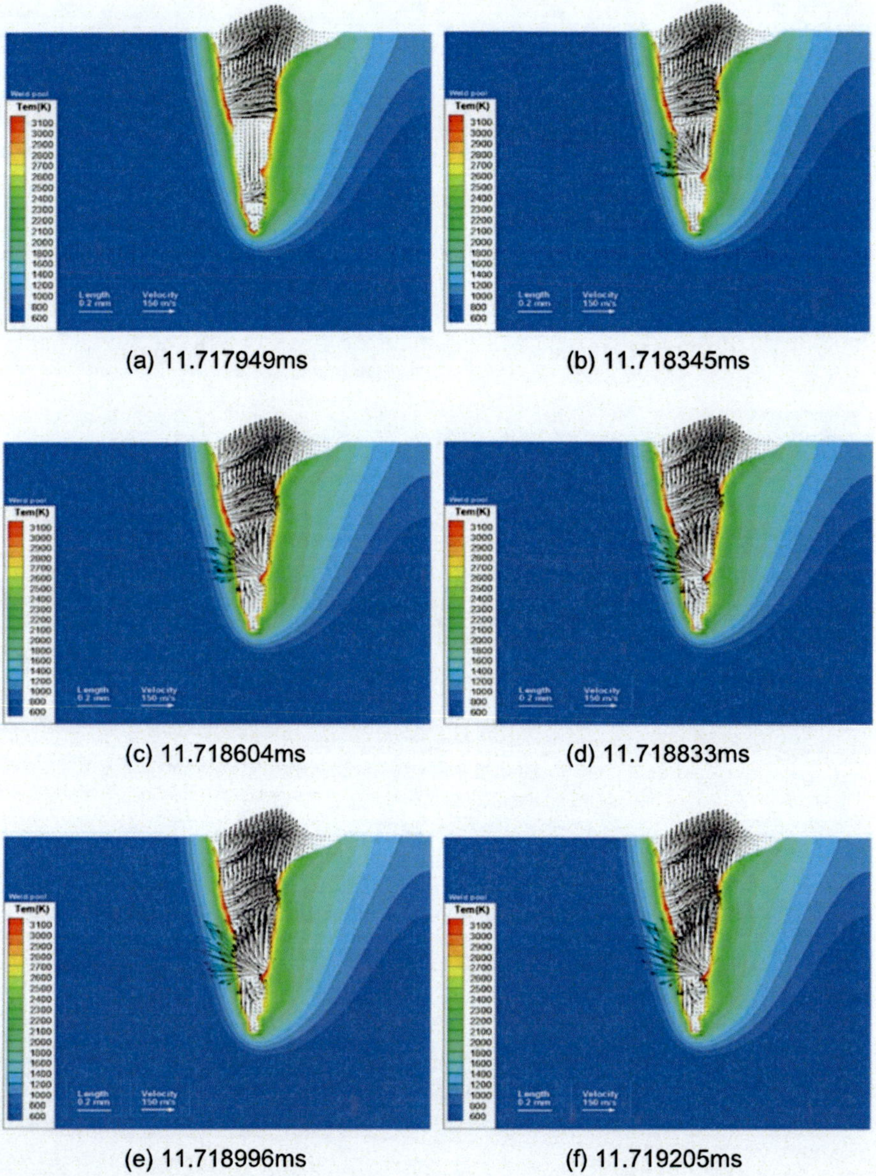

Fig. 5.7 Velocity distribution of metal vapor in the keyhole during the transient period ($P = 1.5$ kW, $V = 3$ m/min)

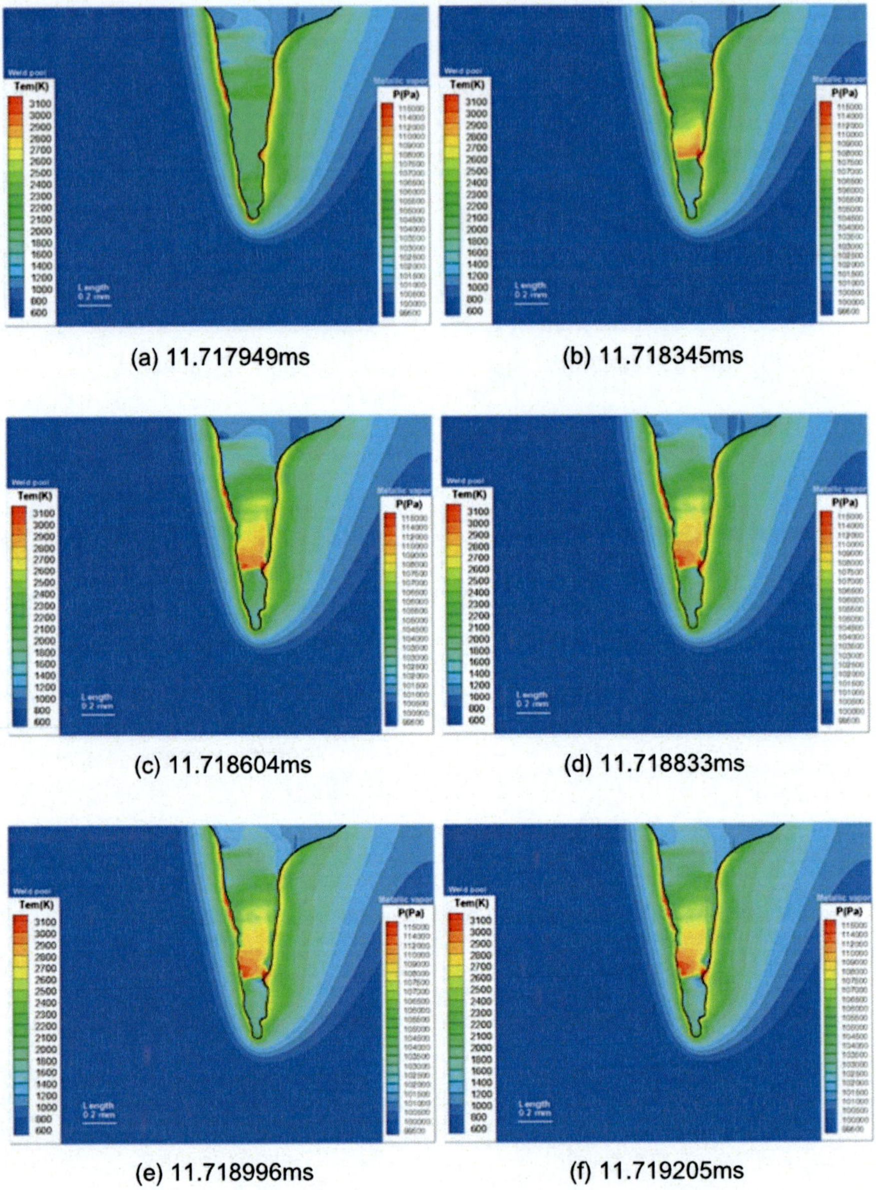

Fig. 5.8 Pressure distribution of metal vapor in the keyhole during the transient period ($P = 1.5$ kW, $V = 3$ m/min)

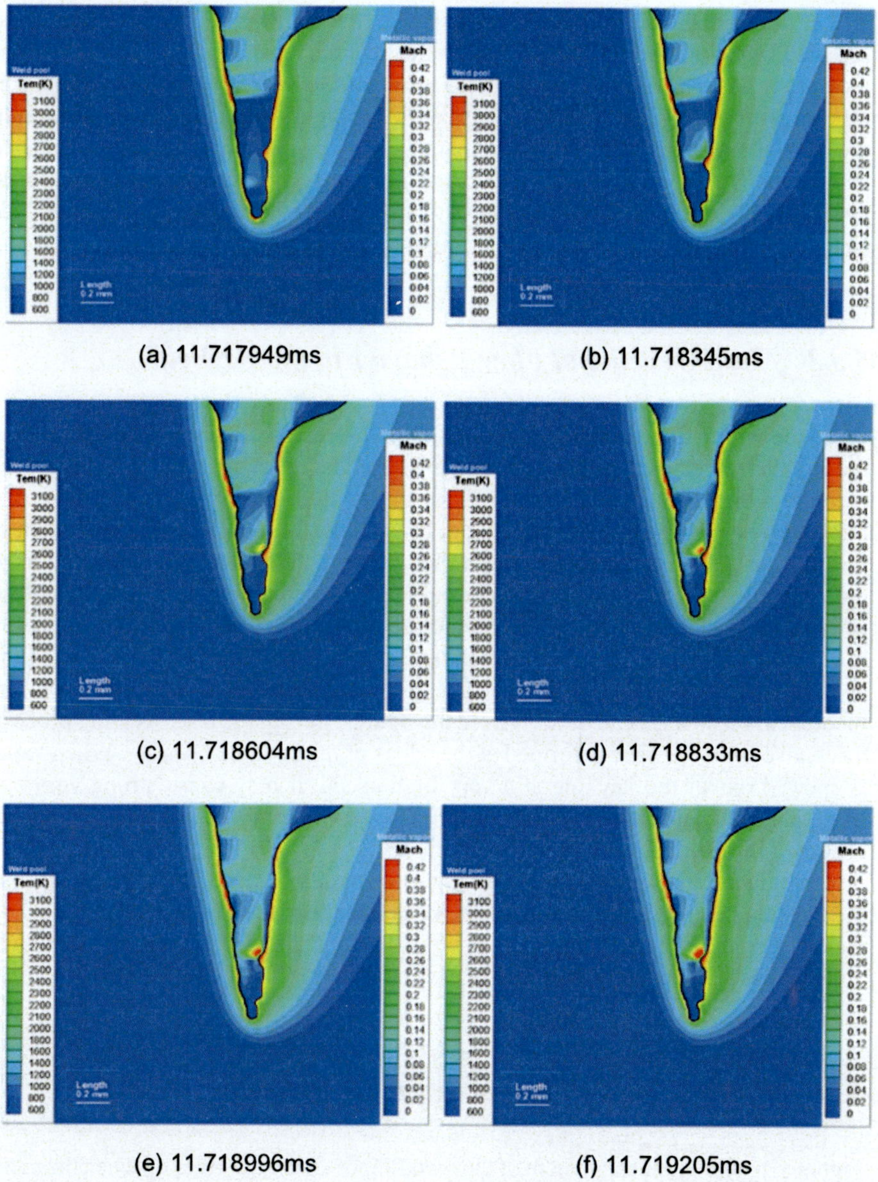

Fig. 5.9 Mach number distribution of metal vapor in the keyhole during the transient period (P = 1.5 kW, V = 3 m/min)

at 3 m/min. As shown in Figs. 5.8 and 5.9a, b, the significant changes of the pressure and Mach number of metal vapor in the keyhole experience are also due to the violent evaporation of metal vapor at the hump of the back wall of the keyhole. It is also shown from Figs. 5.8 and 5.9c–f that the Mach number at the local position of the back hump keeps increasing, and that the nearby vapor pressure state is changed simultaneously. Therefore, the variation of metal vapor velocity, pressure and Mach number in a short period of time shows that the state of metal vapor in the keyhole changes obviously with time and has a high transient property.

5.3.2 Multi-directional Flow Behavior in the Keyhole

From the simulation results in Fig. 5.10, a high-pressure region will be formed in the middle lower bottom part of the keyholes. If the keyhole depth is large, the pressure inside the keyhole, especially at the middle lower part, cannot be easily released, and a high-pressure region can be easily formed. The main component of the gaseous phase in the bubble caused by the keyhole's instability is proved to be ambient gas. It is known that recondensation due to rapid cooling is due to the flow of protective gas into the keyhole during the pulsed laser of liquid zinc. However, how environmental gases enter the bubble is still not well understood, especially in continuous laser welding processes. This problem can be explained by our current theoretical results. As can be seen from the figures, during the laser welding process, a negative pressure zone is formed at the edges of the keyhole opening. When vapor is spewed out of the keyhole at a very high speed, it may cause strong vortex in those places. Under the influence of this flow, some atmospheres may be quickly absorbed into the keyhole, resulting in the metal vapor inside the keyhole possibly flowing toward the bottom while flowing toward the opening. These gases entering the keyhole can reach the bottom of the keyhole, and if the keyhole is split into two parts at this time, the separate part at the bottom of the keyhole will be filled with ambient gases, eventually resulting in cavities.

5.3.3 Violent Oscillation and Swinging Behavior

Figure 5.11 shows the movement of the metal vapor above the opening when the metal vapor is ejected. It can be clearly seen that under the current process conditions, the metal vapor is not spewed out in a fixed direction but oscillates back and forth, and this is consistent with the simulated swing angle of the metal vapor at the opening. The oscillation of the metal vapor means that the direction of the component of the metal vapor velocity parallel to the workpiece is changing. Due to constant oscillation of the wall surface of the keyhole during the laser welding process, the local evaporation on the wall surface of the keyhole is dramatic, and changes with position and time. The recoil pressure generated during evaporation causes the metal vapor to move

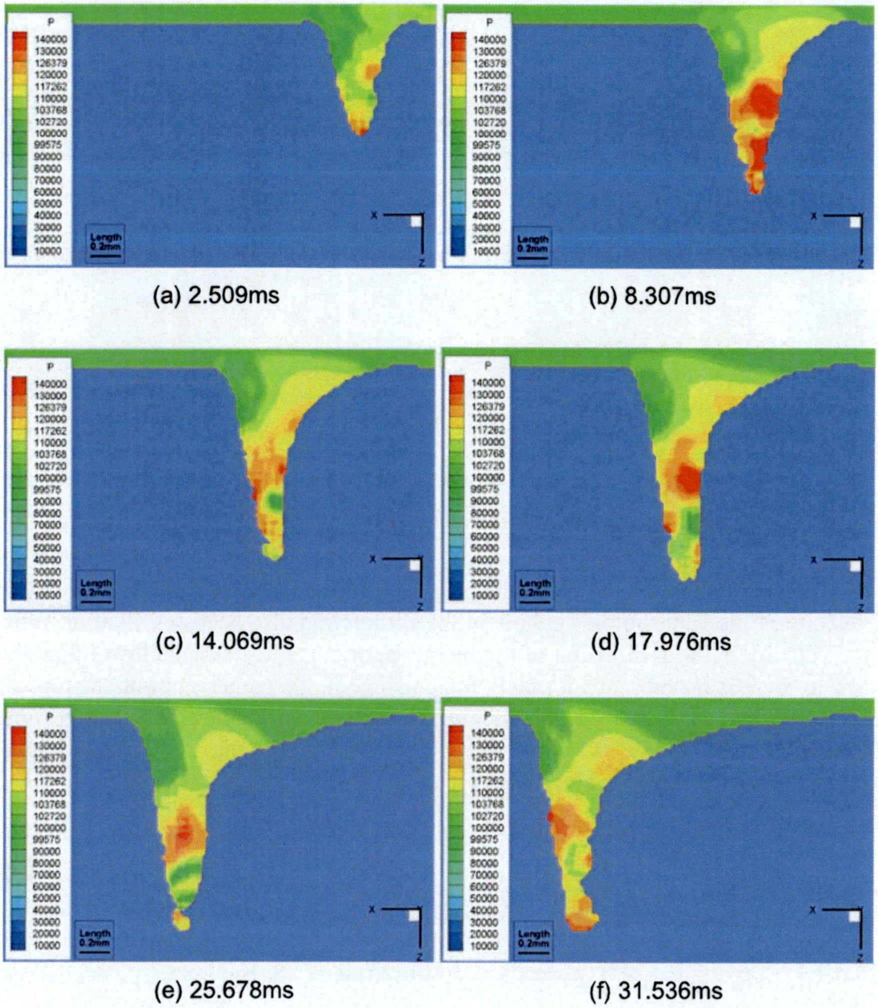

Fig. 5.10 Pressure distribution of metal vapor inside the keyhole

away from the surface of the liquid toward the inner of the keyhole, and the direction of this movement is perpendicular to the wall surface, so the strong local evaporation and inclination of the wall of the keyhole determine the direction of movement of the metal vapor. Due to the constant oscillation of the wall of the keyhole, the direction of the metal vapor changes continuously within a certain range, and finally after interaction inside keyhole, the metal vapor spews out of the keyhole, resulting in swinging behavior. There is a close relationship between the swinging of the vapor outside the keyhole and the oscillation of the wall of the keyhole. As shown in Fig. 5.11b–g, not all metal vapor spewed out of the keyholes is continuous, and some will separate. Due to continuous opening and closing of the keyhole and constant

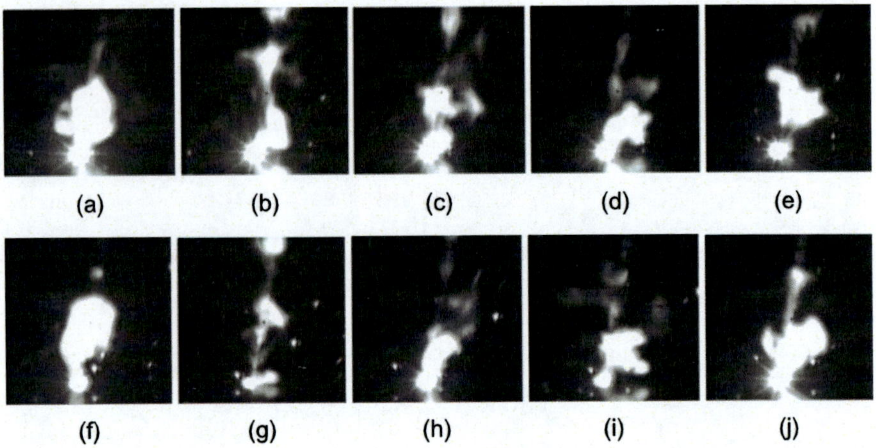

Fig. 5.11 High-speed images of vapor spewed out in laser welding process (P = 1.5 kW, V = 3 m/min)

oscillation of the free interface of the hole wall during the laser welding process, the velocity of the vapor in the keyhole in the Z-direction also keeps oscillating. When the velocity in the Z direction of the metal vapor at $t + t_0$ is much lower than that at t, as there is no subsequent timely replenishment, the metal vapor on the top will be cut off and separate from the underneath metal vapor connected to the keyhole. Therefore, separation that occurs when metal vapor is spewed out is also closely related to oscillation of the keyhole.

5.3.4 Compressibility at Local Evaporation

5.3.4.1 Strong Local Evaporation on the Wall of the Keyhole

Figure 5.11 shows the velocity evolution of the vapor in the transient keyhole during deep penetration laser welding with laser power at 1.5 kW and welding speed at 3 m/min. In the figure, it can be found from the temperature distribution outside the keyhole that the temperature of the wall of the keyhole is mostly between 2800 and 3100 K, and the distribution is not uniform. The local temperature of the keyhole is usually high due to direct or multiple reflection Fresnel absorption of the laser and can reach above 3100 K. The high temperature area may appear at the bottom for a shallow keyhole (Fig. 5.11a), or on the front wall (Fig. 5.11e, f) and the back wall (Fig. 5.11b–d) for a deep keyhole, and violent local evaporation may occur at these locations, which is consistent with the experimental results of Zhang Mingjun et al. The local evaporation phenomenon is also reflected in the spewing out velocity of the metal vapor at the opening, which first increases and then decreases. In the initial stage of deep penetration laser welding, the keyhole has not formed or is shallow, the laser directly heats the surface of the workpiece. The laser spot is on local position of the workpiece, and the metal vapor violently evaporates along the Z-axis direction to produce a very large spewing out speed. After the keyhole is formed, the front wall of the keyhole is directly heated by the laser energy, and the back wall has a local high temperature region due to multiple reflections, and the evaporation is strong. With periodic oscillation of the keyhole morphology, the degree of inclination of the local evaporation position changes, the evaporating direction of the spewed out metal also changes continuously, and the velocity of the spewed out metal vapor in the Z-axis direction is reduced accordingly. This is consistent with the variation pattern of average velocity of the vapor at the opening simulated vapor obtained by simulation.

5.3.4.2 Metal Vapor Compressibility

As shown in Fig. 5.4, it is known that there is a high pressure inside the transient keyhole much larger than 1 atm[1] and the density of the metal vapor is much smaller than the air density. So, it is very possible that highly compressed low density vapor exists in a millimeter keyhole. The compressibility is an important physical property of the vapor plume generated in the deep penetration laser welding process. The degree of compressibility of the metal vapor in the keyhole can, to some extent, reflect the intensity of the metal vapor movement in the keyhole. Figure 5.12 illustrates the peak Mach number evolutions with different welding speeds of 2, 3 and 4 m/min and laser power at 1.5 kW. It can be easily found from Fig. 5.12 that the moment Mach number can easily reach 0.8, and can even exceed 1.0, which indicates the metal vapor may be violently compressed (Fig. 5.13).

5.4 Conclusion

(1) In consideration of complex phenomena such as heat and mass transfer and phase transformation in the weld pool, with the Level Set method employed in tracking the evolution process of keyhole's free interface, and effects of recoil pressure, environmental pressure, and the compressibility of vapor taken into account, a mathematical model depicting dynamic behavior of metal vapor in transient keyhole during deep penetration laser welding is established.

(2) A novel method for setting the boundary temperature of the wall of the keyhole temperature-dependent compressible gas–liquid interface is proposed. The evaporation boundary and reflection boundary are respectively set according to evaporation and condensation caused by the temperature difference on the wall surface of the keyhole. The temperature, pressure, density, and velocity boundary conditions associated with vapor flow are derived.

[1] 1 atm = 101.325 kPa.

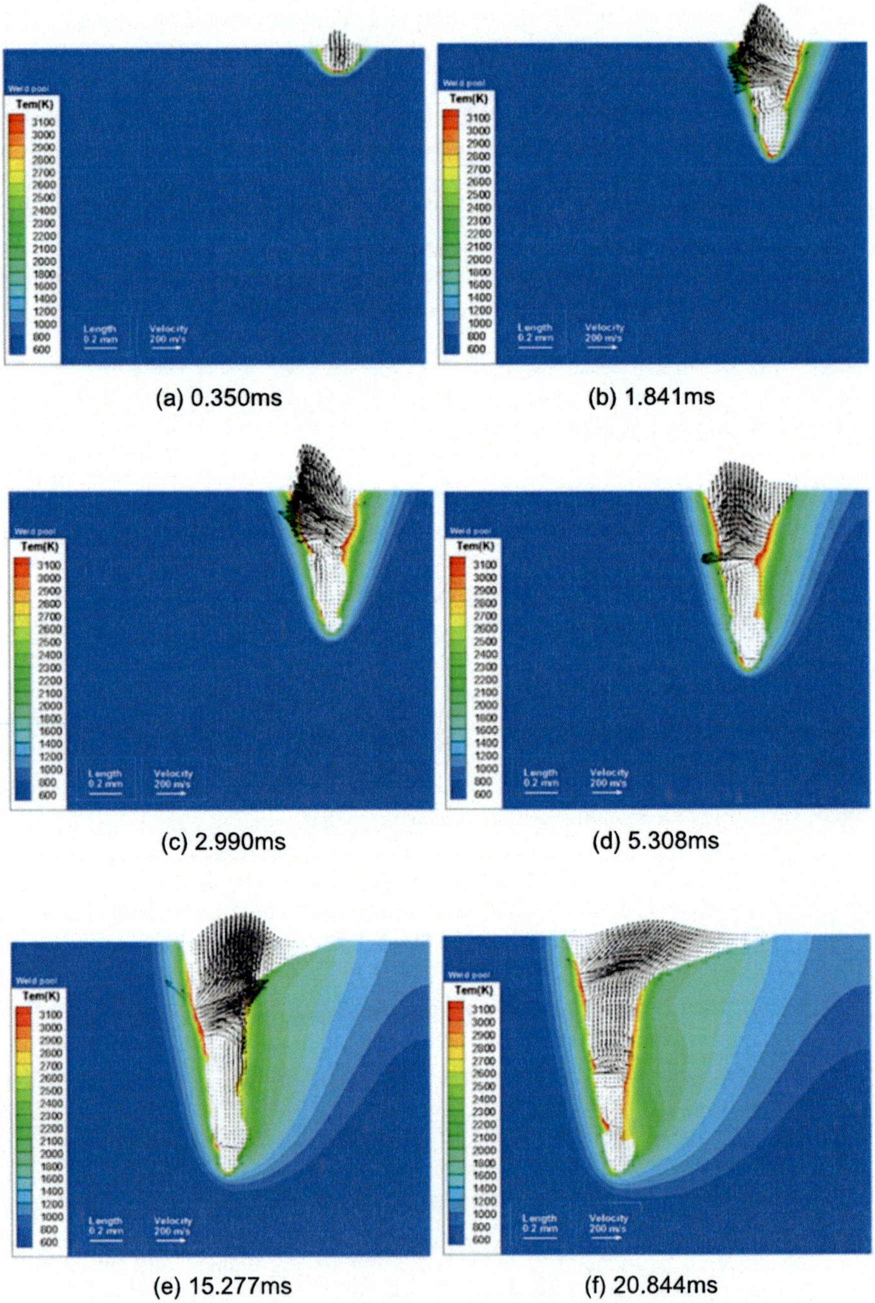

Fig. 5.12 Strong evaporation of metal vapor at local position on the wall of the keyhole (P = 1.5 kW, V = 3 m/min)

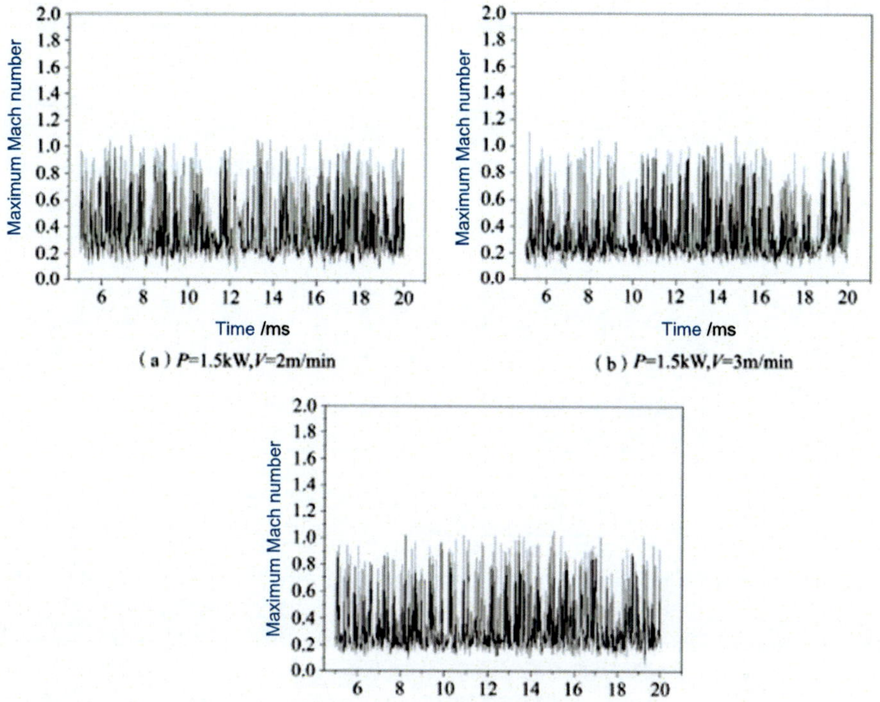

(a) P=1.5kW,V=2m/min

(b) P=1.5kW,V=3m/min

(c) P=1.5kW,V=4m/min

Fig. 5.13 Variation of peak Mach number with welding time under different welding speeds in laser welding process

(3) A highly efficient numerical solution algorithm is proposed based on the dynamic mathematical model of compressible vapor in the keyhole during the deep penetration laser welding keyhole process established in this chapter; the dynamic behavior of metal vapor in the keyhole for 304 stainless steel welding is analyzed under the actual welding process conditions; the distribution non-uniformity, compressibility and high transient of metal vapor inside the keyhole are analyzed; and local evaporation on the free surface of the keyhole and continuous swinging of the vapor at the opening of the keyhole are discussed.

Chapter 6
Behaviors of Keyhole and Weld Pool Under the Effect of Side-Blown Gas

Abstract This chapter introduces the behaviors of keyhole and weld pool under the effect of side-blown gas. Flow patterns of the side-blown gas in laser welding of a Titanium alloy are presented. The influences of side-blown gas on the weld pool profiles and the transient keyhole and weld pool dynamics at the present of keyhole plume are discussed. Moreover, stable mechanisms of side-blown gas on keyhole and welding oscillations are presented.

6.1 Introduction

In deep penetration laser welding, the heat source exerts an effect on the laser beam, plasma, keyhole, weld pool, and other several sub-systems, and these subsystems influence each other during welding. Figure 6.1 illustrates the dynamic coupling of sub-systems in deep penetration laser welding. As can be seen in the figure, (1) The plasma can produce refraction and inverse bremsstrahlung absorption to incident beams, and the fluctuation of the plasma makes power-density distribution on workpiece surface fluctuate. (2) The laser beam provides energy required for producing and maintaining plasma, and the fluctuation of laser power density can result in the fluctuation of plasma. (3) The keyhole and the weld pool offer mass sources for plasma and define the boundary of plasma flow fields. Therefore, the fluctuation of keyhole and weld pool undulates plasma. (4) Fluctuation of velocity and pressure fields in plasma in the keyhole can cause the fluctuation of keyhole and weld pool. (5) The fluctuation of power-density distribution of laser beams changes surface ablation pressures in keyhole and undulates the keyhole and the weld pool. (6) The variation of keyhole shape changes path for multiple reflections of the laser beam in the keyhole, thus resulting in surface laser power-density distribution fluctuating in the keyhole. In addition, the laser welding process is generally protected by using auxiliary gas flow which can directly affect the plasma, keyhole and weld pool, so the introduction and parameters of auxiliary gas flow play an important role in maintaining the stability of welding process.

At present, there are many researches on the influence laws and mechanisms of side-blown gas on laser-induced plasmas and reports on influence laws of side-blown

S. Gong et al., *Weld Pool Dynamics in Deep Penetration Laser Welding*,
https://doi.org/10.1007/978-981-16-0788-2_6

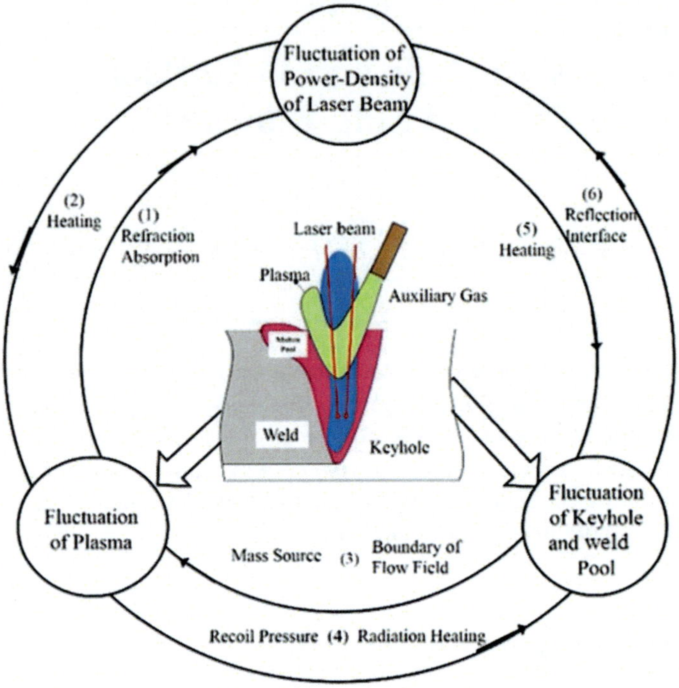

Fig. 6.1 Dynamic coupling of sub-systems in deep penetration laser welding and the influences of side-blown gas

gas on behavior of keyhole and the weld pool in laser welding process. However, very little research has been done on how side-blown gas influences the behaviors of weld pool and keyhole. Based on the previous experimental results, the paper examines the CO_2 full-penetration laser welding process of titanium alloy in an effort to explore the influence mechanisms of auxiliary gas flow on behaviors of keyhole and weld pool in deep penetration laser welding through numerical simulation of fluid dynamics.

6.2 Flow Field of Side-Blown Shielding Gas (for Titanium Alloy)

Figure 6.2 demonstrates the typical calculation results of shielding gas flow fields without considering the generation of metallic vapors in a keyhole. In the figure, the upper left side of the red dashed line shows top-blown gas released from the CO_2 laser head, while the lower right side displays side-blown gas released from the side-blowing nozzle. As presented in the figure, the top-blown gas flow and the side-blown gas flow change their flow directions after they meet and it is obvious that their confluence direction (indicated by the red dashed line in Fig. 6.2) mainly

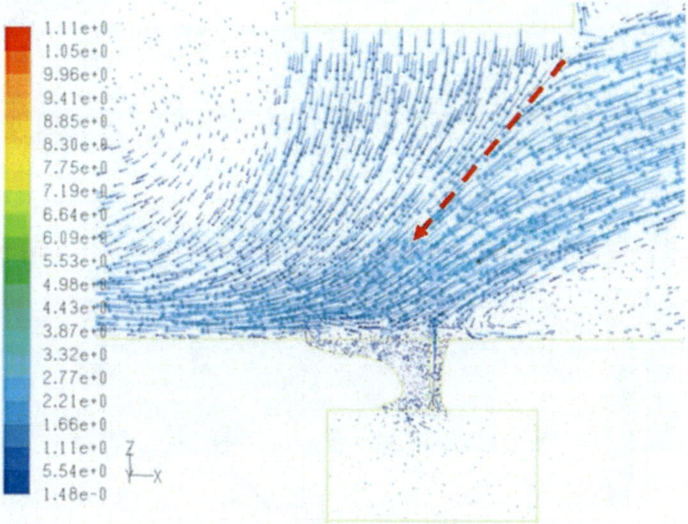

Fig. 6.2 Typical results of velocity vector diagram of auxiliary gas flow fields on symmetry plane without considering jet flows of metallic vapors (welding proceeds from left to right)

depends on the momentum ratio between the two flows. Therefore, the direction of the top-blown gas flow with a small velocity changes greatly, while that of the side-blown gas varies slightly. It can be seen from the figure that the angle between confluence direction and horizontal plane increases in comparison with the angle between the direction of side-blown gas and horizontal direction. Finally, it is the confluence direction rather than the direction of side-blown gas that directly affects gas flow direction on the surface of the weld pool.

Figures 6.3, 6.4 and 6.5 present calculation results of auxiliary gas flow fields at $t_0 + 0.1$ ms in the initial stage with metallic vapors in the keyhole taken into account. Figure 6.3 shows the distribution of pressures on one side of gas phase on symmetry plane at $t_0 + 0.1$ ms in the initial stage. The round areas removed from the figure are molten metal droplets that spatter under the strong impacts of metallic vapors and pressures in these droplets are significantly larger than those in surrounding atmosphere. In order to avoid interferences and misunderstanding, the spattering area of droplets and its adjacent areas are removed from the distribution pattern of gas-phase pressure. Figures 6.4 and 6.5 present the velocity vector diagram and velocity distribution on the side of gas phase on symmetry plane at $t_0 + 0.1$ ms.

From the results in Fig. 6.3, it is clear that the confluence of the side-blown and the top-blown gas flow produces a tilted, bell jar-shaped pressure distribution on the upper area of the workpiece surface and the axis of the bell jar is the flow direction of confluence. Since the calculation results shown in Fig. 6.3 consider the outflow of high-velocity metallic vapors from the keyhole, the ambient pressure distribution above the keyhole is complex. As the figure shows, two local high-pressure areas are formed in the surrounding areas above the keyhole opening. The first high-pressure

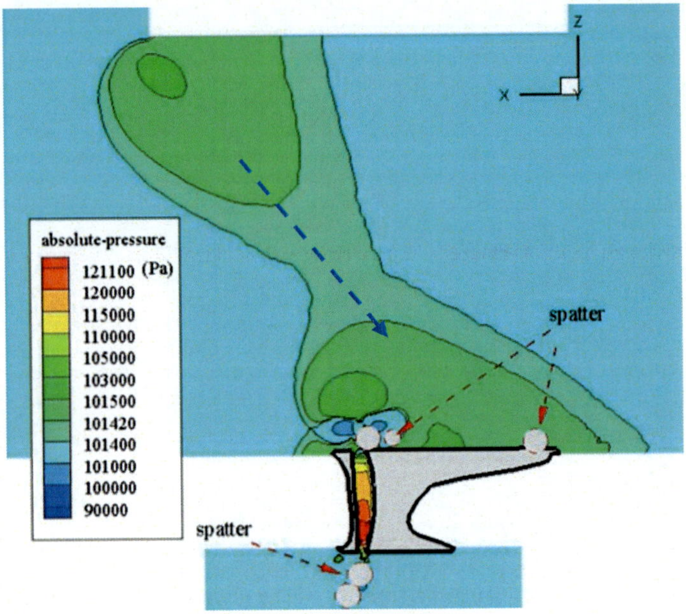

Fig. 6.3 Distribution of gas-phase pressures on symmetry plane at $t_0 + 0.1$ ms in the initial stage with metallic vapors in the keyhole considered

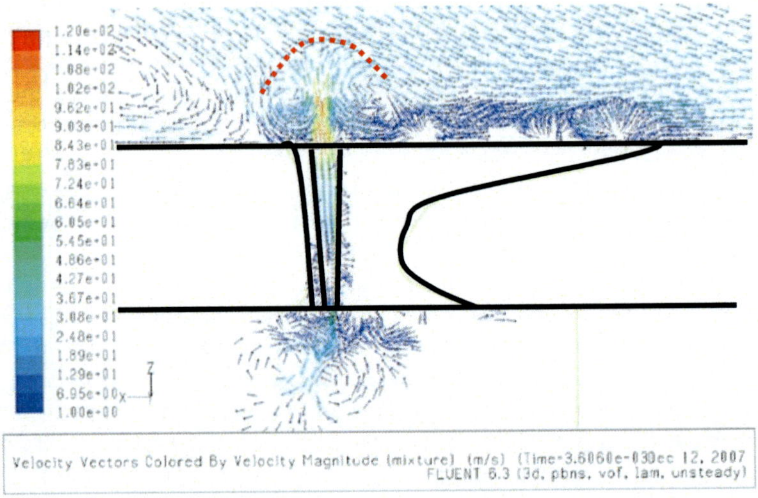

Fig. 6.4 Velocity vector diagram of gas phase on symmetry plane at $t_0 + 0.1$ ms in the initial stage with metallic vapors in the keyhole considered

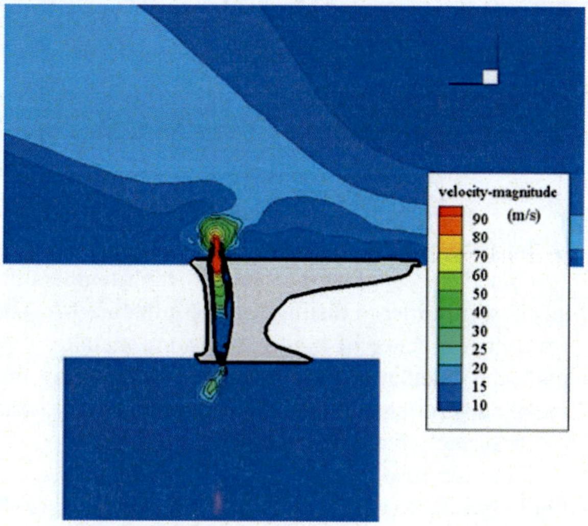

Fig. 6.5 Velocity distribution of gas phase on symmetry plane at t_0 + 0.1 ms in the initial stage with metallic vapors in the keyhole considered

area is located near the intersection point between the axis of the "bell jar" and the surface of the weld pool. The pressure in this area exceeds 101,500 Pa. The other one is over the keyhole, also with a pressure more than 101,500 Pa. Between this high-pressure area and keyhole opening, there are two local low-pressure areas with a pressure smaller than 101,000 Pa. The velocity vector diagram on the side of gas phase on symmetry plane at $t_0 + 0.1$ ms presented in Fig. 6.4 shows that the flow directions of auxiliary gas and metallic vapors suddenly change at the position indicated by the red dashed line, demonstrating that auxiliary gas flow effectively suppresses metallic vapors flowing upwards. By comparing Fig. 6.3 with Fig. 6.4, the local high-pressure area over the keyhole is located above the red dashed line in Fig. 6.4. It can be observed from Fig. 6.4 that the area below the red dashed line shows two vortexes on both sides of the axis of the keyhole. In Fig. 6.3, two local low-pressure areas with pressure lower than 101,000 Pa over the keyhole are located exactly at the positions of the two vortexes. Figure 6.5 presents the velocity distribution of gas phase on symmetry plane at $t_0 + 0.1$ ms. It can be seen from Fig. 6.5 that the gas flow velocity is large around the keyhole opening at $t_0 + 0.1$ ms. On the upper surface of the weld pool, the flow velocity of metallic vapors in the vicinity of the keyhole opening is 90 m/s, and on the lower surface, the flow velocity is 30 m/s. However, the flow velocity of vapors in the keyhole far away from the opening is small. According to Fig. 6.3, the maximum pressure in the keyhole at $t_0 + 0.1$ ms is higher than 121,100 Pa. This shows that the pressure near the keyhole opening is smaller than that in the keyhole. Moreover, the pressure adjacent to the upper opening with a large gas flow velocity is smaller than that around the lower opening with a small gas flow velocity.

6.3 Behaviors of Keyhole and Weld Pool Under the Effect of Side-Blown Gas

6.3.1 Deformation of Weld Pool Surface Under the Effect of Side-Blown Gas

In the actual welding process, the surface of the weld pool was deformed under the combined effects of surface tension, high-temperature metallic vapors and auxiliary gas flow. In order to distinguish the influence of auxiliary gas flow on the weld pool from the influence of high-temperature metallic vapors on the surface of the weld pool, the calculations are simplified by ignoring the effects of temperature on the metal surface tension of the weld pool and the deformation of the keyhole and the lower surface of the weld pool.

From the results in Fig. 6.6, it is clear that after the calculation starts, the gas–liquid interface presents the following changes: (1) the junction point of gas, liquid and solid near the welding line moves to the center of the weld pool; (2) the area around the center of the upper surface of the weld pool bulges, exceeding the height of the upper workpiece surface. After that, under the effects of side-blown auxiliary gas flow, the surrounding liquid metal on the surface of weld pool transfers to the rear of the weld pool. Finally, the gas–liquid interface around the melt front at the front of the weld pool is lower than the workpiece surface, while the gas–liquid interface near the solidification front at the rear of the weld pool is higher than the workpiece surface.

Figure 6.7 shows the surface shapes of the weld pool at 7 ms at different observation angles. Figure 6.7c presents the weld reinforcement and undercut on the surface of the weld pool. According to the calculation results, the weld undercut defects occur mainly because the gas–liquid-solid phase boundary point near the welding line moves toward the center of the weld pool. Since for the convenience of calculations, many complicated factors are neglected, it is easy to see that it is the surface tension that drives the three-phase junction to move towards the center of the weld pool. Figure 6.8 shows the velocity vector diagram of the surface of the weld pool at 0.1 ms. The calculation results in Fig. 6.8 demonstrate that the liquid metal near the solid–liquid interface on the upper surface of the weld pool flows to the center of the weld pool at 0.1 ms. As displayed in Fig. 6.7c, weld reinforcement mainly results from the fact that the liquid metal in the weld pool moves towards the rear of welds, and then accumulates and solidifies. Figure 6.9 shows the velocity field on the weld pool surface at $t = 7$ ms. It is apparent that the metal liquid on the weld pool surface flows towards the rear part of the weld pool. Since the calculation neglects the surface tension gradient and rapidly moving heat source, it is the side-blown flow that forces the metal liquid to move towards the rear of the weld pool.

Figure 6.10 shows the pressure distribution of the gas phase on symmetry plane at 7 ms. When auxiliary gas is applied to the workpiece, a local high-pressure region appears near the gas–liquid surface which is locally concave at the front of the weld

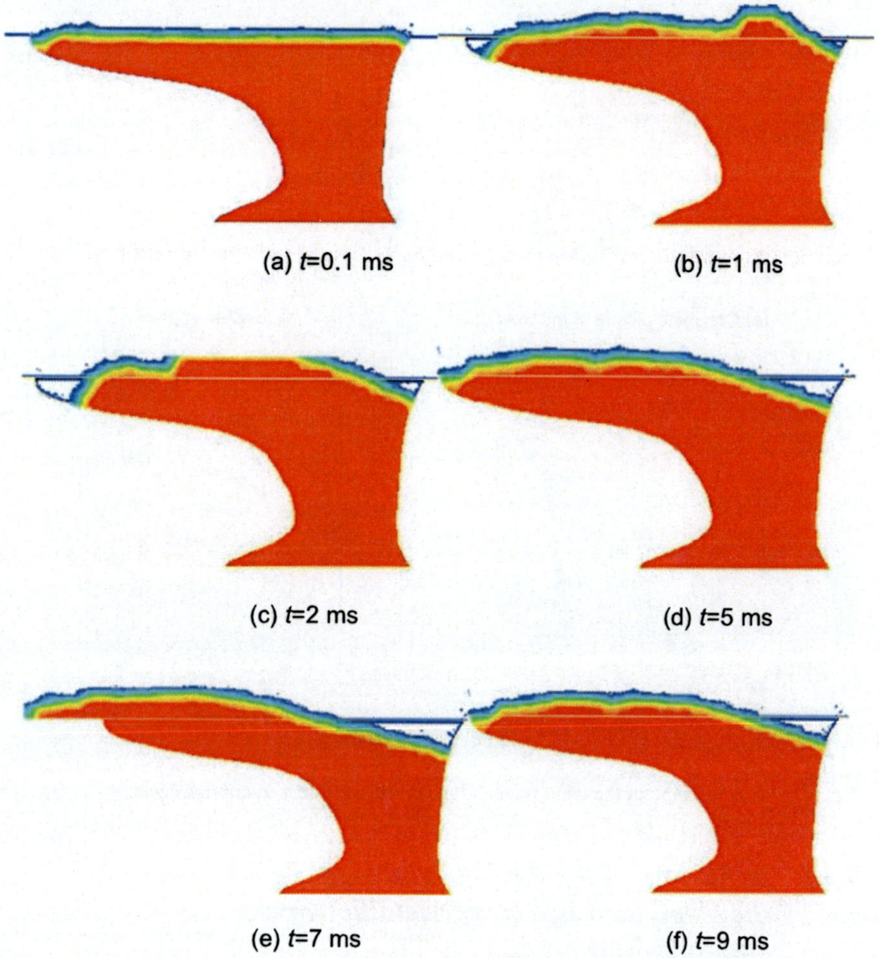

Fig. 6.6 Change process of weld pool surface in the symmetry plane under the effect of auxiliary gas

pool. This pressure distribution will surely cause the gas–liquid surface at the front of the weld pool to sag further.

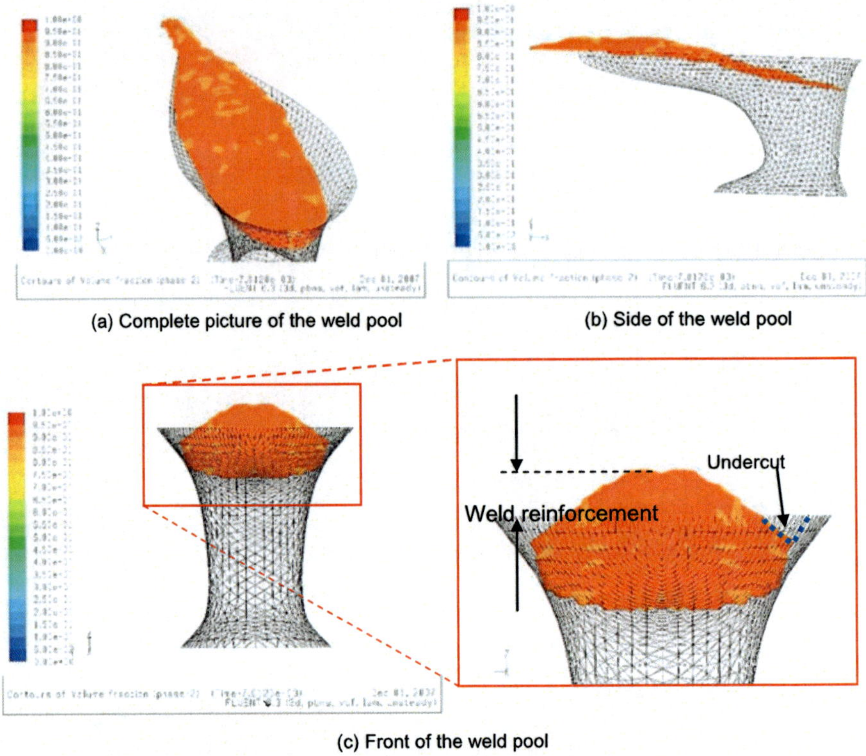

Fig. 6.7 Surface shape of the weld pool at 7 ms under the effect of auxiliary gas

6.3.2 Behaviors of the Keyhole and Weld Pool Under the Combined Effects of Metallic Vapors and Side-Blown Gas

Figure 6.11 shows the transient evolution of the keyhole and weld pool when the side-blown gas is applied. In the figure, the red region indicates liquid metal, and the blue region is a gas phase region. The initial time $t_0 = 3.5$ ms refers to the moment when metal vapor is generated at the beginning. Figure 6.11 also shows the transient behavior of the keyhole and weld pool under the action of metallic vapor in 2.7 ms is given in chronological order. As can be seen from Fig. 6.11, the evolution of keyhole and weld pool consists of three typical stages: initial stage, expansion stage and refilling stage. It is assumed that the keyhole is formed instantaneously in the welding process and the whole welding process is the constant cycles of these three stages.

In the initial stage of the formation of metallic vapor in the keyhole, the liquid metal surrounding the keyhole flows radially outwards due to the impact of metallic vapor generated by the strong evaporation. Figure 6.12 shows the velocity vector

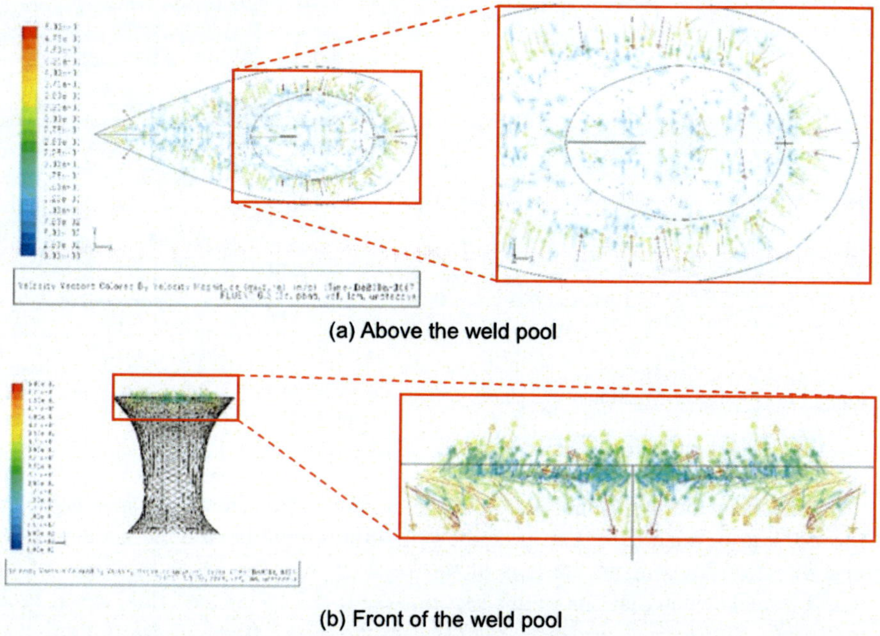

Fig. 6.8 Velocity vector distribution on the weld pool surface at 0.1 ms under the effect of auxiliary gas

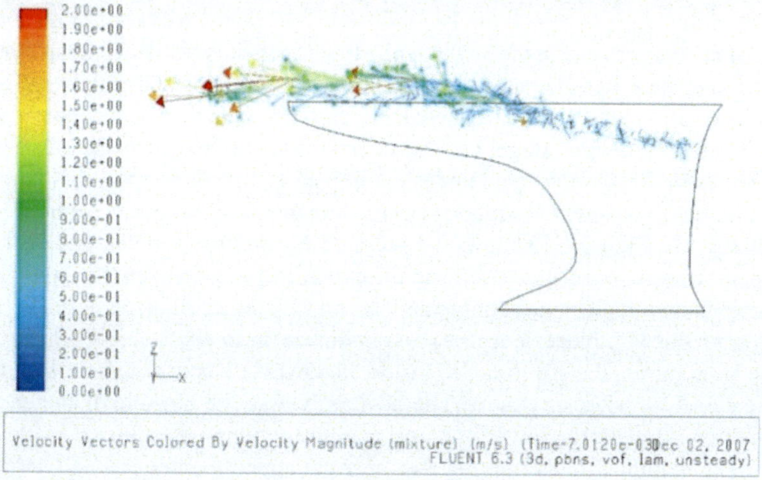

Fig. 6.9 Velocity vector distribution on the weld pool surface at 7 ms under the effect of auxiliary gas

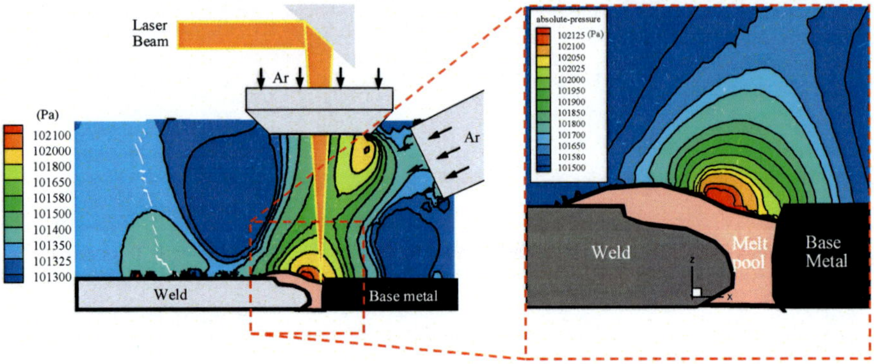

Fig. 6.10 Pressure distribution of gas phase on symmetry plane at 7 ms under the effect of auxiliary gas

distribution of the liquid metal around the keyhole when metallic vapor begins to form. It can be seen from Fig. 6.12 that the liquid metal behind the keyhole flows in a direction away from the axis of the keyhole over the entire thickness of the workpiece. The calculations in this section take 0–0.3 ms as the initial stage. From the results in Fig. 6.11, it can be seen that in the initial stage, no local hump near the keyhole opening is significantly higher than the initial upper surface of the weld pool. In addition, the diameter of the keyhole shows little change along the entire thickness of the workpiece.

The initial stage in which the keyhole and weld pool vary lasts for a short time. Then, since the pressure at the keyhole opening is considerably lower than that inside the keyhole, the gap in diameter between the keyhole opening and middle will become more and more significant. The increasing gap indicates that the keyhole and weld pool come to the second stage, i.e., the expansion stage. In the expansion stage, with the increasing keyhole volume under the impact of metallic vapor, the diameter of the keyhole middle becomes much larger, while the diameter of the keyhole opening varies slightly. Figure 6.13 shows the pressure distribution and velocity distribution on the symmetry plane at $t_0 + 1$ ms in the expansion stage. Given the large diameter of the keyhole middle, a local hump appears on the surface of metal liquid behind the keyhole on the workpiece under the pressure inside the keyhole. During the formation of the local hump, the gas pressure inside the keyhole has to withstand not only the surface tension, but also gravity. The position where the keyhole diameter reaches its peak is closer to the lower opening than to the upper opening. It is clear from Fig. 6.13b that the rapidly moving area near the upper opening is small, largely due to the suppression effect of the auxiliary gas above the workpiece.

Figure 6.14 shows the velocity vector distribution on the symmetry plane at $t_0 +$ 1 ms in the expansion stage. As can be seen from Fig. 6.14d, in the region near the half-depth of the keyhole, the metal liquid surrounding the keyhole flows radically outwards. Figure 6.14 and Fig. 6.14e show the velocity vector distribution of the molten metal behind the keyhole opening near the lower surface and the upper surface,

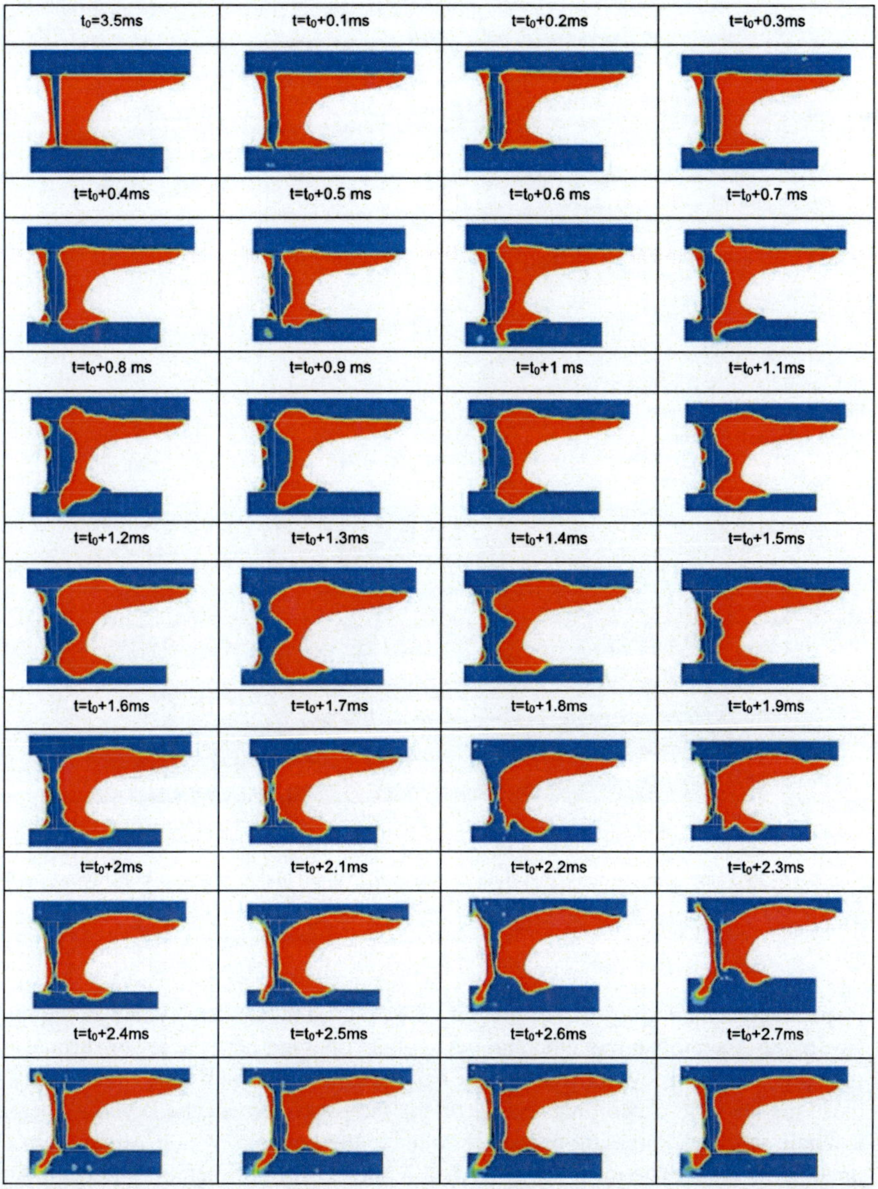

Fig. 6.11 Sequences of evolution of liquid–gas interfaces when side gas is used

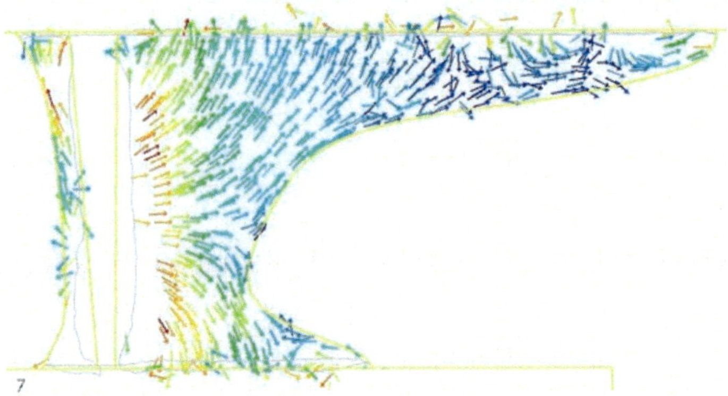

Fig. 6.12 Velocity vector distribution of liquid phase in symmetry plane at $t_0 + 0.1$ ms under the effect of auxiliary gas

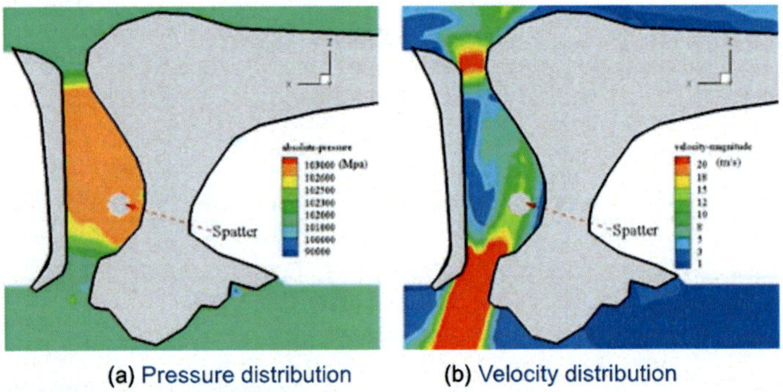

Fig. 6.13 Pressure and velocity distribution (metallic vapor inside keyhole considered) in the symmetry plane at $t_0 + 1$ ms under the effect of auxiliary gas in the expansion stage: **a** pressure field and **b** velocity field

respectively. Under the effect of pressure distribution inside the keyhole as shown in Fig. 6.13b, the molten metal behind the keyhole opening near the lower surface and the upper surface moves toward the axis of the keyhole, which shows a tendency to close the keyhole. According to Fig. 6.13a, the pressure near the keyhole opening is small, while the pressure of the areas inside the keyhole, which are far from the opening, is large. The local hump formed on the surface of metal liquid behind the keyhole opening in the expansion stage allows the side-blown gas to exert a greater effect in the area. Figure 6.14b shows the velocity vector of some metal liquid of a local hump behind the keyhole opening on the upper surface. At the end of the expansion stage, the pressure required to continue the expansion of the keyhole is not met, which restrains the maximum diameter of the keyhole. The period from $t_0 + 0.4$ ms to $t_0 + 1.2$ ms in this calculation result is the expansion stage.

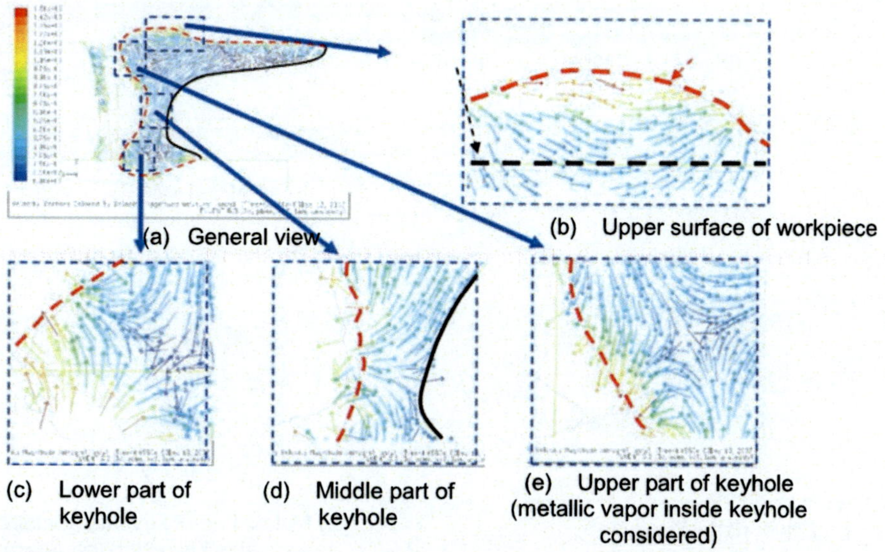

Fig. 6.14 Velocity vector distribution in symmetry plane at $t_0 + 1$ ms under the effect of auxiliary gas in the expansion stage

The period from $t_0 + 1.3$ ms to $t_0 + 2.3$ ms in Fig. 6.11 is the refilling stage, in which the surface tension is an important factor in the behavior of the keyhole and the weld pool. Typical velocity vector distribution of the refilling stage on the symmetry plane is presented in Fig. 6.15 at $t_0 + 2$ ms. As shown in Fig. 6.15c, the metal liquid behind the lower keyhole opening flows in the direction of the keyhole axis. This bottom-up and keyhole-facing flow behavior is obviously not caused by metallic vapor and gravity. In addition, the lower surface of the weld pool is free from the effect of auxiliary gas. It is evident, therefore, that the bottom-up and keyhole-facing flow behavior is driven by surface tension, which is now a leading cause of the behaviors of the keyhole and weld pool. According to Fig. 6.15d, driven by surface tension and gravity, the metal liquid behind the upper keyhole opening shares a similar flow behavior to the metal liquid behind the lower keyhole opening. Specifically, they both flow in the direction of keyhole axis. At $t_0 + 2$ ms, the refilling process comes its later period. Logically, surface tension and gravity are supposed to drive the metal liquid in the rear of the weld pool, which is far from the keyhole opening and near the upper surface, to flow inwards in the direction of keyhole axis. From the results in Fig. 6.15b, however, that metal liquid flows away from the keyhole, and its maximum speed near the liquid surface reaches 0.5 m/s. This suggests that the flow of the metal liquid in the rear of the weld pool away from the keyhole is the result of the auxiliary gas flow overcomes the effects of surface tension and gravity. The results in Fig. 6.15 further demonstrate the role of auxiliary gas flow in controlling the laser weld pool.

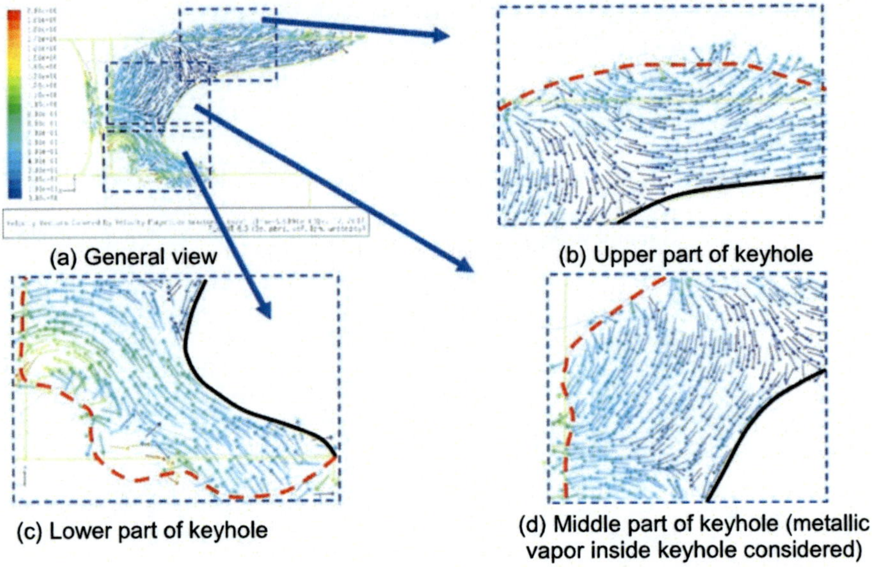

Fig. 6.15 Velocity vector distribution in symmetry plane at $t_0 + 2$ ms under the effect of auxiliary gas in the refilling stage

From the calculation results at time $t_0 + 2.4$ ms in Fig. 6.11, it can be seen that after the completion of the refilling stage, the keyhole starts to expand radially outwards under the action of metallic vapor.

6.4 Influences of Side-Blown Gas on the Stability of the Keyhole and Weld Pool

The transient evolution behavior of a keyhole and weld pool without applying side-blown gas is calculated for comparison with the physical mechanism of keyhole and weld pool evolution when side-blown gas is applied. Figure 6.16 shows the calculation results obtained when side-blown gas is not used, with the red region indicating liquid metal and the blue region indicating gas phase region.

Figure 6.17 compares the transient evolution behavior of the average pressure of a keyhole when side-blown gas is applied and when it is not applied. From 0.8 to 1.5 ms, the average pressure in both cases is relatively stable. At the same time, the average pressure with side-blown gas applied is about 1200 Pa higher than that without applying side-blown gas. This will surely lead to the difference in keyhole and weld pool behaviors between the two cases. As can be seen in Figs. 6.11 and 6.16, during the period from 0.8 to 1.5 ms, the top surface of the weld pool swells and the melt flows towards the rear part of the weld pool when side-blown gas is considered, which is quite different from the situation for the case without side-blown gas. The

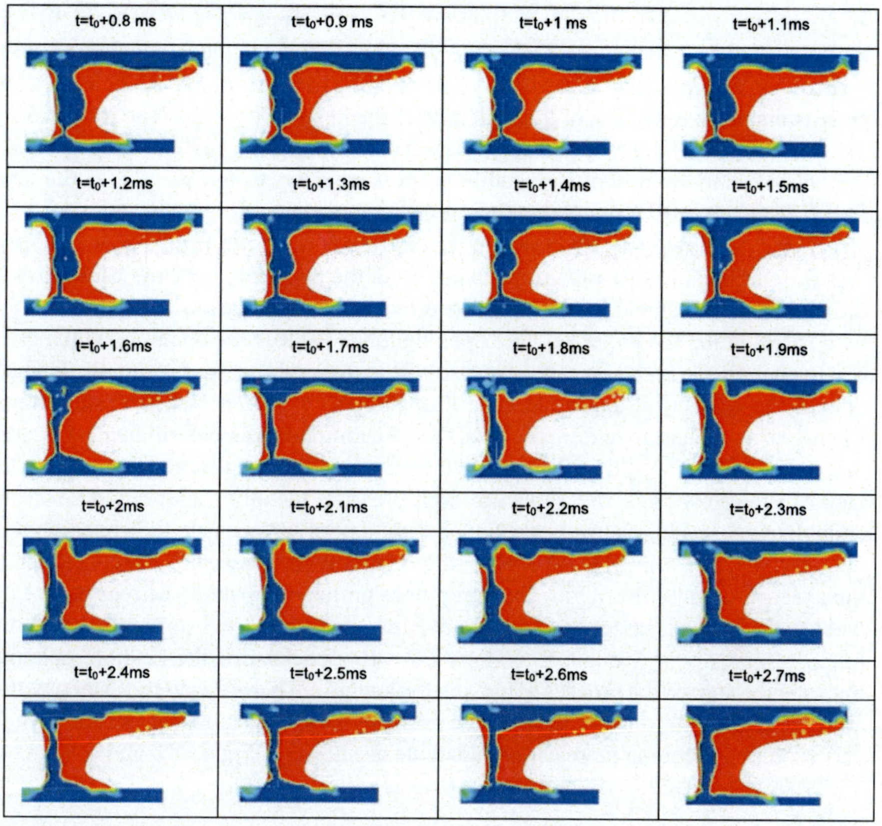

Fig. 6.16 Calculation results of transient behavior of keyhole and weld pool without auxiliary gas (metallic vapor inside keyhole considered)

Fig. 6.17 Variation of average pressure of keyhole (metallic vapor inside keyhole considered)

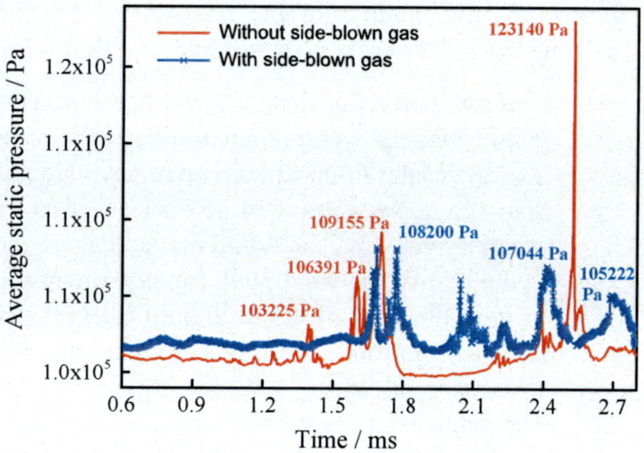

applied side-blown gas pushes the liquid metal on the upper surface of the weld pool to flow backwards to the rear part of the weld pool. This enables a concave shape to form on the liquid surface near the keyhole opening in the front part of the weld pool, and results in the reduction of the melt around the keyhole opening. The formation of the concave shape and reduction of the melt greatly help the keyhole keep open and the keyhole and weld pool stay stable in laser welding. Consequently, as the laser action lasts, the peak average pressure of the keyhole decrease step by step under the effect of side-blown gas, as shown in the result of the period from 1.5 to 2.8 ms in Fig. 6.17. When the side-blown gas is not used, the melt does not flow backwards to the rear part of the weld pool. Under the effect of surface tension, a large amount of melt surrounding the upper keyhole opening tends to close the opening, which gives rise to an increase in the average pressure within the keyhole. As a result, the keyhole opening becomes wide again under the strong impact of high-pressure metallic vapor. During an actual laser welding process, this phenomenon might periodically occur so that the weld pool would be disturbed violently. It is should be noted that despite the lack of side-blown gas, the impact of high-pressure metallic vapor could be strong enough to cause a hump to emerge on local weld pool surface near the upper keyhole opening, as shown in Fig. 6.16 at $t_0 + 2.3$ ms. In this case, however, because of the absence of side-blown gas, the hump does not move towards the rear part of the weld pool; instead, under the combined effects of surface tension and gravity, the hump reserves energy enough to allow it to flow backwards the keyhole opening, marking the start of the process of closing the keyhole. Therefore, as the laser welding continues, the peak value of the average pressure increases stepwise, and the keyhole and weld pool become increasingly unstable as shown in Figs. 6.16 and 6.17.

6.5 Conclusions

Below are the findings of the research on the influences of side-blown gas on deformation of the weld pool surface.

(1) Under the effect of surface tension, the junction point of gas, liquid and solid phases near the welding line moves to the center of the weld pool and the area near the center of upper surface of the weld pool bulges, with the hump higher than the upper surface of workpiece. Then, under the effect of side-blown auxiliary gas flow, the liquid metal near the surface of the weld pool transfers to the rear of the weld pool. Finally, the gas–liquid interface around the melt front in the front of the weld pool is lower than the workpiece surface, while that near the solidification front in the rear of the weld pool is higher than the workpiece surface.

(2) The main reason for the appearance of weld undercut defects is that the junction point of gas, liquid and solid phases near the welding line moves to the center of the weld pool driven by surface tension. The weld reinforcement is mainly caused by the movement of liquid metal in the weld pool to the rear of welds,

and then accumulation and solidification there. The side-blown gas obviously helps liquid metal flow to the rear of the weld pool.

(3) A local high-pressure area can be formed in the vicinity of a locally depressed gas–liquid interface in the front of the weld pool after auxiliary gas is blown to the workpiece surface. Backwards from the area, with the increase of distance from the front of the weld pool, pressure constantly decreases.

The following are the findings of the research on behaviors of the keyhole and weld pool under the effects of metallic vapors.

(1) Confluence formed by side-blown gas and top-blown gas flow shows a tilted, bell jar shaped pressure distribution in the upper area of the workpiece surface and the axis of the "bell jar" is the flowing direction of confluence. There is a local high-pressure area in the vicinity of the intersection point of the axis of the "bell jar" and the workpiece surface. In addition, due to the inhibition of auxiliary gas flow on high-speed gas flow ejected from the keyhole, a local high-pressure area can be found in the surrounding areas above the keyhole opening.

(2) Under the effect of metallic vapors, the changes of the keyhole and the weld pool can be divided into three stages, i.e. initial stage, expansion stage and refilling stage. It is assumed that the keyhole is formed instantaneously in the welding process and the whole welding process is the constant cycles of these three stages.

In the initial stage, liquid metal around the keyhole flows in the direction departing the axis of the keyhole. The local uplift obviously exceeding the initial upper surface of the weld pool is not formed close to the keyhole opening, and the diameter of the keyhole changes slightly in the direction of whole thickness of the workpiece. In the expansion stage, molten metal around the wall in the middle of the keyhole constantly expands and moves to the direction far away from the axis of the keyhole. However, molten metal in the rear of keyhole opening on the lower and upper surfaces moves towards the axis of the keyhole and shows the trend of closing the keyhole opening. In addition, local uplift formed on the metal surface in the rear of the keyhole opening can enhance the effects of side-blown gas on the part, so that a part of molten melt in the local uplift behind the keyhole opening on the upper surface flows to the rear of the weld pool. In the refilling stage, molten melt in the weld pool behind the lower keyhole opening flows towards the keyhole and from the bottom up under the effect of surface tension. Driven by surface tension and gravity, surrounding liquid metal behind the upper keyhole opening shows a similar flowing trend to the liquid metal behind the lower keyhole opening, that is, liquid metal flows towards the axis of the keyhole. However, molten melt in the rear of the weld pool around the upper surface flows to the direction far away from the keyhole. Such a flowing behavior is formed under the effects of auxiliary gas flow after overcoming the influences of surface tension and gravity, which proves the control effects of auxiliary gas flow on the weld pool during laser welding.

(3) Under the pressures of metallic vapors, the diameter of the keyhole on the cross section first increases with the expansion of the keyhole, and then decreases with the refilling of molten melt after reaching the maximum; Moreover, after completing the refilling stage of the molten melt, the diameter increases again with the expansion of the keyhole. The expansion of the keyhole in the radial direction can make the wall of the keyhole closer to the solid–liquid interface. In this way, the energy carried by the beam directly acts at the position closer to the solid–liquid interface and is more effectively and directly applied in the workpiece far away from the axis of the keyhole in the radial direction. Therefore, this facilitates the formation of a larger weld width.

Chapter 7
Keyhole and Weld Pool Dynamics in Dual-Beam Laser Welding

Abstract This chapter introduces the time-dependent keyhole and weld pool dynamics in dual beam laser welding. Mathematical model of self-consistent keyhole and weld pool dynamics in dual beam welding is presented. The coupled keyhole and weld pool dynamics in dual beam welding are discussed by numerical simulations. The stabilization mechanisms of a dual beam mode laser arrangement in welding process are given.

7.1 Introduction

The laser welding features small beam spot diameter, high thermal efficiency and small thermal input of the welding line, thus the deformation and residual stress of laser welding are very small. Meanwhile, the thermal influence zone of welding is narrow and the metal grain coarsening tendency of the weld joint is small, endowing laser welding typical application advantages. Due to the small beam spot diameter, high precision is demanded for weldment with single beam laser welding. At present, for some sheet metal forming thin-wall structural components, it is difficult to ensure the forming precision as there is a large gap of weldment and certain misalignment, thus it is often difficult to achieve the assembly precision required by ordinary single beam laser welding technology. This posed certain challenges for the application of laser welding technology on the thin-wall structural component. In order to break through the limitation of single beam laser welding, the dual beam laser welding technology is applied to the welding of thin-wall components. It has become an important technical approach to realize the optimization and high adaptability welding of the thin-wall structural component.

Dual beam laser welding separates the same laser into two separate laser beams by an optical method, or uses a combination of two different types of laser for welding. By changing the energy ratio, beam spacing and arrangement mode of two laser beams, the adaptability of laser welding to assembly accuracy can be improved, the welding temperature field and flow field can be adjusted flexibly, and the evolution behavior of keyhole and flow form of weld pool can be changed, improving the stability of the welding process. Therefore, dual beam laser welding is featured with

S. Gong et al., *Weld Pool Dynamics in Deep Penetration Laser Welding*,
https://doi.org/10.1007/978-981-16-0788-2_7

small residual stress deformation and good joint quality. Moreover, it can solve the problems such as poor gap adaptability of conventional laser welding, large porosity tendency and serious burning loss of alloying elements. According to the arrangement of the beam, the dual beam laser welding can be divided into parallel and tandem arrangement. If the lining of the focus between two beams is parallel to the welding direction, it is called tandem dual beam welding. If the lining of the focus between two beams is perpendicular to the welding direction, it is called parallel dual beam welding.

Taking dual beam welding as the research object, this chapter studies the establishment of transient coupling model of keyhole and weld pool in dual beam laser welding, focusing on the dynamics behavior of the keyhole and weld pool in dual beam welding, and exploring the stability mechanism in dual beam laser welding.

7.2 Transient Coupling Model of Keyhole and Weld Pool in Dual-Beam Welding

7.2.1 Control Equation of Transient Coupling Model

7.2.1.1 Basic Assumption

The dual beam laser welding process includes the very complex polyphase transition of plasma—gas—liquid—solid, free interface evolution and heat transfer flow coupling in the weld pool. Therefore, the current numerical calculation needs to be simplified. The metal vapor plume and fluid mechanics effect of protective gas on the keyhole are ignored, and the condensation process of metal vapor is not considered. For high power CO_2 laser welding, laser beam scattering and inverse bremsstrahlung absorption of plasma matter a lot. In numerical simulations, these effects are characterized by artificially increasing the radius of the laser beam. However, in Nd: YAG laser welding, due to the low ionization of metal vapor, inverse bremsstrahlung absorption has little effect, thus scattering and refraction effect is simply considered. But for general low power dual beam laser welding process, the effect can be neglected. What's more, according to the method proposed by Ki et al., heat loss due to evaporation near the keyhole wall is addressed. The model also considers the thermal convection and thermal radiation of workpiece surface.

7.2.1.2 Control Equation

Assume that the metal liquid in the welding weld pool is incompressible fluid, the density of liquid metal does not change during the solid-liquid phase transition. Therefore, the conservation equation of mass, momentum and energy is expressed as follows

$$\nabla \cdot \vec{U} = 0 \tag{7.1}$$

$$\rho\left(\frac{\partial \vec{U}}{\partial t} + \left(\vec{U} \cdot \nabla\right)\vec{U}\right) = \nabla \cdot \left(\mu_l \nabla \vec{U}\right) - \nabla p - \frac{\mu_l}{K}\vec{U} - \frac{C\rho}{\sqrt{K}}\left|\vec{U}\right|\vec{U} + \rho\vec{g}\beta\left(T - T_{ref}\right) \tag{7.2}$$

$$\rho C_p\left(\frac{\partial T}{\partial t} + \left(\vec{U} \cdot \nabla\right)T\right) = \nabla \cdot (k\nabla T) \tag{7.3}$$

where: $\vec{U}$—three-dimensional velocity vector;

μ_l—dynamic viscosity of fluid;

ρ—density;

p—pressure;

$\vec{g}$—three-dimensional gravity acceleration vector;

β—coefficient of thermal expansion;

T_{ref}—reference temperature;

K—Carman-Kozeny coefficient in mixed phase model, i.e., transmission coefficient, which is closely related to liquid mass fraction f_l of the current grid cell.

K can be determined by the following formula

$$K = \frac{f_l^3 d^2}{180(1 - f_l)^2} \tag{7.4}$$

where, d-closely related to the size of dendritic arms, which is usually a constant. For laser welding, the order of magnitude of its size is 10^{-2} cm.

In addition, C in Formula (7.2) is a nonreactivity coefficient related to liquid mass fraction, which is calculated by the following formula

$$C = 0.13 f_l^{-3/2} \tag{7.5}$$

where, f_l—liquid mass fraction

In order to improve the value calculation speed, the liquid mass fraction may well have a linear relation with the temperature, thus

$$f_l = \begin{cases} 1 & T > T_l \\ \frac{T - T_s}{T_l - T_s} & T_l \geq T \geq T_s \\ 0 & T < T_s \end{cases} \tag{7.6}$$

where, T_l and T_s—liquidus temperature and solidus temperature of welded alloy material.

Based on Level Set approach, the kinematic equation describing the transient keyhole interface in laser welding can be expressed as:

$$\frac{\partial \varphi}{\partial t} + \vec{U} \cdot \nabla \varphi = 0 \tag{7.7}$$

where, $\vec{U}$—kinematic velocity of keyhole interface.

7.2.1.3 Fresnel Absorption Formula

According to the relative position of two laser beams and welding direction in dual beam laser welding, it can be classified into parallel dual beam welding and tandem dual beam laser welding. By the ray-tracing method, the laser energy density q absorbed at any position on the free interface of the weld pool in dual beam welding can be calculated. The formula is expressed as

$$\begin{aligned} q = {} & I_1(r_1, z_1)\left(\vec{I}_1 \cdot \vec{n}\right)\alpha_{Fr}(\theta_1) + \sum_{m=1}^{N} I_m(r_1, z_1)\left(\vec{I}_m \cdot \vec{n}_m\right)\alpha_{Fr}(\theta_m) \\ & + I_2(r_2, z_2)\left(\vec{I}_2 \cdot \vec{n}\right)\alpha_{Fr}(\theta_2) + \sum_{k=1}^{N} I_k(r_2, z_2)\left(\vec{I}_k \cdot \vec{n}_k\right)\alpha_{Fr}(\theta_k) \end{aligned} \tag{7.8}$$

$$\alpha_{Fr}(\theta) = 1 - \frac{1}{2}\left(\frac{1 + (1 - \varepsilon\cos\theta)^2}{1 + (1 + \varepsilon\cos\theta)^2} + \frac{\varepsilon^2 - 2\varepsilon\cos\theta + 2\cos^2\theta}{\varepsilon^2 + 2\varepsilon\cos\theta + 2\cos^2\theta}\right) \tag{7.9}$$

where, $I_1(r_1, z_1)$ and $I_2(r_2, z_2)$—energy distribution function of initial two laser beams;

$I_{\mathrm{m}}(r_1, z_1)$—density distribution of the residual energy of the first laser beam after the mth reflection;

$I_{\mathrm{k}}(r_2, z_2)$—density distribution of the residual energy of the second laser beam after the kth reflection;

θ—angle between the incident laser beam and the normal vector of the keyhole wall;

α_{Fr} (θ)—Fresnel absorption coefficient;

N—number of laser beam incidence considering multiple reflections;

$\vec{I}$—normalized laser beam direction;

$\vec{n}$—normalized normal vector of the keyhole wall;

ε—constant related to the laser device and material.

The diagram for tracking and calculating the light of tandem dual beam is shown in Fig. 7.1.

In this study, the approximate distribution of laser energy density is assumed to be a Gaussian distribution, and then the distribution function equation of laser energy can be written as

$$I_0(r, z) = 3Q/\left(\pi R^2\right)\exp\left(-3\left(r^2\right)/R^2\right) \tag{7.10}$$

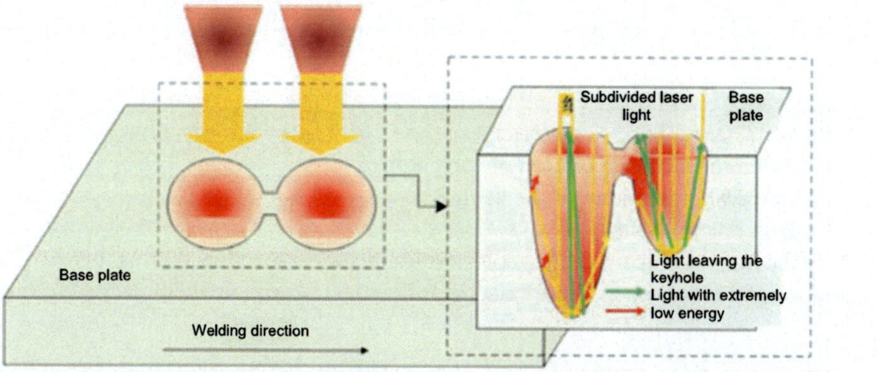

(a) Without considering recoil pressure (b) With recoil pressure considered

Fig. 7.1 Diagram for tracing and calculating the light in tandem dual beam laser welding

where, R—light spot radius;
Q—laser power density.

7.2.2 Boundary Conditions of Transient Coupling Model

To make the model easy to deal with, the frictional effect of the vapor jet on the hole wall is ignored. The motion of the free interface of the keyhole is mainly affected by surface tension, thermal capillary force, recoil pressure, fluid shear flow stress, hydrostatic pressure and fluid dynamic pressure in the weld pool. Based on the sharp interface boundary condition method, the boundary condition of the free interface is expressed as:

$$
\begin{aligned}
\left(\mu\nabla\vec{U}\right)_f = {} & \mu\left(\vec{n}\ \vec{t}_1\ \vec{t}_2\right)\left(\vec{n}\ \vec{0}\ \vec{0}\right)^T\left(\nabla\vec{U}\right)\left(\vec{n}\ \vec{0}\ \vec{0}\right)\left(\vec{n}\ \vec{t}_1\ \vec{t}_2\right)^T \cdot \mu\left(\vec{n}\ \vec{t}_1\ \vec{t}_2\right)\left(\vec{0}\ \vec{t}_1\ \vec{t}_2\right)^T\left(\nabla\vec{U}\right) \\
& - \mu\left(\vec{n}\ \vec{t}_1\ \vec{t}_2\right)\left(\vec{n}\ \vec{0}\ \vec{0}\right)^T\left(\nabla\vec{U}\right)^T\left(\vec{0}\ \vec{t}_1\ \vec{t}_2\right)\left(\vec{n}\ \vec{t}_1\ \vec{t}_2\right)^T \\
& + \left(\vec{n}\ \vec{t}_1\ \vec{t}_2\right)\begin{pmatrix} 0 & \nabla_s\sigma\cdot\vec{t}_1 & \nabla_s\sigma\cdot\vec{t}_2 \\ 0 & 0 & 0 \\ 0 & 0 & 0 \end{pmatrix}\left(\vec{n}\ \vec{t}_1\ \vec{t}_2\right)^T
\end{aligned} \tag{7.11}
$$

$$
p_f = p_r + \sigma\kappa + 2\mu\vec{n}\cdot\nabla\vec{U}\cdot\vec{n} \tag{7.12}
$$

where, f—free interface;
μ—density of the metal liquid in weld pool;
p_r—recoil pressure.
Based on the recoil pressure model proposed by Semak, et al., p_r is expressed as

$$p_r = 0.54 A B_0 (T)^{-1/2} \exp(-\frac{U}{kT}) \tag{7.13}$$

where, A—constant related to material;

B_0—constant related to material;

U—latent heat of evaporation of each atom;

T—surface temperature of the keyhole;

k—Boltzmann's constant.

On the free interface of the keyhole, due to Fresnel absorption, thermal convection, radiation and evaporation, there exist the following temperature boundary conditions

$$k\frac{\partial T}{\partial \vec{n}} = q - h(T - T_\infty) - \varepsilon_r \sigma \left(T^4 - T_\infty^4\right) - \rho V_{evp} T_v \tag{7.14}$$

where, q—laser energy density absorbed by Fresnel effect, determined by Formula (7.8);

T_v—evaporating temperature;

V_{evp}—receding speed of the keyhole interface caused by evaporation, determined in this study by the formula proposed by Ki, et al.

On other boundaries of the calculated region, there exist the following temperature boundary conditions

$$k\frac{\partial T}{\partial \vec{n}} = -h(T - T_\infty) - \varepsilon_r \sigma \left(T^4 - T_\infty^4\right) \tag{7.15}$$

where, h—convection coefficient;

ε_r—black body radiation coefficient;

σ—Boltzmann constant.

7.3 Coupling Behavior of Keyhole and Weld Pool in Dual-Beam Welding

7.3.1 *Evolution Behavior of Dynamic Keyhole in Welding*

Figures 7.2 and 7.3 show the temperature field distribution and morphological evolution of the keyhole in typical parallel dual beam laser welding respectively. In the study, the material is 304 stainless steel, the welding speed is 2 m/min, the power of each beam is 2 kW, the beam spot spacing is 0.5 mm, and the laser spot radius is 0.25 mm. As shown in Figs. 7.2a, b and 7.3a, b, when the welding starts, the workpiece absorbs laser energy, with the temperature rising rapidly, and starts to evaporate. The recoil pressure generated by metal evaporation causes the metal liquid surface to dip, forming two relatively independent keyholes. At 0.83 ms, two separate keyholes are connected to form a single keyhole, and the center of the keyhole opening is concave,

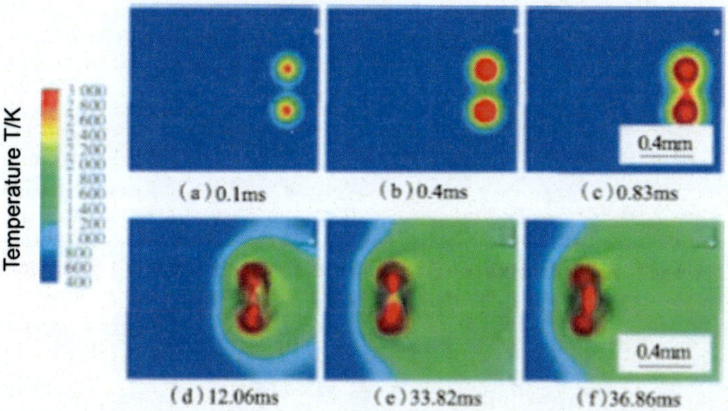

Fig. 7.2 Top view of the evolution process of keyhole shape in parallel dual beam welding

Fig. 7.3 Evolution process of keyhole cross-section shape in parallel dual beam welding

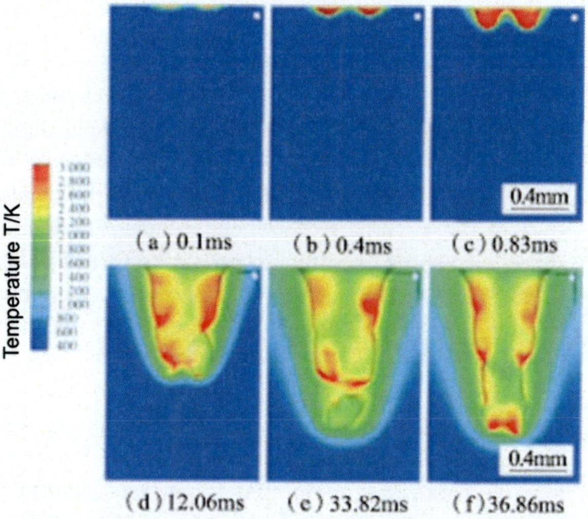

with the lower end appearing an inverted hump shape, as shown in Figs. 7.2c and 7.3c. At 12.06 ms, the melting zone at the center of the beam spot spacing expands, and the keyhole opening takes on an oval shape, with the lower ends connected, as shown in Figs. 7.2d and 7.3d. After that, the keyhole continues to absorb laser energy, and it keeps growing and deepening until the depth of the keyhole reaches a quasi-steady state, with its shape shown in Figs. 7.2f and 7.3f. The keyhole may collapse during the evolution process. At 33.82 ms, the keyhole collapses obliquely, and is broken into two parts, as shown in Fig. 7.3e. Therefore, it can be seen that in parallel dual beam laser welding, the evolution process of keyhole may experience the following stages: two separate keyholes—keyholes connected—keyholes merging. The interaction of keyholes makes the evolution more complicated.

Figure 7.4 shows the typical evolution process of a keyhole in tandem dual beam

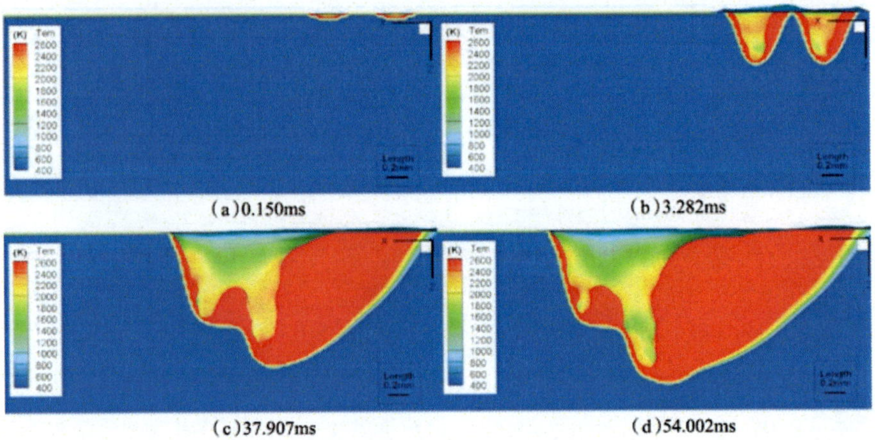

Fig. 7.4 Evolution process of keyhole shape in tandem CO_2 laser beam welding

laser welding of 5052 aluminum alloy. The openings of the two keyholes formed at the beginning stage gradually merge together, and the tips of the two holes remain, but the keyhole at the rear is obviously deeper than the one at the front. Moreover, the shape of the keyhole is not smooth, with many folds.

In tandem dual beam laser welding, with the increase of the beam spacing, the state of the keyhole experiences a change from sharing a same opening and tip, sharing the same opening but not tip, to forming two completely separate keyholes. In the state of keyhole sharing the same opening but not tip, the keyhole at the rear is obviously deeper than that at the front, and there exist many folds on the free interface of the keyhole, as shown (see Fig. 7.5).

Moreover, the instability of the keyhole in welding results in oscillation of the keyhole depth. According to the evolution curve of keyhole depth shown in Fig. 7.6, as the welding speed increases from 2.54 to 6.25 m/min, the oscillation amplitude of the keyhole depth in tandem dual beam laser welding decreases gradually. Therefore, it can be seen that the increase in welding speed is conducive to improving the keyhole stability. According to the quantitative analysis of the depth oscillation, the frequency of the depth oscillation in typical tandem double beam laser welding is about 1.4 kHz (see Fig. 7.7), while the frequency of the depth oscillation in single beam laser welding is about 1.7 kHz (see Fig. 7.8). Compared to single beam laser welding, it can be found that the oscillation amplitude of the keyhole depth in tandem dual beam welding decreases significantly with the same heat input. As can be seen from Fig. 7.9, that the oscillation amplitude of the rear keyhole depth is significantly higher than that of the front keyhole. The oscillation frequency of the rear keyhole is about 1.4 kHz, while that of the front keyhole is between 2 and 3 kHz (Fig. 7.9).

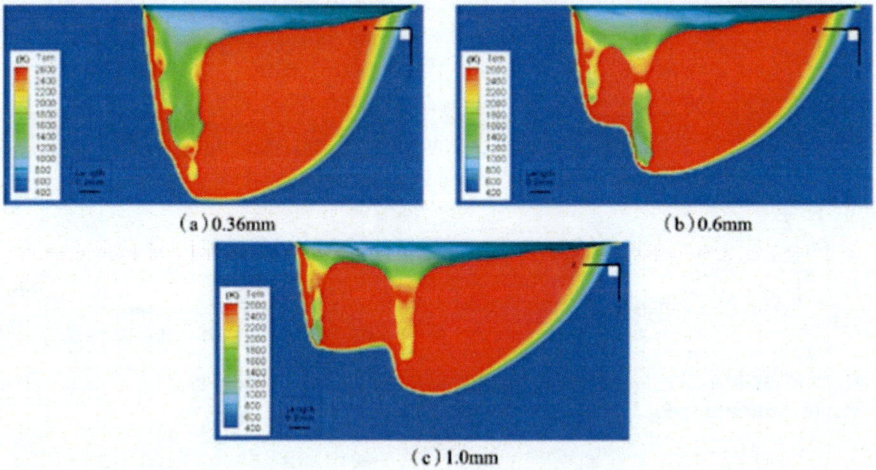

Fig. 7.5 Shape of the longitudinal section of keyholes in tandem dual beam laser welding with different beam spacing at 30.71 ms

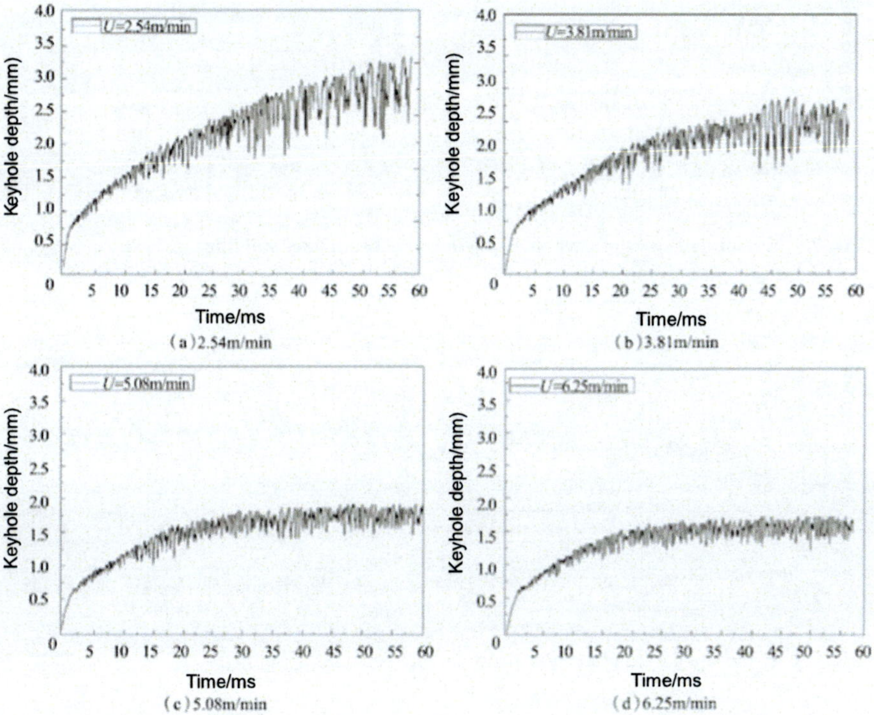

Fig. 7.6 Evolution curves of keyhole depth with different dual beam laser welding speeds

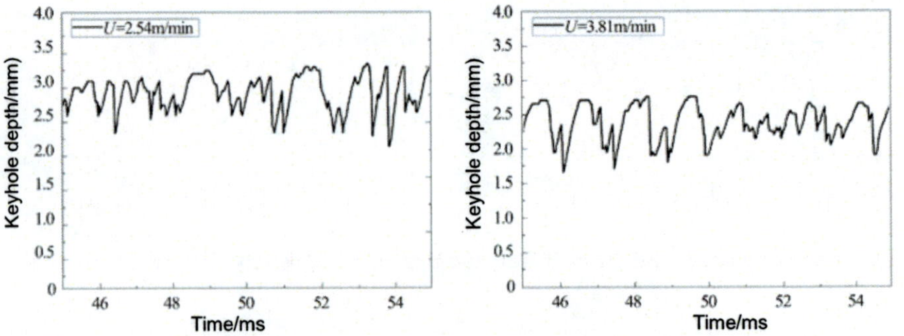

(a) 2.54m/min (b) 3.81m/min

Fig. 7.7 Oscillation curves of keyhole depth between 45 and 55 ms in dual beam laser welding with two different speeds

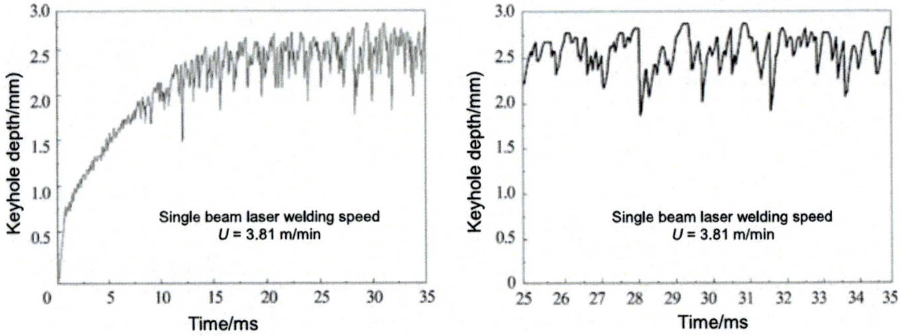

Fig. 7.8 Evolution curves of keyhole depth in single beam laser welding

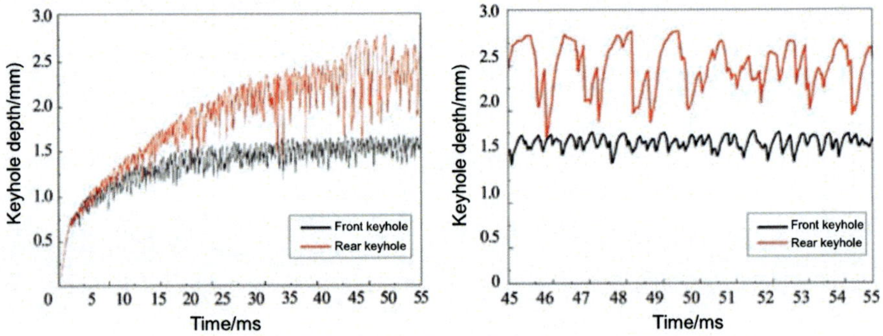

Fig. 7.9 Evolution curves of keyhole depth in dual beam laser welding

7.3.2 *Flow Behavior of Moving Weld Pool in Welding*

In tandem double beam welding, when the two keyholes merge together, the flow velocity on the surface of the keyhole opening is more intense and complex than that on the surface of other areas of the weld pool. There exists an intense downward flow near the front wall of the keyhole. Besides, there are some characteristic flows at the rear wall of the bottom of the rear keyhole, as shown in Fig. 7.10. The specific performance is as follows: first, on the upper part of the rear wall of the keyhole there exists an upward flow along the wall surface; second, there is a downward flow near the rear wall of the bottom of the rear keyhole, which leads to vortices of flow behind the keyhole.

Compared to single beam laser welding, under the same heat input conditions, the flow in the weld pool in tandem dual beam laser welding is apparently different. For example, the flow in dual beam welding is slower and gentler, and the downward flow near the wall surface of the keyhole is weaker than that in single beam laser welding. However, there are similarities between dual beam laser welding and single laser welding, e.g., the downward flow and vortices behind the keyhole.

In conclusion, the flow behavior of the weld pool in tandem dual beam laser welding is very complicated. In the current model, the influence of two main dynamic factors of recoil pressure and thermal capillary force is considered. Some flow behaviors, such as the surface flow at the keyhole opening and the upward flow at the rear wall of the keyhole, are mainly driven by recoil pressure and thermal capillary force. However, the high-speed downward flow at the front of the other keyhole and the flow at the rear wall of the keyhole tip are mainly affected by recoil pressure (Fig. 7.11).

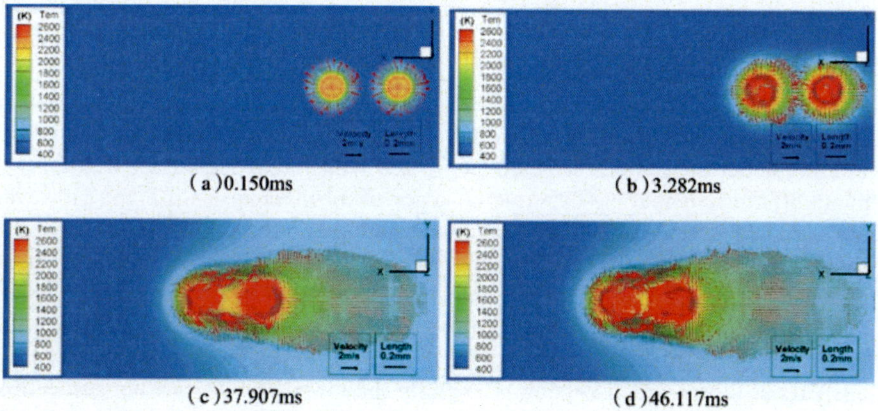

Fig. 7.10 Top view of the flow state on the free interface of the keyhole in tandem dual beam welding

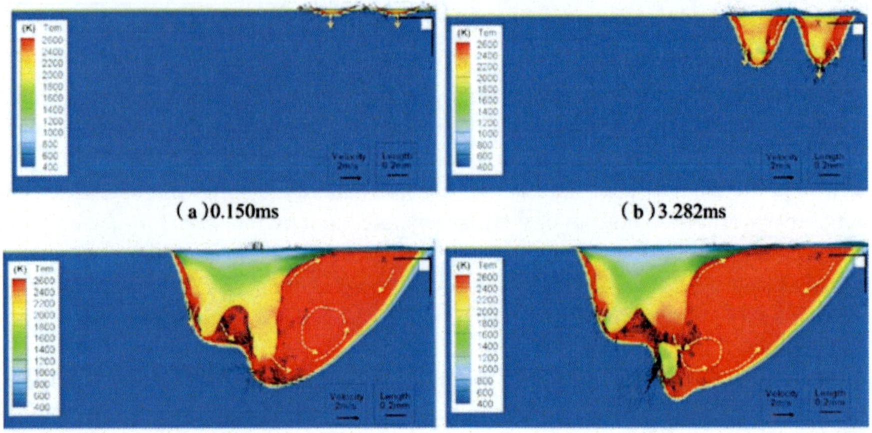

(a)0.150ms (b)3.282ms (c)37.907ms (d)46.117ms

Fig. 7.11 Flow state of a weld pool in tandem dual beam welding

7.4 Mechanism and Influencing Factors of Stability of Dual-Beam Welding

7.4.1 Stability Mechanism of Dual-Beam Welding

The experimental results by Xie et al., have shown that tandem dual beam laser welding can significantly improve the stability of the welding process compared with single beam laser welding. Some studies suggest that the reason for the better stability of tandem dual beam laser welding is that a bigger opening of the keyhole can allow gas to escape more easily, thus providing a better degassing effect.

Previous quantitative experimental studies have verified that the flow of weld pool in tandem dual beam laser welding is more stable than that in single beam laser welding under the same heat input. Compared with single beam laser welding, the main difference is that the dynamic behavior of metal vapor is different. However, the frictional effect of metal vapor is not considered in the simulation study in this chapter. Therefore, besides the frictional effect of metal vapor, the simulation results in this chapter imply that there must be some other stability mechanism for tandem dual beam laser welding.

The results of this study show that the double beam welding technology can obviously reduce the depth of keyhole. Generally shallower keyholes are not susceptible to the surface tension and hydrostatic pressure of the metal liquid, so a more stable process can be obtained. Moreover, the flow in the weld pool is more stable during the dual beam welding process, because the tandem arranged dual beams can reduce the power density and increase the surface area of the weld pool, thus reducing the effect of surface tension. All of these physical factors can help stabilize the laser welding process. Therefore, the stability mechanism of dual beam welding, compared with single beam welding process, is not only due to the degassing effect, but also the combined action of the above-mentioned factors.

7.4.2 *Influence of Process Parameters on Stability of Dual-Beam Welding*

7.4.2.1 Influence of Welding Speed on the Evolution Process of Keyhole Shape

Figure 7.12 shows the keyhole shape in parallel dual beam welding with different welding speeds at 35 ms. According to the figure, at the current welding speeds, the keyhole shape is very irregular, and protrusions are generated in many places on the wall surface of the keyhole. Figure 7.13 shows the variation curve of the keyhole depth with the above-mentioned welding speeds at 35 ms. At this point, the

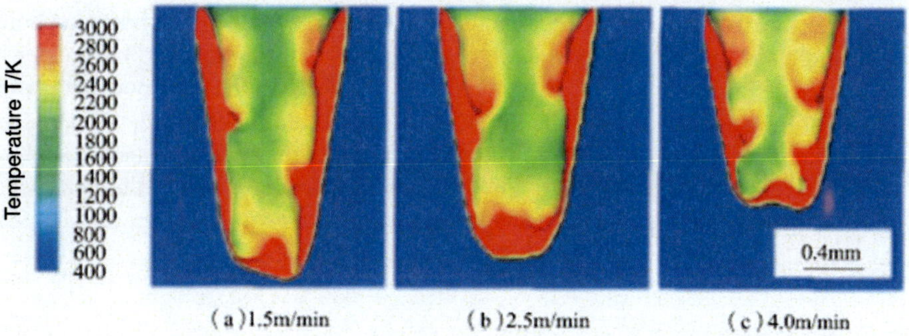

Fig. 7.12 Shape of keyhole in parallel dual beam welding with different welding speeds at 35 ms

Fig. 7.13 Curve of the keyhole with different welding speeds at 35 ms

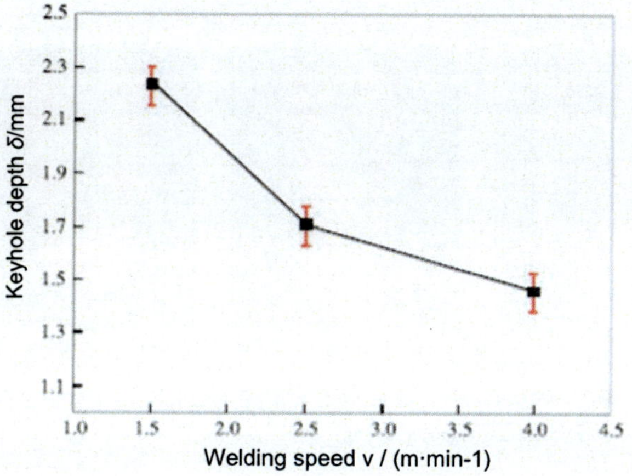

keyhole depth is 2.23 mm, 1.70 mm and 1.45 mm, respectively. It can be seen that the keyhole depth decreases as the welding speed increases. Besides, in parallel dual beam welding, the keyhole shape is generally irregular, and protrusions are generated in many places on the wall surface of the keyhole.

7.4.2.2 Influence of Beam Spot Spacing on the Evolution Process of Keyhole Shape

Figure 7.14 shows the keyhole shape in parallel dual beam welding with beam spot spacing of 0.25 mm, 0.5 mm and 1.0 mm respectively at 35 ms. It can be seen that: ① when the beam spot spacing is relatively small, two beams will produce a keyhole together in the same weld pool, as shown in Fig. 7.14a, and the keyhole has a high depth and a small opening; ② as the spot spacing increases, the two beams create a keyhole respectively in the same weld pool, with the front wall of the keyhole created by the first beam closely connected with the rear wall of the keyhole created by the second beam, as shown in Fig. 7.14b, and at this point, the depth of the keyhole decreases and the radius of the opening increases; ③ when the beam spot spacing is too large, the two keyholes will be separated, and two relatively independent keyholes will be formed in the same weld pool, or two separate weld pools and keyholes will be formed, as shown in Fig. 7.14c. At this point, the depth of the weld pool decreases, as well as the depth of keyhole and the radius of the opening. Therefore, it can be seen that in parallel dual beam welding, with the increase of the spot spacing, the keyhole depth decreases, while the radius of the opening increases. When the beam spot spacing is too large, the two beams will form two relatively independent keyholes in the weld pool, and the welding mechanism will also change.

7.4.2.3 Stages of Keyhole Depth Oscillation

Figure 7.15 shows the variation curves of keyhole depth with time in parallel dual

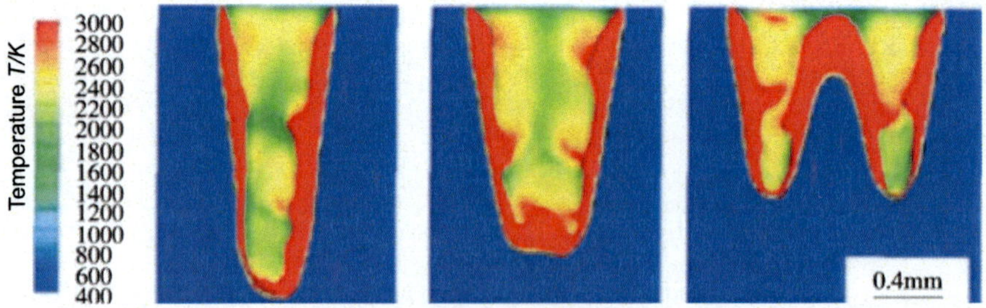

(a) Beam spot spacing of 0.25 mm (b) Beam spot spacing of 0.5 mm (c) Beam spot spacing of 1.0 mm

Fig. 7.14 Shape and temperature distribution of keyhole created by parallel dual beams with different beam spot spacing

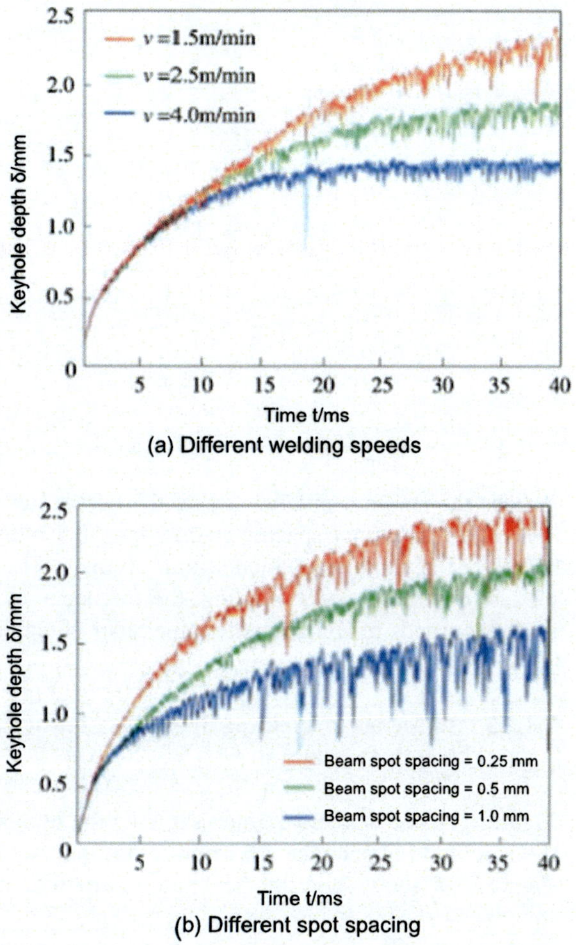

Fig. 7.15 Evolution curves of keyhole depth-time with parallel dual beams in different processes

beam welding with different welding speeds and different spot spacing. It can be seen that in certain welding process, the keyhole oscillates violently in welding, with its depth roughly divided into three stages: ① linear increase stage, in which the keyhole depth grows rapidly in a linear way; ② oscillatory increase stage, in which the growth rate of keyhole depth slows down, accompanied by small oscillation; ③ stable oscillation stage, in which the keyhole depth remains at a relatively fixed value and oscillates violently up and down.

7.4.2.4 Periodicity of Keyhole Depth Oscillation

Figure 7.16 shows the oscillation curves of keyhole depth within 30–35 ms in the welding process with different spot spacing. It can be seen that after the keyhole

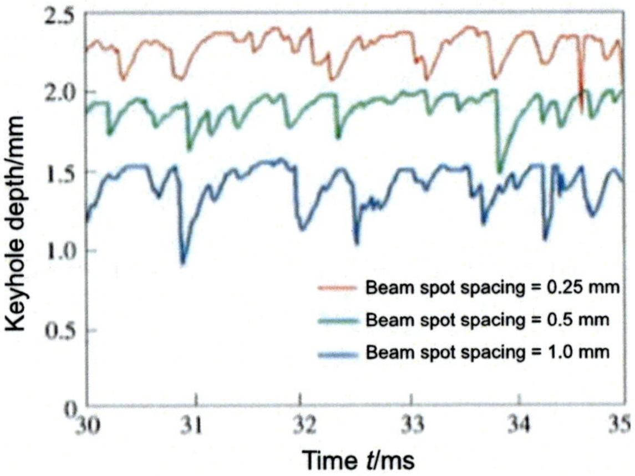

Fig. 7.16 Oscillation curves of keyhole depth within 30–35 ms in parallel dual beam welding with different spot spacing

reaches the stable oscillation stage, the oscillation takes on a certain periodicity. When the beam spot spacing is 0.25 mm, 0.5 mm and 1.0 mm respectively, the number of keyhole oscillation within 30 ms is 5–8 times, which indicates that in parallel dual beam laser welding, the frequency of keyhole depth oscillation can reach thousands of Hertz, on the same order of magnitude as single laser welding.

7.4.2.5 Influence of Welding Speed on Characteristics of Keyhole Depth Oscillation

Figure 7.17 is a statistical graph showing the number of keyhole oscillations with the amplitude larger than 0.15 mm at different welding speeds within 35–40 ms. Figure 7.18 shows how the maximum amplitude of the keyhole varies with the

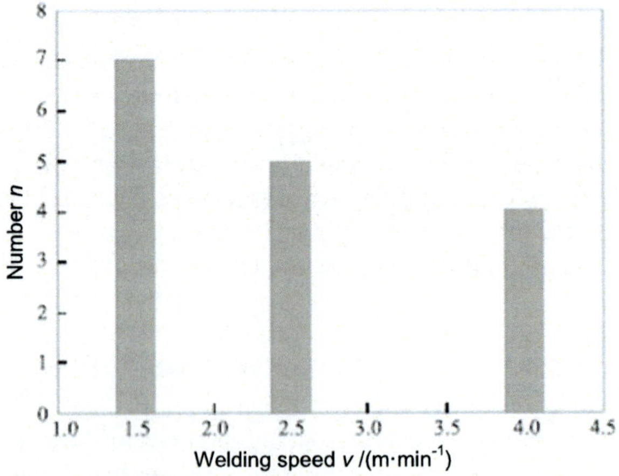

Fig. 7.17 Number of keyhole oscillations with the amplitude larger than 0.15 mm at different welding speeds within 35–40 ms

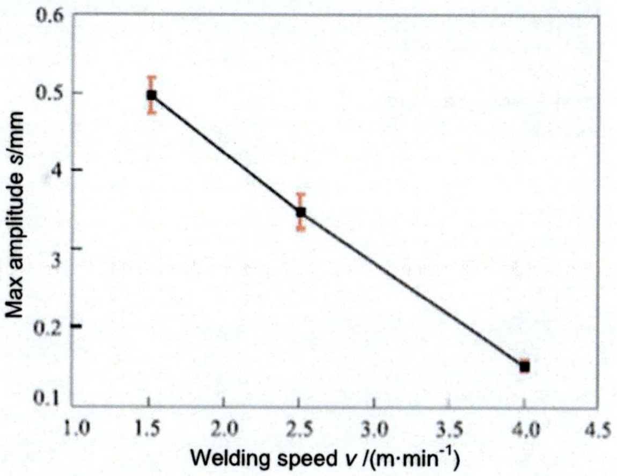

Fig. 7.18 Curve of the maximum keyhole amplitude varying with the welding speed

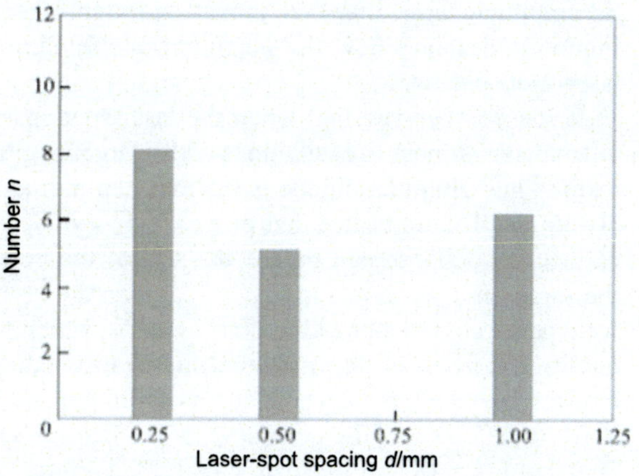

Fig. 7.19 Statistics of number of keyhole deep oscillations with the amplitude larger than 0.20 mm within 30–35 ms

welding speed. It can be observed that, in simulated welding, with the increase of the welding speed, the number of keyhole oscillations with the amplitude larger than 0.15 mm and the maximum amplitude decrease; in particular, when the welding speed reaches 4.0 m/min, the maximum amplitude of the keyhole decreases greatly to only 0.15 mm. It can be inferred that the keyhole may be in a stable state when the welding speed is increased to a certain value. It can be seen that in the parallel dual beam welding process, increasing the welding speed appropriately can improve the stability of the keyhole.

7.4.2.6 Influence of Laser-Spot Spacing on Characteristics of Keyhole Depth Oscillation

Figure 7.19 is a statistical graph showing the number of keyhole oscillations with

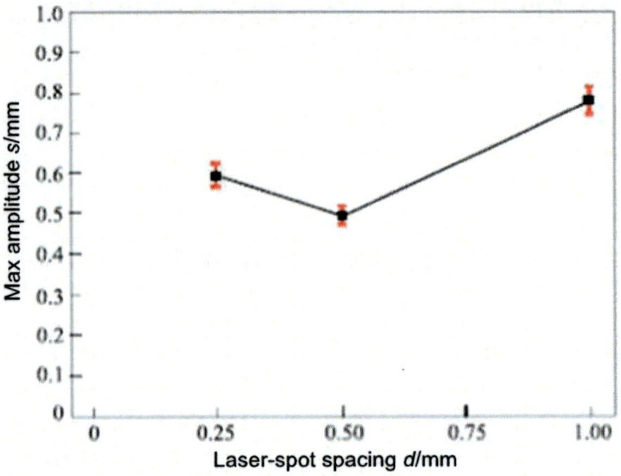

Fig. 7.20 Curve of the maximum amplitude of keyhole depth oscillation varying with laser-spot spacing

the amplitude larger than 0.20 mm at a different laser-spot spacing within 30–35 ms. Figure 7.20 shows how the maximum amplitude of the keyhole varies with the laser-spot spacing.

It can be observed that when the laser-spot spacing is 0.25 mm, 0.5 mm, and 1.0 mm, the number of oscillations with a larger amplitude is 8, 5, and 6, respectively, and the maximum amplitude is 0.6 mm, 0.5 mm and 0.8 mm, respectively, within 30–35 ms. This indicates that proper laser-spot spacing can reduce the likelihood of inclined collapse and reduce the size of the pores that might be generated. At the same time, as shown in Fig. 7.20, when the laser-spot spacing is too large, the weld penetration depth will fluctuate greatly, which will seriously affect the welding quality. It can be seen that in parallel dual beam welding, if the laser-spot spacing is too large or too small, the keyhole depth oscillation amplitude will increase, and severe porosity defects may be generated; and when the spacing is excessively large, the penetration depth will fluctuate greatly. Therefore, appropriate spot spacing should be employed when the weld penetration depth permits.

7.5 Summary

(1) In the parallel dual beam welding process, the variation of keyhole depth with time can be divided into three stages, namely linear growth stage, oscillatory growth stage and stable oscillation stage. In the stable oscillation stage, the oscillation of the keyhole exhibits a certain periodicity, and its depth oscillation frequency can reach thousands of Hertz, which is in the same order of magnitude as that for single beam laser welding.

(2) Under a specific parallel dual beam welding process, with the increase of the welding speed, the keyhole depth decreases and the oscillation amplitude

decreases too. When the welding speed increases to a certain value, the keyhole is even to be in a stable status.

(3) In the process of parallel dual beam welding, with the increase of the laser-spot spacing, the keyhole depth decreases but the radius of the opening increases generally. When the laser-spot spacing is too large, two beams will form two relatively independent keyholes in the weld pool, and the welding mechanism will also change. In addition, if the laser-spot spacing is too large or too small, the keyhole depth oscillation amplitude will increase, and when the spacing is too large, the weld penetration depth will fluctuate greatly.

(4) In tandem dual beam laser welding, with the increase of the beam spacing, the welding process experiences the change from a single keyhole, to one opening but different hole bottoms, and to two independent keyholes. In the general process of welding with one opening but different hole bottoms, the keyhole in the rear is larger than the one in the front due to the effect of preheating.

(5) The frequency of keyhole depth oscillation in tandem dual beam laser welding is several kilohertz, which is equivalent to that for single beam laser welding. The keyhole depth oscillation amplitude is smaller than that for single beam laser welding, and the keyhole depth oscillation gradually becomes stable with the increase of the welding speed.

(6) Under the same heat input conditions, the flow in the weld pool for tandem dual beam laser welding is more stable than the flow in the weld pool for single beam laser welding. For dual beam laser welding, the power density in the weld pool is lower, the superficial area of the weld joint larger, the surface tension effect weaker, and the weld pool during the welding process is more stable.

(7) Compared with single beam laser welding, the process stability mechanism of dual beam laser welding is not only caused by the degassing effect of the larger keyhole, but also by the combined effect of several conducive physical factors.

Chapter 8
Keyhole and Weld Pool Dynamics in Laser Welding with Filler Wires

Abstract This chapter introduces the time-dependent keyhole and weld pool dynamics in laser welding with filler wires. Mathematical models of the welding process including wire feeding model, boundary conditions, and keyhole and weld pool coupling model are presented. Keyhole and weld pool behaviors under different wire transfer modes including droplet and continuous transfer are discussed in detail. The oscillation mechanisms of keyhole and weld pool in laser welding with filler wires are investigated. The time dependent dilution behaviors of weld pool by the filler wire are also studied.

8.1 Introduction

Currently, the common laser welding is basically self-fluxing welding. In self-fluxing welding, the parent metal is heated and molten under laser action and solidified to form a weld joint without the use of wire or welding flux. This welding method can produce excellent weld joints for most materials like steel. Materials like aluminum alloy and magnesium alloy, however, feature high thermal conductivity, high coefficient of thermal expansion, low surface tension of the weld pool, and volatility of alloy compositions like Mg, Zn, and Li. When they are subject to laser self-fluxing welding, evaporation of alloy composition with low-boiling point will degrade the property of a weld joint and cause cracks, pores, poor welding formation, and other defects. Moreover, many high-strength aluminum alloys have bad weldability. As laser self-fluxing welding does not need additional materials, the chemical compositions of the weld pool and weld joint metal are hard to control. As a result, it is difficult to form a satisfactory weld joint. In addition, the self-fluxing welding beam of the ordinary laser is extremely small in diameter, setting high requirements on gap joints and welding centering (allowance <0.1 mm generally). These factors have restricted the application of laser self-fluxing welding in practice to some extent.

To address the above limitations of the laser self-fluxing welding and better expand application of the laser welding, the technique of laser welding with filler wires is developed based on the laser self-fluxing welding. The welding technique is similar to MIG welding in principle; the wire feeder is used to drive the wire to enter the

S. Gong et al., *Weld Pool Dynamics in Deep Penetration Laser Welding*,
https://doi.org/10.1007/978-981-16-0788-2_8

laser beam action area at certain speed, the wire melts after being heated, the molten metal enters the parent metal weld pool area, and finally the weld joint is formed after solidification of the weld pool. Under the appropriate technical condition, by adding the wire, the weld joint compositions and organization can be controlled, any welding defects can be restrained, and the tolerance of pre-weld assembly clearance can be increased.

With the laser welding with filler wires—non self-fluxing welding process as a study object, this chapter introduces the simulation mathematical model for keyhole and weld pool dynamics in laser welding with filler wires to reproduce the heat transfer, flow, free interface motion, and other transient coupling behaviors between wire, keyhole, and moving weld pool under different technical conditions, focusing on interpretation of keyhole and weld pool stability, and related mechanisms during laser welding with filler wires, followed by preliminary exploration of the motion law of wire element in the weld pool, as well as quantitative prediction of the distribution characteristics of wire chemical elements in the weld joint.

8.2 Multiphase Transient Coupling Model in Laser Welding with Filler Wires

8.2.1 Mathematical Model of Wire Melting

In the chapter, addition of droplet is processed into interval increase of enthalpy and wire composition in the weld joint. Meantime, the enthalpy increment carried by the droplet is regarded as a volume heat source which has the following characteristic parameters: radius (R_v), height (d), and energy density (S_v), including the interaction between molten droplet and weld pool under different welding conditions. Assumed that the radius of the volume heat source is 2.7 times the radius of the droplet, the height is calculated through the following equation based on conservation of energy:

$$d = h_v - x_v + D_d \tag{8.1}$$

where, h_v—The estimated height of the cavity formed by impact of the molten droplet;

x_v—The moving distance of the center of the latter droplet during collision of two continuous falling droplets;

D_d—Diameter of the droplet.

The effective input heat of the molten droplet Q_a is defined to be:

$$Q_a = \rho\pi r_w^2 w_f H_d \tag{8.2}$$

where, ρ—Density;

r_w—Wire radius;
w_f—Wire feed rate;
H_d—Total enthalpy of the droplet.

It should be noted that the heat input Q_d carried into the weld pool by the droplet will keep the additive at the liquidus temperature. Therefore, the effective heat carried into the weld pool by the droplet is defined to be as follows:

$$Q_d = \rho \pi r_w^2 w_f c_{pl} (T_d - T_l) \tag{8.3}$$

where, c_{pl}—Specific heat capacity of liquid metal;
T_d—Temperature of the droplet;
T_l—Liquidus temperature.

Based on the conservation of energy, values of both *hv and xv* in Eq. (8.1) are calculated as follows:

$$h_v = \left(-\frac{2\gamma}{D_d \rho g} + \sqrt{\left(\frac{2\gamma}{D_d \rho g}\right)^2 + \frac{D_d v_d^2}{6g}} \right) \tag{8.4}$$

$$x_v = \left(h_v + \frac{2\gamma}{D_d \rho g} \right) \left\{ 1 - \cos\left[\left(\frac{g}{h_v}\right)^{1/2} \Delta t \right] \right\} \tag{8.5}$$

where, γ—Surface tension of the molten metal;
g—Gravitational acceleration;
v_d—Droplet impact velocity;
Δt—Time interval of the two continuous droplets ($\Delta t = 1/f$, f is the metal-droplet transition frequency).

The average time energy density on the grid point of the volume heat source S_d is calculated as per the following equation:

$$S_d = \frac{Q_d}{\pi D_d^2 d} \tag{8.6}$$

The solute increment carried into the weld pool by the wire is correlated to the mean time volume mass source S_m in the solute conservation equation, and the volume mass source size S_m is assumed the same as the volume heat source, so the net solute mass Q_t in the droplet is calculated as follows:

$$Q_t = \rho \pi r_w^2 w_f (C_f - C) \tag{8.7}$$

where, C_f—Concentration of the solute in the droplet;
C—Concentration of the solute in the parent metal.

Furthermore, the mean time volume mass source S_m on the grid point of the mass source is given by the following equation :

$$S_m = \frac{Q_t}{\pi D_d^2 d} \tag{8.8}$$

8.2.2 *Transient Coupling Model of Keyhole and Weld Pool*

Irrespective of the impact of wire and parent metal density on the process of laser welding with filler wires, the conservation of mass equation during the welding with filler wire can be described as :

$$\nabla \bullet \vec{U} = 0 \tag{8.9}$$

The method similar to deal with the solid–liquid mushy region in the mathematical model of the single-beam laser welding process is adopted, and the conservation of energy equation for describing the laser welding with filler wires is:

$$\rho\left(\frac{\partial \vec{U}}{\partial t} + \left(\vec{U} \bullet \nabla\right)\vec{U}\right) = \nabla \bullet \left(\mu_l \nabla \vec{U}\right) - \nabla p - \frac{\mu_l}{K}\vec{U} - \frac{C\rho}{\sqrt{K}}\left|\vec{U}\right|\vec{U} + \rho \vec{g} \beta(T - T_{ref}) \tag{8.10}$$

In this equation, the related variable and symbols are consistent with Eq. (3.2) in Chap. 3. Considering the effect of overheating factor of the wire, convection and conduction in the weld pool on coupling, the conservation of energy equation for describing the work piece in the welding process is:

$$\rho c_p\left(\frac{\partial T}{\partial t} + \left(\vec{U} \bullet \nabla\right)T\right) = \nabla \bullet (k \nabla T) + \tilde{S}_d \tag{8.11}$$

where, $\tilde{S}_d$—Energy carried into the weld pool by the wire, to be determined according to Eq. (8.6).

To track the free surface motion of the droplet during metal-droplet transition and free surface motion of the keyhole, Level Set Method is also adopted in the chapter to track the free interface. Level Set Control Equation for the free interface is:

$$\frac{\partial \varphi}{\partial t} + \vec{U} \bullet \nabla \varphi = 0 \tag{8.12}$$

8.2.3 Boundary Conditions

On the droplet, liquid bridge, and keyhole free interfaces, considering the impact of surface tension, recoil pressure, and other forces on the free interface motion, there exist the pressure boundary conditions consistent with the single-beam laser welding on the free interfaces.

$$p_f = p_r + \sigma\kappa + 2\mu \vec{n} \bullet \nabla\vec{U} \bullet \vec{n} \tag{8.13}$$

Meantime, on the droplet, liquid bridge, and keyhole free interfaces, there also exist viscous stress boundary conditions caused by the thermal capillary force, which are similar to single-beam laser welding.

$$\begin{aligned}\left(\mu\nabla\vec{U}\right)_f &= \mu(\vec{n}\ \vec{t}_1\ \vec{t}_2)(\vec{n}\ \vec{0}\ \vec{0})^T\left(\nabla\vec{U}\right)(\vec{n}\ \vec{0}\ \vec{0})(\vec{n}\ \vec{t}_1\ \vec{t}_2)^T \\ &\mu(\vec{n}\ \vec{t}_1\ \vec{t}_2)(\vec{0}\ \vec{t}_1\ \vec{t}_2)^T\left(\nabla\vec{U}\right) \\ &-\mu(\vec{n}\ \vec{t}_1\ \vec{t}_2)(\vec{n}\ \vec{0}\ \vec{0})^T\left(\nabla\vec{U}\right)^T(\vec{0}\ \vec{t}_1\ \vec{t}_2)(\vec{n}\ \vec{t}_1\ \vec{t}_2)^T \\ &+(\vec{n}\ \vec{t}_1\ \vec{t}_2)\begin{pmatrix}0 & \nabla_s\sigma\bullet\vec{t}_1 & \nabla_s\sigma\bullet\vec{t}_2 \\ 0 & 0 & 0 \\ 0 & 0 & 0\end{pmatrix}(\vec{n}\ \vec{t}_1\ \vec{t}_2)^T\end{aligned} \tag{8.14}$$

In addition, on the free interface, considering the effect of multiple reflection absorption, thermal radiation, convection, and evaporation on the welding process, there exist the energy boundary conditions consistent with the single-beam laser welding.

$$k\frac{\partial T}{\partial \vec{n}} = q - h(T - T_\infty) - \varepsilon_r\sigma\left(T^4 - T_\infty^4\right) - \rho V_{evp} T_v \tag{8.15}$$

In Eq. (8.12) through (8.15), the meanings of the involved symbols are consistent with those in Eqs. (3.59) through (3.62) of Chap. 3. At last, due to existence of the wire feed rate $\vec{U}_{filler}$, during the falling process of the droplet every time, its initial speed is

$$\vec{U}_0 = \vec{U}_{filler} \tag{8.16}$$

In addition, during the liquid bridge transition, the initial speed of the liquid bridge entering the moving weld pool is also equal to the wire feed rate.

8.3 Keyhole and Weld Pool Dynamics in Welding with Filler Wires

8.3.1 Behaviors of Transient Keyholes and Weld Pools Under Free Transition Conditions

Free transition refers to the droplet separating from the wire end and freely falling to the weld pool, without any contact between wire end and weld pool. This case is shown in Fig. 8.1. In favor of numerical simulation, it's assumed that the droplet falling position is always in front of the keyhole. Figure 8.2 shows the initial free interface in the numerical calculation.

In Fig. 8.2, the block is parent metal, and the arrow is welding speed direction. The position of the droplet during welding is shown in Fig. 8.1. According to test results and welding process, the droplet falling interval during numerical simulation is set to be 9.64 ms, initial temperature of the droplet is 2500 K, the diameter is 0.9 mm, and the wire feed rate is 0.1 m/s. What's more, the falling position of the droplet is not stationary but moving in the welding speed direction at the speed equal to the welding speed. The technical parameters for the simulated laser welding with filler wires are shown in Table 8.1.

For numerical simulation, it's proposed to take the uniform cube grid, with the specific calculating parameters shown in Table 8.2.

Fig. 8.1 Schematic diagram for the droplet free transition

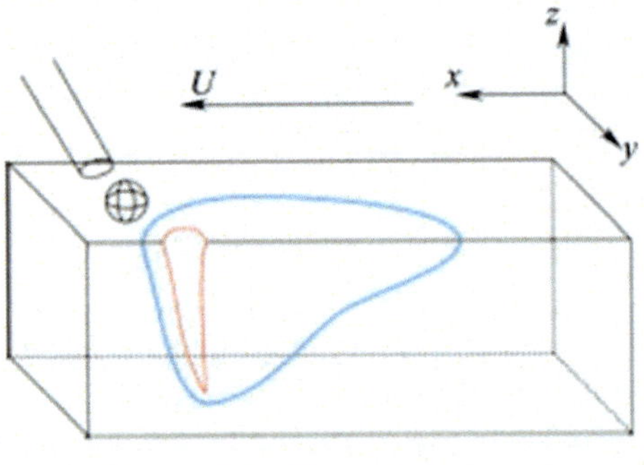

Fig. 8.2 Initial free interface

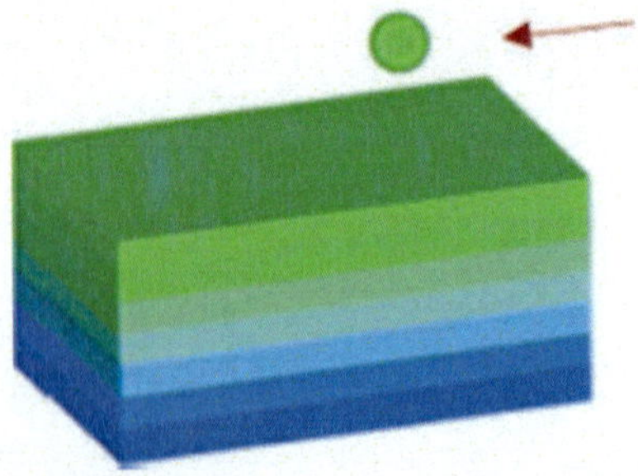

Table 8.1 Technical parameters of laser welding with filler wires

Laser power/kw	Laser spot radius/mm	Defocusing amount/mm	Welding speed/ (m/min)
1.5	0.2	0.0	3

Table 8.2 Numerical simulation parameters

Grid quantity	Spatial step/m	Time step/s	Initial temperature of parent metal/K	Ambient temperature/K
160 × 60 × 100	3.0×10^{-5}	3.0×10^{-6}	300	300

The welding parent metal is TC4, and the physical property parameters are shown in Table 8.3. Figure 8.3 is the dynamic free interface morphology during droplet free transition process under the current technical conditions. In the figure, the welding direction is from right to left, the red part represents gas, and the blue part represents droplet ad parent metal. Figure 8.3a through f show the free interface morphology of both droplet and parent metal respectively at 0 ms, 2.09 ms, 3.94 ms, 7.08 ms, 9.64 ms, and 13.59 ms.

As it can be seen from Fig. 8.3, at about 7.08 ms, the droplet is falling to the front edge of the weld pool. After falling, on one hand, the wire fluid carried by the

Table 8.3 Thermophysical parameters for parent metal

Density /(kg/m^3)	Specific heat capacity /[J/(kg•K)]	Thermal conductivity /[W/(m•K)]	Liquidus temperature /K	Solidus temperature /K	Latent heat of melting /(J/kg)	Latent heat of evaporation /(J/kg)	Boiling point /K
4000	660	25	1928	1878	3.7×10^5	8.9×10^6	3315

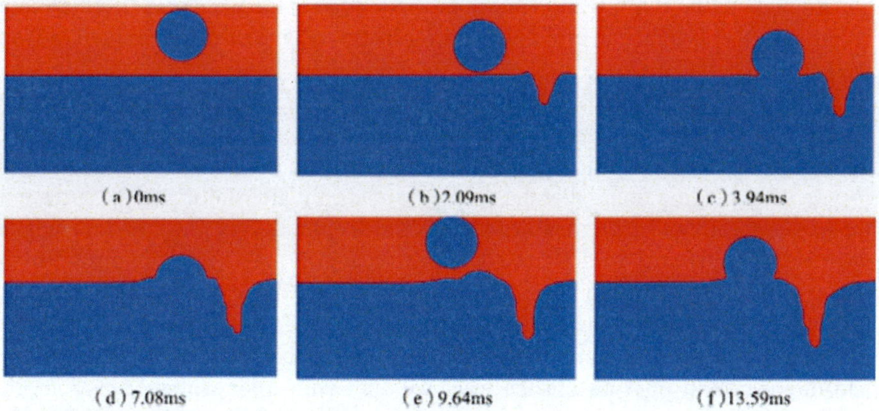

Fig. 8.3 Transient keyhole morphology on the longitudinal section during droplet free transition

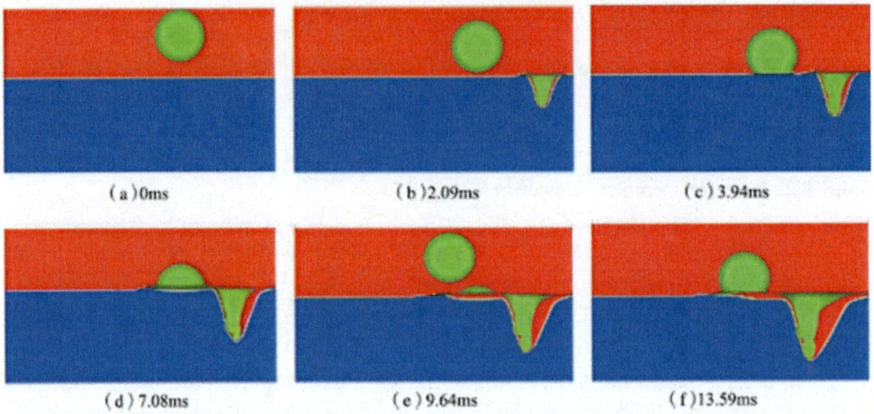

Fig. 8.4 Weld pool morphology of longitudinal section during the droplet free transition process

droplet will flow into the keyhole along the front wall of the keyhole. On the other hand, due to overheating of the droplet, the carried heat will melt the parent metal in the front part of the keyhole, hence to thicken the weld pool close the front wall of the keyhole. The above two factors tend to aggravate the front boss phenomenon of the keyhole wall, as shown in Fig. 8.3d through f. Since the droplet falling interval is 9.64 ms, the second droplet appears in Fig. 8.3e.

Figure 8.4 is the weld pool transient morphology on the longitudinal section during the droplet free transition. In the figure, the green part (other than the droplet part) represents the keyhole, and the red part between parent metal and keyhole is the weld pool.

As it can be seen from Fig. 8.4, the weld pool is also gradually forming while the keyhole is forming. Over time, the liquid in the weld pool at the back edge of the keyhole gradually increases by accumulation, and therefore, it can be inferred that the weld pool finally reaches the quasi-steady state when the welding speed is consistent with the solidification rate at the back edge of the weld pool. Furthermore, addition of the droplet can effectively increase the volume of the weld pool in front wall of the keyhole.

Figure 8.5 is the transient evolution process of the 3D free interface morphology during the metal-droplet transition process. In the figure, the color represents temperature, and the blue plane represents the surface of the parent metal. According to Figs.8.3 and 8.4, it's found that the droplet spreads out after falling to the surface of the parent metal, and scatters all round. As the droplet is very close to the keyhole, some droplets are easy to flow into the keyhole, which may aggravate instability of the keyhole. It shows from analysis and discussion of the above simulation results that, the metal-droplet transition mathematical model developed in the chapter can reasonably simulate the complicated transient interaction between droplet, keyhole, and moving weld pool during the laser welding with filler wires. Meantime, the above-mentioned results also show that, under the technical condition of the wire

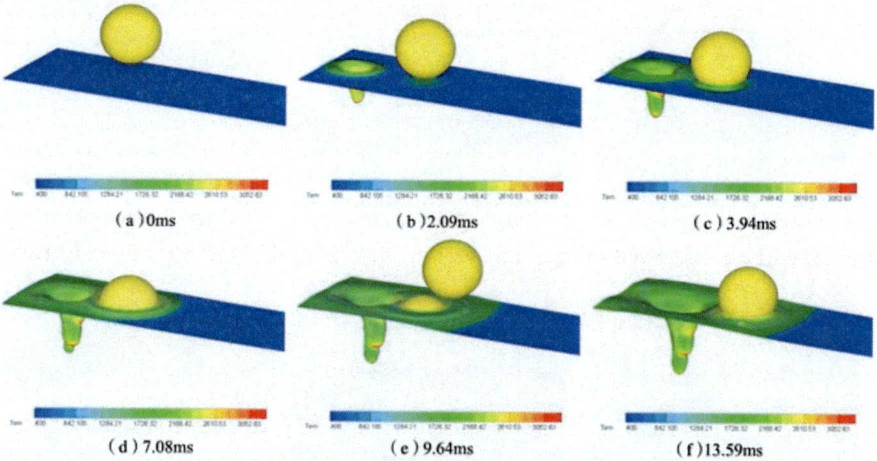

Fig. 8.5 3D morphology of the droplet free transition

transitioning to the weld pool in the form of droplet with the wire keeping in front, part of metal liquid carried by the droplet may flow into the keyhole, which will aggravate the boss phenomenon in the front wall of the keyhole, hence to aggravate the instability of the keyhole.

8.3.2 Influences of Wire Feed Speed on Transient Keyhole and Weld Pool Behaviors in Free Transition

Unless otherwise noted, the numerical simulation parameters mentioned in the section are consistent with the corresponding parameters in Sect. 8.3.1.

Figure 8.6 is the morphology comparison of the free interface, moving weld pool, 3D droplet, and keyhole free interface at the two wire feed rates of 0.1 and 0.06 m/s. Figure 8.6a through Fig. 8.6c are respectively the 2D sectional drawing of keyhole, 2D sectional drawing and 3D morphology of weld pool after 13.59 ms of the free transition transient laser welding at the speed of 0.1 m/s; Fig. 8.6d through f are respectively the longitudinal section of keyhole, longitudinal section and 3D morphology of weld pool after 13.59 ms of the transient laser welding of the droplet free transition at the speed of 0.06 m/s. It is found through comparison that, when the simulation result is at 13.59 ms at the speed of 0.1 m/s, part of the second droplet has entered the weld pool, while for the simulation at the speed of 0.06 m/s, the first droplet has not completely entered the weld pool at the same time. It shows that the higher the wire feed rate is, the shorter the time of the droplet liquid entering the keyhole, and the deeper the impact on the keyhole stability will be, as shown in Fig. 8.6b, e.

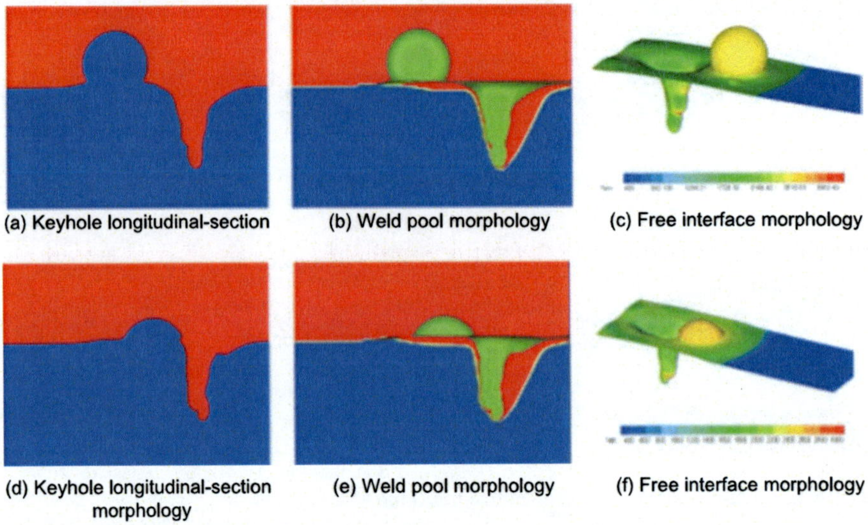

Fig. 8.6 Comparison of the results at the welding time of 13.59 ms at different wire feed rates with the droplet in the free transition mode.(Wire feed rate from (**a**) to (**c**) is 0.1 m/s, while that from (**d**) to (**f**) is 0.06 m/s)

Figure 8.7 is the changing curve for the absolute values of maximum speeds over time in the welding pool when the wire feed speed is at 0.1 and 0.06 m/s. In the figure, the horizontal axis is the time, and the longitudinal axis is the absolute value of maximum speed in the weld pool. Where, the red line represents the changing curve for the maximum speeds of the laser welding pool over time at the wire feed rate of 0.1 m/s, while the green line is the curve at the wire feed rate of 0.06 m/s. It can be seen from the figure that, when the welding time is at 0–6 ms, changes of the two curves have no distinct difference, because the droplet gradually spreads on the surface of the parent metal before impacting the transient behaviors of both

Fig. 8.7 Curve of maximum speed absolute value-time function of the weld pool at different wire feed rates

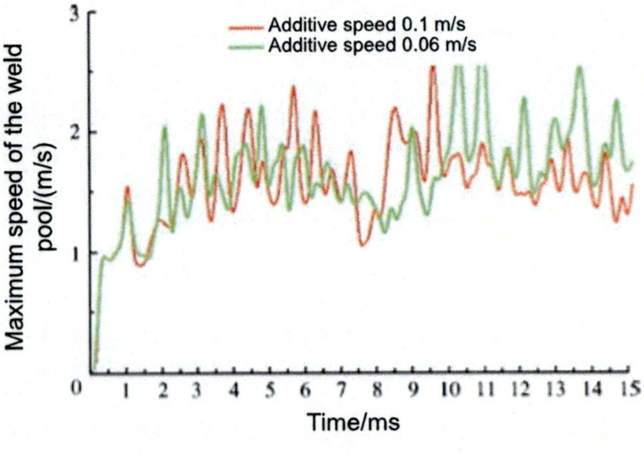

weld pool and keyhole. At the interval of 7–11 ms, it can be found that when the wire feed rate is low, the time reaching the peak of maximum speed is slightly later than the process at the high wire feed rate, mainly because when the wire feed rate is relatively low, the time affecting the weld pool is slower than that when the wire feed rate is high. At the interval of 11–15 ms, when the wire feed rate is low, the change of maximum speed of the weld pool is slightly greater than that when the wire feed rate is 0.1 m/s, so it can be assumed that the welding pool mobility is more unstable. It is suggested through analysis that, the reason for the above result is that when the wire feed rate is high, there are already two droplets entering into the weld pool within the time under study, so that the volume is larger than that when the wire feed rate is low. Therefore, at the same laser power, change of the maximum speeds of the weld pool is gentler than that when the wire feed rate is low.

It shows from the above simulation results that, when the wire enters into the weld pool in the free transition mode, increasing the wire feed rate may cause instability of the keyhole. But under certain condition, reasonably raising the wire feed speed can increase the volume of the weld pool, hence to improve the stability of the weld pool to certain extent.

8.3.3 *Influences of Wire Diameter on Transient Keyhole and Weld Pool Behaviors in Free Transition*

Unless otherwise noted, the numerical simulation parameters mentioned in the section are consistent with the corresponding parameters in Sect. 8.3.1.

Figure 8.8 shows the comparison of the simulation results at the transient welding moment of 13.59 ms under different wire diameters in free transition mode. The figures (a) through (c) show the keyhole cross-section morphology, weld pool morphology and free interface morphology under the wire diameter of 1.2 mm. The figures (d) through (f) show the corresponding results under the wire diameter of 0.9 mm.

The larger the molten droplet diameter, the more intense the protrusion of the front wall of the keyhole, as shown in the figure. It can be speculated that excessive large diameter of wire will bring adverse effects on the stability of the keyhole. In addition, in case of fixed wire feed rate, the momentum of the metal-droplet transition to the weld pool increases due to the increased radius, so the addition of wire has an intensified disturbance effect on the weld pool. To further illustrate this issue, the fluctuation of maximum speed absolute value of the weld pool under the two conditions is analyzed.

Figure 8.9 shows the comparison of maximum speed absolute values of the weld pool under different wire diameters in free transition mode. Where, the red line is the variation curve of the maximum speed absolute value of the weld pool at a wire diameter of 1.2 mm as a function of time, and the green line is the variation curve of the maximum speed absolute value of the weld pool at a wire diameter of 0.9 mm as a function of time. It can be seen from the figure that in the 0–5 ms stage, the two curves change in a very similar way, but in the 5–15 ms stage, the maximum speed of the weld pool changes dramatically in the event of large wire diameter. It can be concluded that excessive large diameter of the wire will weaken the stability of the weld pool in the process of laser welding with filler wires.

The foregoing simulation results indicate that when the wire enters the weld pool in free transition mode, the excessive large diameter of the wire will weaken the stability of the transient keyhole and the moving weld pool.

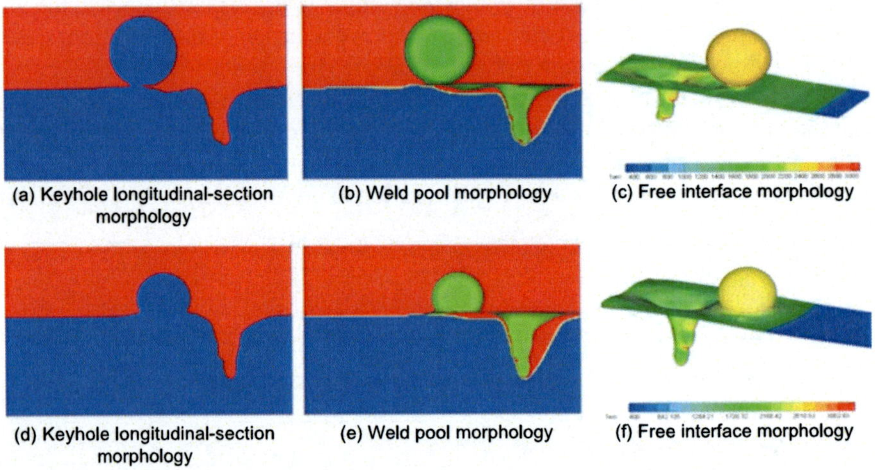

Fig. 8.8 Comparison of the results at the welding moment of 13.59 ms under different wire diameters in free transition process

Fig. 8.9 Curve of maximum speed absolute value-time function of the moving weld pool under different wire diameters in free transition mode

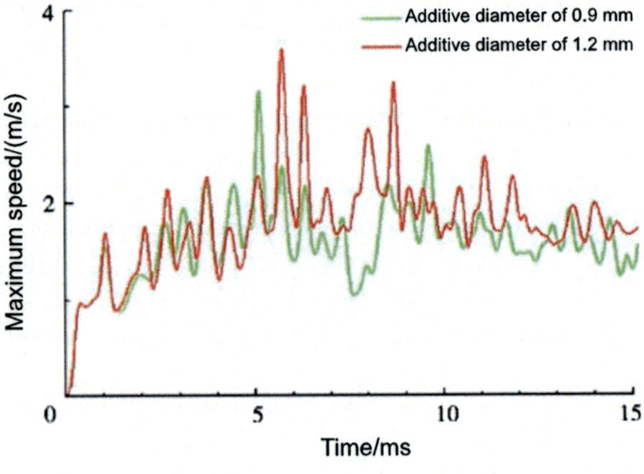

8.3.4 Transient Keyhole and Weld Pool Behaviors Under Contact Transition Conditions

Contact transition refers to the contact between the metallic liquid at the end of the wire and the surface of the weld pool for the purpose transition. In this transition condition, the connection between the wire and the weld pool is normally called liquid bridge in studies. The case is shown in Fig. 8.10. The end of the liquid bridge formed through the melted wire is simplified as hemispherical shape at the initial stage in the numerical simulation process (see Fig. 8.11). The position of the wire entering the weld pool (such as the laser welding position) is shown in Fig. 8.10.

It is assumed that the inlet diameter of the liquid bridge is 0.9 mm and the falling speed is 0.1 m/s in the simulation process according to the technical parameters. Besides, the liquid is set to enter from the inlet boundary in the numerical simulation process. The laser, located behind the inlet boundary, moves forward together with the inlet boundary at the welding speed. Relevant welding technical parameters are listed in the Table 8.4. For numerical simulation grid, it's proposed to use the uniform cube grid, with the specific calculating parameters shown in the Table 8.5. The welding parent metal is TC4 in the welding process, with the physical property parameters shown in the Table 8.6.

Figure 8.12 shows the longitudinal-section morphology of the transient keyhole in the contact transition process under the current technical condition. In the figure,

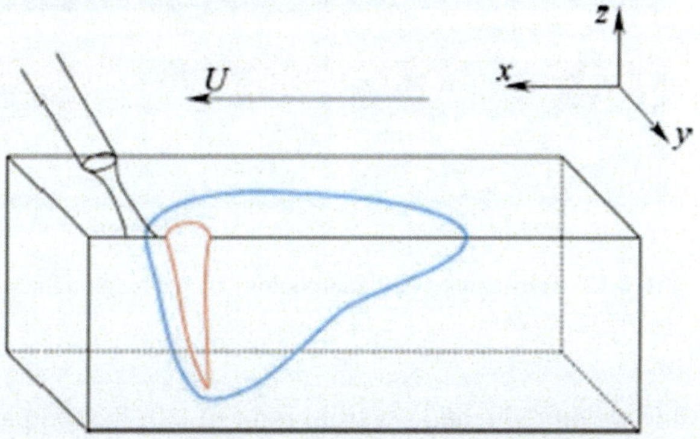

Fig. 8.10 Contact transition diagram

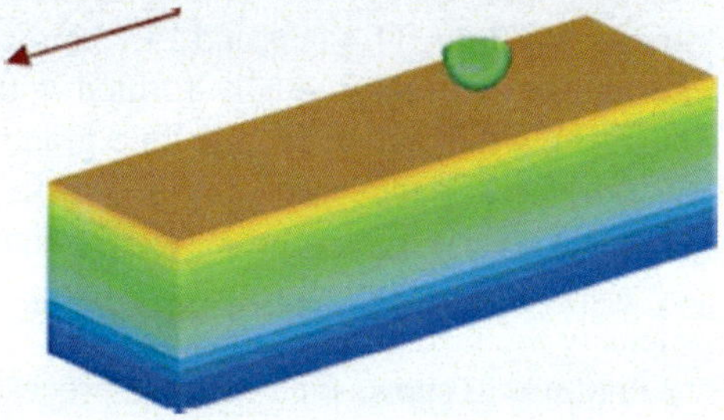

Fig. 8.11 Initial free interface morphology

Table 8.4 Technical parameters of laser welding with filler wires

Laser power/kw	Laser spot radius/mm	Defocusing amount/mm	Welding speed/ (m/min)
1.5	0.2	0	3

Table 8.5 Numerical simulation parameters

Grid quantity	Spatial step/m	Time step/s	Initial temperature of parent metal/K	Ambient temperature/K
160 × 60 × 80	3.0×10^{-5}	3.0×10^{-6}	300	300

Table 8.6 Thermophysical parameters for parent metal

Density /(kg/m^3)	Specific heat capacity /[J/(kg•K)]	Thermal conductivity /[W/(m•K)]	Liquidus temperature/K	Solidus temperature/K	Latent heat of melting /(J/kg)	Latent heat of evaporation /(J/kg)	Boiling point /K
4000	660	25	1928	1878	3.7×10^5	8.9×10^6	3315

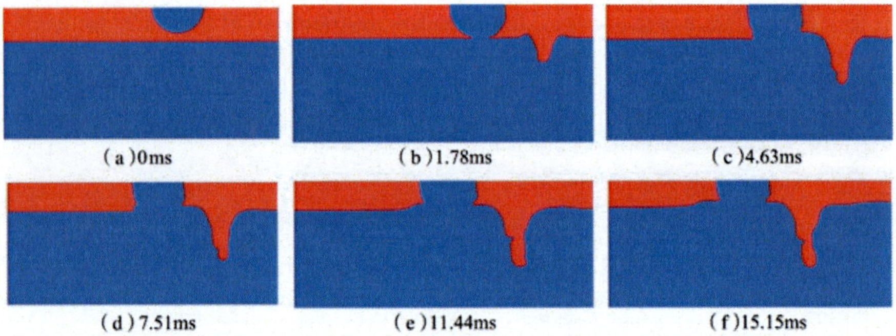

Fig. 8.12 Transient keyhole morphology on the longitudinal section of the molten droplet contact transition process

the welding direction is from right to left, the red part represents gas, and the blue part represents droplet ad parent metal. Figure 8.12a through Fig. 8.12f show the free interface morphology on the longitudinal section at the moment of 0 ms, 1.78 ms, 4.63 ms, 7.51 ms, 11.44 ms and 15.15 ms, respectively. It can be seen from the figures that the keyhole is gradually formed with the advancing of the welding process; meanwhile, the molten droplet fluid gradually flows into the keyhole along the wall surface of the keyhole, leading to a tendency of keyhole closing at the moment of 15.15 ms. Therefore, similar to free transition, the flow of the molten droplet to the weld pool in contact transition mode under given technical condition may also intensify the instability of the keyhole.

Figure 8.13 shows the evolution process of the weld pool. In the figure, the green

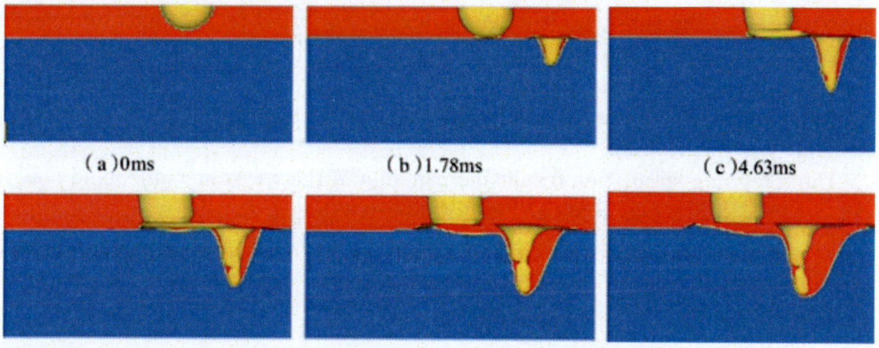

Fig. 8.13 Transient formation process of the moving weld pool in contact transition

part (except the molten droplet part) represents the keyhole, and the red part between the blue part and the keyhole represents the weld pool. Figure 8.12a through Fig. 8.12f show the cross-section views of the weld pool in contact transition after 0 ms, 1.78 ms, 4.63 ms, 7.51 ms, 11.44 ms and 15.15 ms respectively. It can be seen from the figures that with the advancing of the welding process, the liquid in the weld pool gradually increases along the keyhole back edge and the weld pool volume also increases; the molten droplets not only accumulate in the weld pool front edge but also melt partial parent metal. As the molten droplets continuously enter the weld pool, a large protrusion emerges on the front wall of the keyhole, which is very evident at the moment of 15.15 ms.

Figure 8.14 shows the three-dimensional free interface morphology in contact transition, where color represents temperature and blue plane represents the parent metal surface. Figure 8.14a through Fig. 8.14f show three-dimensional free interface morphology at 0 ms, 1.78 ms, 4.63 ms, 7.51 ms, 11.44 ms and 15.15 ms, respectively. The temperature field indicates that the temperature at the corresponding position of the parent metal starts to rise when the high-temperature liquid at the end of the

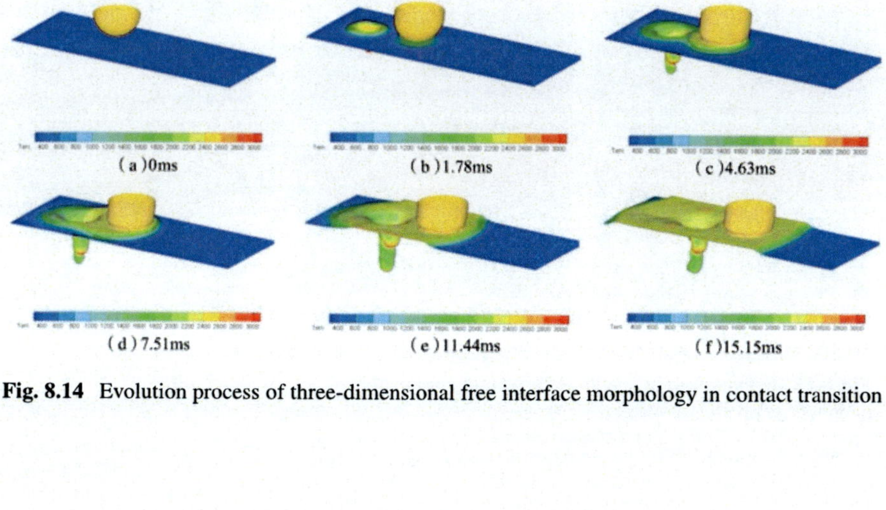

Fig. 8.14 Evolution process of three-dimensional free interface morphology in contact transition

wire transits to the surface of the parent metal. The front edge temperature of the keyhole increases as the metal liquid spreads out. The higher the temperature of the keyhole front wall, the greater the recoil pressure. Therefore, the wire solution will flow quickly into the keyhole under the action of recoil pressure, resulting in instability of the keyhole.

The foregoing simulation results indicate that, if the wire enters the weld pool in contact transition mode in the process of laser welding with filler wires, the molten droplet liquid may flow into or be pushed into the keyhole by the recoil pressure to reduce the stability of the keyhole.

8.3.5 *Influences of Wire Feed Speed on Transient Keyhole and Weld Pool Behaviors in Contact Transition*

The keyhole and weld pool behaviors under the two wire feed rates of 0.1 and 0.06 m/s are discussed below. Figure 8.15 shows the comparison of simulation results at different additive speeds in contact transition.

Figure 8.15a through Fig. 8.16c show the keyhole longitudinal-section morphology, the weld pool morphology and the free interface morphology at the welding transient moment of 15.15 ms under the wire feed rate of 0.06 m/s respectively. Figure 8.15d through Fig. 8.15f show the keyhole longitudinal-section morphology, the weld pool morphology and the free interface morphology under the wire feed rate of 0.1 m/s. It can be seen from the figures that the larger the wire feed rate, the larger the weld pool near the keyhole front and back edge.

To better analyze and compare the two technical conditions, the variation curve of the maximum speed absolute value of weld pool with time at different contact transition speeds is created, as shown in Fig. 8.16. The red line represents the maximum speed variation curve of the weld pool with time at the contact transition speed of

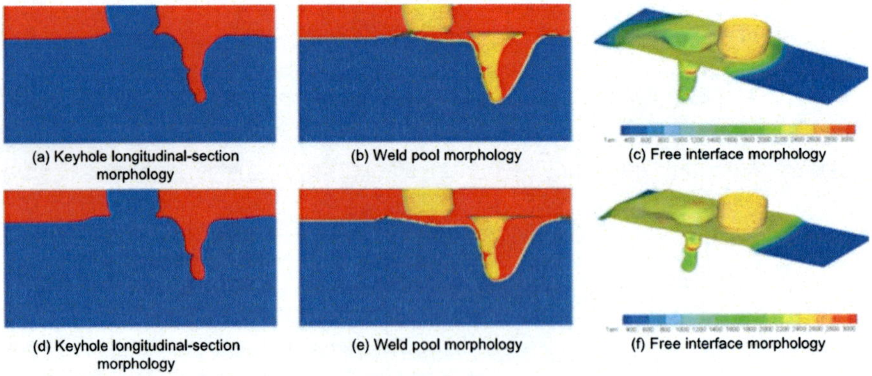

Fig. 8.15 Comparison of results at the welding moment of 15.15 ms under different wire feed rates in contact transition mode

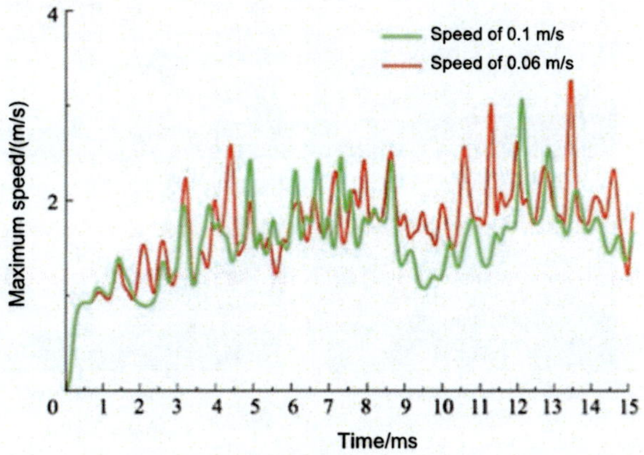

Fig. 8.16 Curve of maximum speed absolute value-time function of the weld pool at different wire feed rates in contact transition

0.06 m/s, while the green line represents the maximum speed variation curve of the weld pool with time at the contact transition speed of 0.1 m/s. As can be seen from the figure, the little difference in the wire feed rate under the two conditions leads to small fluctuation in the maximum speed tendency of the weld pool. However, the larger the wire feed rate, the smaller the maximum amplitude of the weld pool maximum speed fluctuation. The analysis suggests that this is mainly because large wire feed rate improves the stability of the weld pool to some extent.

It can be concluded from the foregoing analysis and discussion that, in the process of laser welding with filler wires, when the wire enters the weld pool in contact transition mode, increasing the wire feed rate to some extent can improve the stability of the weld pool.

8.3.6 Influences of Wire Diameter on Transient Keyhole and Weld Pool Behaviors in Contact Transition

Unless otherwise noted, the numerical simulation parameters mentioned in this section are consistent with the corresponding parameters in Sect. 8.3.4.

Figure 8.17 shows the comparison of the simulation results at the welding moment of 13.59 ms under different wire diameters in free transition mode. The figures (a) through (c) show the keyhole cross-section morphology, weld pool morphology and free interface morphology under the wire diameter of 1.2 mm. The figures (d)–(f) show the corresponding results under the wire diameter of 0.9 mm. According to the figures, it is apparent that the increase in the wire diameter and the flow liquid will generate more impact force on the weld pool, which weakens the stability of the flow in the weld pool; this can also be verified by the comparison with the three-dimensional free interface morphology: the surface of weld pool becomes rougher with the increase in the diameter. To better illustrate this conclusion, the maximum speed values of the weld pool under the two conditions are compared, with the

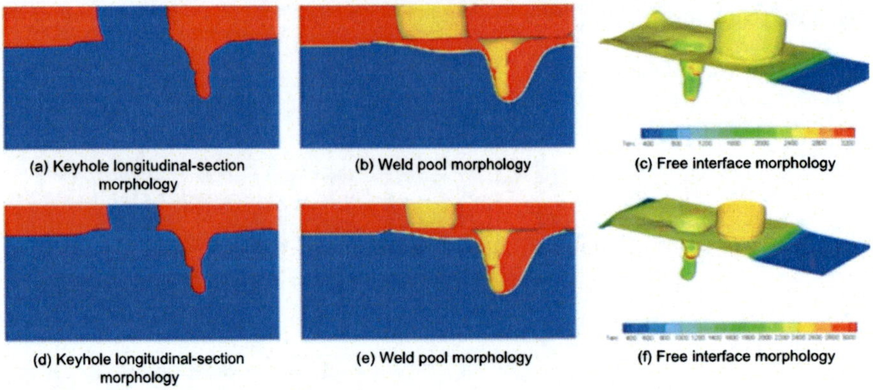

Fig. 8.17 Comparison of the results at the welding moment of 13.59 ms under different wire diameters in free transition mode

statistical results shown in Fig. 8.18. The red line represents the maximum speed variation curve of the weld pool with time at the additive diameter of 1.2 mm in contact transition, while the green line represents the maximum speed variation curve of the weld pool with time at the additive diameter of 0.9 mm in contact transition.

According to the overall tendency of the red curve in Fig. 8.18, big fluctuation exists in the maximum speed of the weld pool with a larger diameter between 3 and 5 ms in contact transition. The analysis shows that this is because of the impact on the weld pool due to the wire that just enters the weld pool. The overall tendency of the two curves reveals that the larger the contact transition radius, the more dramatic the fluctuation. In case of the additive diameter of 0.9 mm, the maximum speed of the weld pool changes between 0.5 and 3.25 m/s in contact transition, in case of the additive diameter of 1.2 mm, the maximum speed of the weld pool changes between

Fig. 8.18 Curve of maximum speed absolute value fluctuation-time function of the moving weld pool under different wire diameters

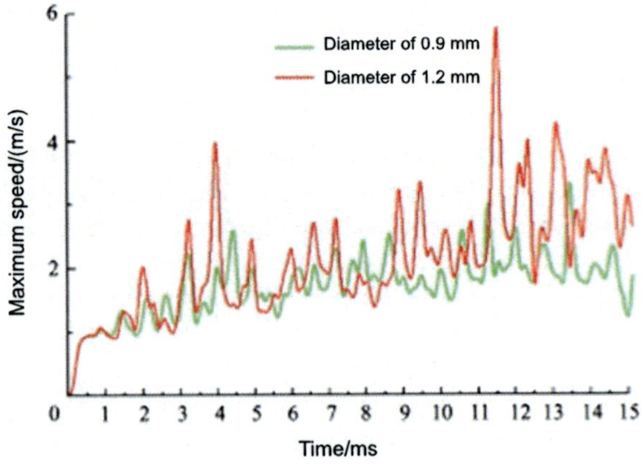

1.0 and 2.0 m/s. When the maximum speed of the weld pool fluctuates violently, the flow stability is relatively poor, meaning poor flow stability of the weld pool.

It can be concluded from the foregoing simulation analysis and comparison that in the process of laser welding with filler wires, when the wire enters the weld pool in contact transition mode, the increase of wire diameter under given conditions will result in the increase of the maximum speed absolute value fluctuation of the weld pool, thus weakening the stability of the weld pool.

8.3.7 Comparison of Transient Keyhole and Weld Pool Behaviors in Free Transition with Those in Contact Transition

The transient keyhole and moving weld pool behaviors in the process of laser welding with filler wires and the process of single-beam laser welding are compared and analyzed in this section. The main technical conditions used during the comparison and analysis are consistent, as shown in Tables 8.1 and 8.4. The same is true for the material of the parent metal and the parameters used in numerical calculation. In the process of laser welding with filler wires, the contact transition and free transition are simulated respectively.

Figure 8.19 shows the comparison of simulation results of non-metal-droplet transition (single-beam laser welding), free transition and contact transition at the same moment under the same conditions. At the same moment, the single-beam laser welding has a greater penetration depth and a smaller weld pool in the keyhole front and back edge, as shown in the figure. The analysis reveals that in the process of welding with filler wire, the addition of wire absorbs some laser energy, which leads to a decrease in the penetration depth.

Figure 8.20 shows the comparison of the maximum speed of the weld pool in non-metal-droplet transition, free transition and contact transition. Where, the red line represents the maximum speed variation curve of the weld pool with time in contact transition, the black line represents the maximum speed variation curve of the weld pool with time in free transition, and the blue line represents the maximum speed variation curve of the weld pool with time in non-metal-droplet transition. According to the variation trend of the three curves, the weld pool in non-metal-droplet transition (blue line) has a significantly greater fluctuation frequency of maximum speed than that in contact transition and free transition, which indicates that the flow stability of the weld pool in non-metal-droplet transition is lower than that in free transition and contact transition. However, the maximum speed variation fluctuation of the weld pool with time in free transition is slightly larger than that in contact transition. In the initial stage, because of the influence of transition droplets, the maximum speed of the weld pool in free transition increases from 0 to 2.2 m/s, and then fluctuates around 1.75 m/s, with the maximum value of 3.0 m/s and the minimum value of 1.5 m/s. However, the fluctuation in contact transition is gentle. It fluctuates around

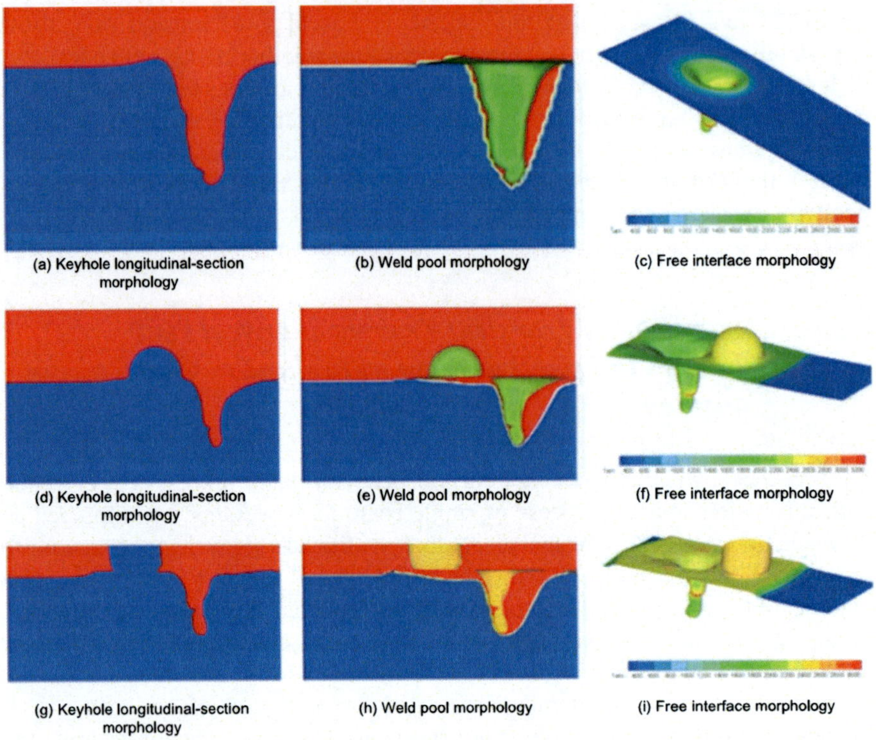

Fig. 8.19 Comparison of simulation results at the welding moment of 15.15 ms in non-metal-droplet transition, free transition and contact transition. **a–c** non-metal-droplet transition; **d–f** free transition; **g–i** contact transition

Fig. 8.20 Curve of the maximum speed absolute value-time function of the weld pool in non-metal-droplet transition, free transition and contact transition

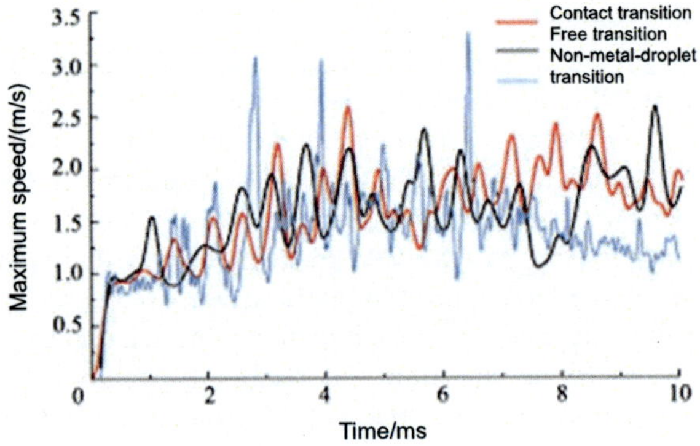

1.75 m/s, with the maximum value of 2.6 m/s and the minimum value of 1.6 m/s. The quantitative analysis indicates that compared with the free transition, the contact transition can better promote the stability of the flow in the weld pool of laser welding during the wire addition process. According to the analysis based on Fig. 8.19, it is mainly because the addition of wire increases the width and length of the weld pool around the keyhole, that is, the flow range around the keyhole, which further reduces the maximum speed fluctuation of the weld pool and improves the stability of the weld pool.

The above simulation results show that under the current research conditions, the process of single-beam laser welding can bring about a greater penetration depth than the process of laser welding with filler wires. However, the weld pool has a better stability in the process of laser welding with filler wires than the process of single-beam laser welding, no matter the wire enters the weld pool in free transition mode or contact transition mode. Relatively speaking, the moving weld pool has the optimal stability in contact transition.

8.4 Instability of Keyhole and Weld Pool in Welding with Filler Wires

8.4.1 Weld Pool Dynamics in Single Beam Laser Welding and Laser Welding with Filler Wires

In this section, the weld pool dynamics in the process of single-beam laser welding and the process of laser welding with filler wires is studied through a high-speed photography test to qualitatively verify some conclusions in Sect. 8.3.1. In the welding test, the equipment used is YLR-4000 type fiber laser, with the material of aluminum alloy. The diameter of wire is 1 mm and the material is SA1-Mg5. To avoid oxidation of aluminum alloy weld pool during welding, argon is adopted for protection.

Figure 8.21 shows typical images of the weld pool surface during single deep penetration laser welding. It can be seen from the figure that the surface of the weld pool is very unsmooth, and the weld pool fluctuates violently. The subsidence on the surface of the weld pool is relatively serious, and there are two serious subsidence occurrences near the opening of the keyhole and the tail of the weld pool. The above test results indicate that the stability of the weld pool is poor during the single-beam laser welding of aluminum alloy.

Figure 8.22 shows typical images of the weld pool surface during laser welding with filler wires. As can be seen from the figure, due to the addition of the wire, the fluctuation of the surface of the weld pool is significantly reduced, and the unevenness of the subsidence on the surface of the weld pool is more moderate compared with that of the single-beam laser welding.

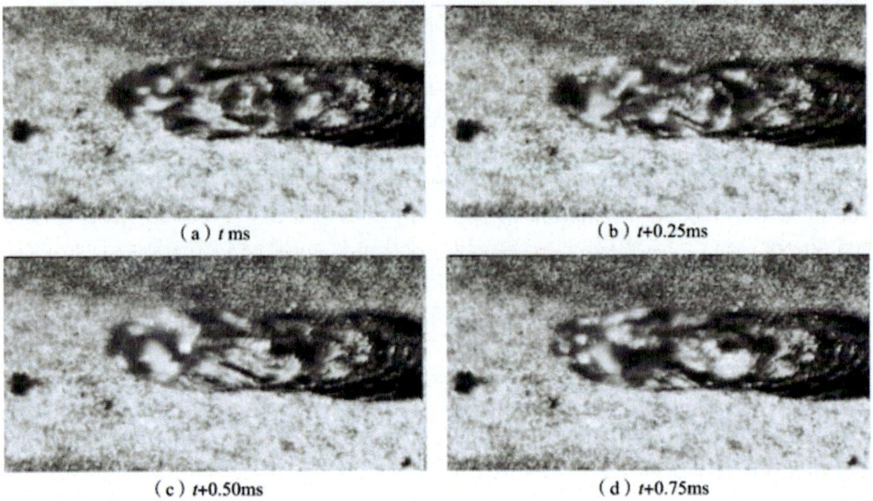

Fig. 8.21 High-speed photography images of the weld pool surface during single-beam laser welding (Power: 3 kW, welding speed 1.5 m/min)

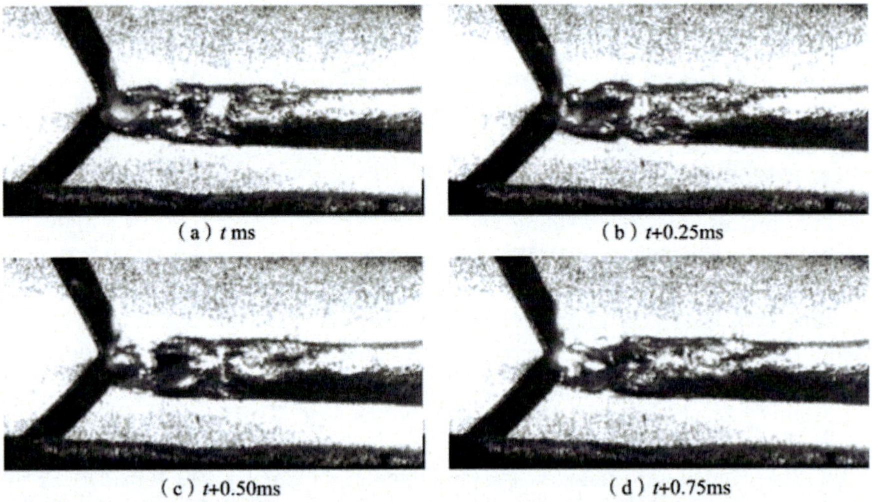

Fig. 8.22 High-speed photography images of the weld pool surface during laser welding with filler wires (Power: 2.6 kW, welding speed 2 m/min, wire feed rate 7 m/min)

Although the welding technical parameters in Figs. 8.21 and 8.22 are not consistent, the above test results prove that laser welding with filler wires can effectively reduce the fluctuation of weld pool and improve the stability of weld pool. The test results are qualitatively consistent with the simulation results (Sect. 8.3.7) of the transient keyhole and the moving weld pool in the welding with filler wire process established in this chapter, which confirms the mechanism, proposed in Sect. 8.3 in this chapter, that the addition of wire weakens the maximum velocity fluctuation in the weld pool and improves the stability of the weld pool is reasonable.

8.4.2 Weld Pool Dynamics in Welding with Filler Wires Under Different Wire Feed Speeds

In this section, the weld pool dynamics during laser welding with filler wires at different wire feed rates is studied through a high-speed photography test, and then the simulation results obtained in Sect. 8.3 are verified qualitatively. Consistent with Sect. 8.4.1, in the welding test, a fiber laser is used as the laser, and the material is aluminum alloy; the diameter of the wire is 1 mm and the material is SA1-Mg5; in addition, argon gas is used to protect the weld pool during the welding process.

Figure 8.23 shows the weld pool morphologies in the process of laser welding with filler wires at the wire feed rate of 2 m/min and 6 m/min, respectively. As can be seen from the figure, when the wire feed rate is 2 m/min, there are many large waves on the surface of the weld pool, especially near the keyhole, which indicates that the surface of the weld pool fluctuates violently under this technical condition; when the wire feed rate is 6 m/min, the fluctuation of the weld pool in other parts is very weak except that there is a small fluctuation in the weld pool near the location of the keyhole. Therefore, the above test can show that increasing the wire feed rate under certain conditions can effectively improve the fluctuation of the weld pool.

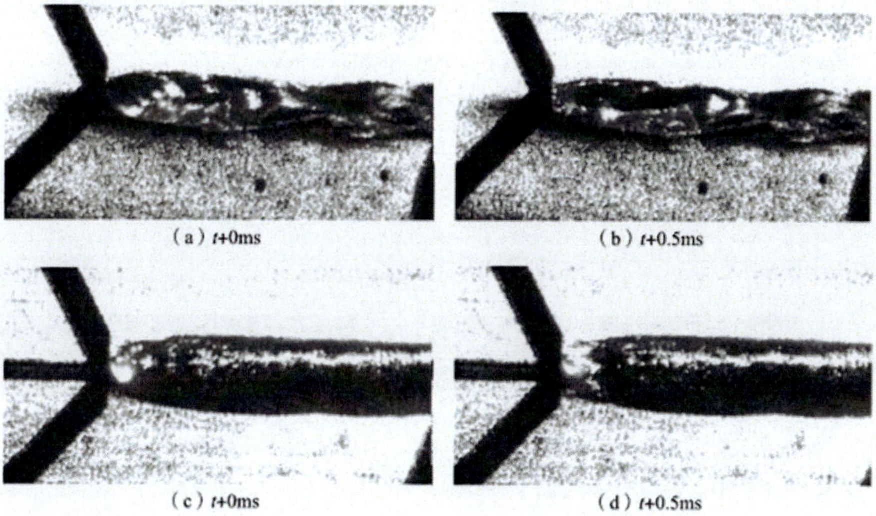

Fig. 8.23 Images of weld pool surface during welding with filler wire at different wire feed rates (power 2.6 kW, welding speed 2 m/min). **a–b** Wire feed rate 2 m/min; **c–d** Wire feed rate 6 m/min

8.4.3 Mechanism of Keyholes Instability in Welding with Filler Wires

According to the numerical simulation results in Sect. 8.3, in the process of laser welding with filler wires without complete penetration, when the wire is located in front of the keyhole, the use of welding with filler wire can stabilize the stability of the moving weld pool; however, because the metallic liquid may flow into the keyhole or be pushed into the keyhole by recoil pressure along the forearm of the keyhole, whether the wire enters the weld pool in the form of molten drop or liquid bridge, there is a tendency to increase the lug boss phenomenon of the front wall of the keyhole, that is to say, the addition of the wire may increase the instability of keyholes.

In order to verify the rationality of the mechanism, the recent test work on laser welding with filler wires by Dr. Yu Yangchun from Huazhong University of Science and Technology (HUST) is quoted for verification, which is shown below. In their welding tests, the parent metal is 4 mm thick 5A06 aluminum alloy plate, and the additive is SA1-Mg5 wire with a diameter of 1 mm. In addition, YLR4000-type fiber laser of IPG company is used for welding. The defocusing amount is set to zero, and single laser welding and laser welding with filler wires tests are performed separately during the welding process. The test parameters and technical parameters used are shown in Table 8.7. Figure 8.24 shows the corresponding pictures of the longitudinal section and cross section of the weld joint.

It can be seen from Fig. 8.24 that the filler wire penetration under the condition of the same power and welding speed is a little shallower than that of single-beam laser welding, which is consistent with the simulation results of Sect. 8.3.7 in this

Table 8.7 Technical parameters of laser welding with filler wires

No	Laser power/kw	Welding speed/ (m/min)	Wire feed rate/(m/min)	Plate thickness/mm
1	3	2	0	4
2	3	2	4	4
3	3.5	1.2	7	4
4	3.8	1.2	7	4

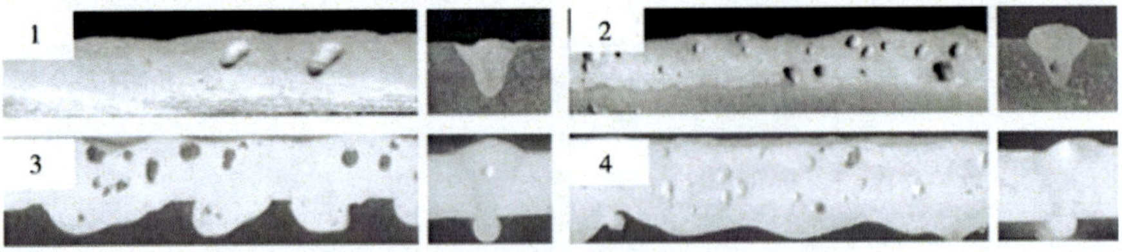

Fig. 8.24 Longitudinal section and cross section (weld joint) of the weld joint under the corresponding process in Table 8.7

chapter; in addition, under the same power and welding speed conditions, there are much more large pores inside the cross section of the weld bead than that during the single-beam laser welding process (see Fig. 8.24 (1) and (2)), which strongly proves that the keyhole may be more unstable in the laser welding with filler wires process than in the single-beam laser welding process when the wire is located in front of the keyhole. In addition, even if the stability of the moving weld pool is improved by increasing the wire feed rate, there are still many large pores on the weld bead, which further indicates that the stability of the keyhole in the laser welding with filler wires process is very poor. According to the mechanism of single-beam laser welding process confirmed in Chapter 3, under general conditions, the shallower the penetration, that is, the shorter the keyhole, the higher the stability of the keyhole. However, the results of Fig. 8.24 (1) and (2) confirm that a keyhole with shallower penetration is more unstable. Obviously, the only possible explanation is that the addition of the wire increases the instability of the keyhole in the laser welding with filler wires process.

In view of the above phenomenon, according to the test results, Dr. Yu Yangchun speculated that the wire may form periodic unstable molten drops in the laser welding with filler wires process, and during the transition into the weld pool, some of the molten drops enter the keyhole, thus increasing the instability of the keyhole. In addition, the jitter of the wire feeder may also increase the instability of the keyhole. In this study, through theoretical simulation, the process of the wire entering the keyhole and increasing the instability of the front wall of the keyhole is reproduced visually, which, in theory, directly confirms that Dr. Yu Yangchun's speculation that the molten drops entering the keyhole may increase the instability of the keyhole is reasonable. In addition, the theoretical study in this chapter also confirms that even if periodic molten drops can exist steadily in the process of welding with filler wire and enter the weld pool under ideal conditions without unstable wire feed structure, it is still likely to increase the instability of the keyhole, and even if the wire enters the weld pool in an ideal and uniform contact transition (liquid bridge transition) way, it is still likely to increase the instability of the keyhole.

8.5 Dynamic Dilution Behavior of Moving Weld Pools in Welding with Filler Wires

8.5.1 Model of Dilution in Weld Pools

In laser welding with filler wires, the chemical compositions of the wire convectively move along with the velocity field of the moving weld pool after entering the moving weld pool. At the same time, the chemical compositions can diffuse inside the weld pool. The calculation by using the thermophysical property calculation software JMatPro shows that the diffusion coefficients of general wire metal elements in a liquid weld pool are small, about 10^{-8}m^2/s orders of magnitude. Despite varying diffusion coefficients depending on the chemical composition in a weld pool, the high-temperature weld pool exists for a very short time (usually tens of milliseconds). In addition, given the high-speed fluid dynamics in the moving weld pool, convection is likely to have a dominant influence on the movement of chemical compositions of a wire. Therefore, the model in this study is established based on assumption that the chemical compositions of wires have the same diffusion coefficients in the moving weld pool. The model ignores solid-phase diffusion, considering small diffusion coefficients of chemical compositions in solid alloys. In addition, in non-self-fluxing laser welding, there is little difference between wire density and the density of parent metal. Therefore, in the numerical simulation process, we assume that wire density and the density of parent metal are always equal. Based on the above assumptions, the two-phase flow equation of the wire fluid and the parent metal fluid in a moving weld pool can be defined by single-phase flow Navier–Stokes equations in combination with a conservation of concentration field equation.

The flow and diffusion of the metallic liquid formed when the wire is melted (different from the metallic liquid formed when the parent metal is melted) in the weld pool is regarded as an indivisible "chemical composition". Comprehensively considering the influence factors of convection and diffusion of the "chemical composition" of the wire in the moving weld pool, the conservation of the concentration field equation of the wire compositions is as follows:

$$\frac{\partial C}{\partial t} + u\frac{\partial C}{\partial x} + v\frac{\partial C}{\partial y} + w\frac{\partial C}{\partial z} = \frac{\partial}{\partial x}(\gamma\frac{\partial C}{\partial x}) + \frac{\partial}{\partial y}(\gamma\frac{\partial C}{\partial y}) + \frac{\partial}{\partial z}(\gamma\frac{\partial C}{\partial z}) \quad (8.17)$$

where, C—Concentration of the element of the wire;

γ—Diffusion coefficient of the element of the wire inside the moving weld pool;

u, v, w—The velocities of metallic liquid in three directions in the moving weld pool. They can be obtained by solving a similar equation from (8.9) to (8.12). To solve Eq. (8.17), the Navier–Stokes equation and the conservation of energy equation need to be solved. If the evolution process of a transient keyhole is considered, the Level Set equation describing free interface motion must be also solved.

After the concentration distribution of the wire in any micro area in the moving weld pool is obtained by solving the Eq. (8.17), the specific distribution of chemical compositions at any time in the weld pool can be determined quickly. The following is an example to illustrate how to calculate the specific distribution of the concentration of chemical compositions in the moving weld pool. It may be assumed that the concentration of wire in a micro area is C, and given an element (assumed to be aluminum) in the wire alloy, if the content of aluminum in the wire is C_{Al}^{filter} and the content of aluminum in the parent metal is C_{Al}^{base}, then the content of aluminum in the micro area is C_{Al}, which is as follows:

$$C_{Al} = C \times C_{Al}^{filler} + (1 - C) \times C_{Al}^{base} \quad (8.18)$$

In order to obtain the information of the flow field, temperature field and free interface of keyholes in the moving weld pool, boundary conditions such as the pressure, viscous stress and energy similar to that in Eqs. (8.13)–(8.16) must be considered in the simulation process. In addition to the above boundary conditions, the initial boundary conditions of the chemical compositions of the wire must also be considered to simulate the dynamic dilution process of the chemical compositions of the wire.

In theory, the simulation results of the keyhole and weld pool in the welding with filler wire in this chapter can provide initial conditions such as composition for dynamic dilution research. However, because the physical mechanism of the laser welding with filler wires is very complicated, if the simulation results in this chapter are used as initial conditions, it will bring great difficulties to the test verification of the simulation results of the dynamic dilution research. In order to facilitate the test verification of the established dynamic dilution simulation of the moving weld pool, an idealized boundary condition is set up and a layer of wire powder is required to be laid on the surface of the moving weld pool. That is, at these locations, the concentration values of the chemical compositions of the wire have the following boundary conditions

$$C = 1.0 \tag{8.19}$$

8.5.2 *Dynamic Dilution Behavior in Quasi-Steady Weld Pools*

As the laser welding process goes on, the moving weld pool will enter a relatively stable state, and the obtained weld joint tends to be in a relatively stable shape, that is, the welding process enters the quasi-steady state. After the welding process reaches the quasi-steady state, on the one hand, the melted wire material will continuously enter the weld pool from the melting front of the front end of the moving weld pool; on the other hand, the chemical compositions carried in the wire will gradually leave the moving weld pool from the solidification front of the moving weld pool. Therefore, studying the dilution process of the chemical compositions of the wire in the moving weld pool when the welding process reaches the quasi-steady state is of great significance to explain the dynamic dilution mechanism of the moving weld pool in the laser welding process and reveal the distribution law of the chemical compositions of wire inside the weld joint.

8.5.2.1 Calculation for Quasi-Steady Moving Weld Pools

By solving the transient coupling model of laser welding and laser welding with filler wires, important information such as the three-dimensional morphology, temperature field, flow field and keyhole morphology of the quasi-steady moving weld pool during

Table 8.8 Technical parameters of laser welding

Laser power/kW	Laser spot radius/mm	Defocusing amount/mm	Welding speed/ (m/min)
2	0.3	0	2

single-beam laser welding and welding with filler wire may be obtained. However, due to current limited computer capabilities, it is difficult to obtain, by calculation in a reasonable time based on the above method, a moving weld pool that reaches the quasi-steady state.

Therefore, in this section, a model proposed by the authors is used to obtain a quasi-steady moving weld pool. In this model, the formation process of keyholes and the formation process of moving weld pools are considered separately; when a keyhole is formed, it is solidified, and the wall temperature of the keyhole is fixed to the evaporation temperature, and then the quasi-steady moving weld pool is calculated by taking into account the thermal capillary force and other factors. The calculation speed may be greatly improved in this way. Although it can be seen from the existing research conclusions that this simple model cannot fully reflect the transient nature of laser welding process, it can still increase our understanding of the dynamic dilution process behavior of the moving weld pool by studying the behavior of the chemical compositions of the wire in this quasi-steady moving weld pool.

The technical parameters used in the laser welding process are shown in Table 8.8. The parent metal is Ti–Al alloy and the wire is Al-Si alloy. The contents of the chemical compositions of the two materials are shown in Table 8.9 and Table 8.10 respectively. Table 8.11 shows the thermophysical properties of the parent metal.

Table 8.9 Chemical compositions of parent metal

Al	V	Ni	Bi	H	Ti
6	4	2	7	0.1	Equilibrium state

Table 8.10 Chemical compositions of wire

Si	Mn	Ti	O	Al
20	8	10	0.2	Equilibrium state

Table 8.11 Thermophysical parameters for parent metal

Density /(kg/m^3)	Specific heat capacity /[J/(kg•K)]	Thermal conductivity /[W/(m•K)]	Liquidus temperature/K	Solidus temperature/K	Latent heat of melting /(J/kg)	Latent heat of evaporation /(J/kg)	Boiling point /K
4000	660	25	1928	1878	3.7×10^5	8.9×10^6	3315

Figure 8.25 shows the calculated flow and heat transfer characteristics of quasi-steady moving weld pool. From Fig. 8.25a, it can be seen that after the moving weld pool reaches the quasi-steady state, the width of the moving weld pool is about 1.8 mm and the length of the weld pool is about 4.5 mm. The length–width ratio is in good agreement with the length–width ratio, observed in general tests, of the weld pool obtained by low-speed deep penetration laser welding; in addition, the temperature distribution on the upper surface of the weld pool is close to the motion Gaussian distribution, which is consistent with the usual welding experience. As can be seen from the results of Fig. 8.25c–f, the absolute value of velocity near both sides of the keyhole in the weld pool is large. In addition, the pressure gradient near the front and the rear of the keyhole is large, and thus the value of velocity is also large. Although it is very difficult to directly verify these theoretical results, the above results are basically reasonable from practical experience. Figure 8.25g and h show the results of the flow field inside the weld joint and inside the longitudinal section of the weld pool, respectively. By comparing the simulation results based on the coupling of the transient keyhole and the weld pool in Chap. 4, it can be seen that the results of the flow field in the upper part of the moving weld pool are basically consistent with the transient simulation results.

8.5.2.2 Dynamic Dilution Process of Wire Compositions in Quasi-Steady Moving Weld Pool

According to the analysis results in Sect. 8.5.2.1, based on the velocity trend of the fluid in the weld pool obtained from the quasi-steady-state model established in this chapter, the results of the flow field in the upper part of the quasi-steady moving weld pool are of practical significance. Therefore, in this section, the dynamic dilution behavior of the wire compositions in the upper part of the quasi-steady moving weld pool is mainly studied.

In order to study the dynamic dilution behavior of chemical compositions of the wire in the upper part of the quasi-steady moving weld pool, in this section, the wire alloy, Si element, Ti element and Al element are respectively used as research objects, and their specific percentage content and characteristics are shown in Table 8.12 (where being diluted by parent metal indicates that the chemical composition of an element in the wire is higher than that in the parent metal, so the wire is diluted by parent metal; that the parent metal is diluted indicates that the chemical composition of an element in the wire is lower than that in the parent metal, so the parent metal is diluted).

According to the boundary conditions in Sect. 8.5.1, the wire initial boundary conditions shown in Fig. 8.26a are adopted in the study: in the red zone, the concentration of the wire composition is set to be 1.0 (red zone), while that in other positions is 0. After the mathematical model built in the chapter is used to figure out the quasi-steady moving weld pool, Eq. (8.17) is solved to calculate the dynamic dilution process of wire composition to the moving weld pool.

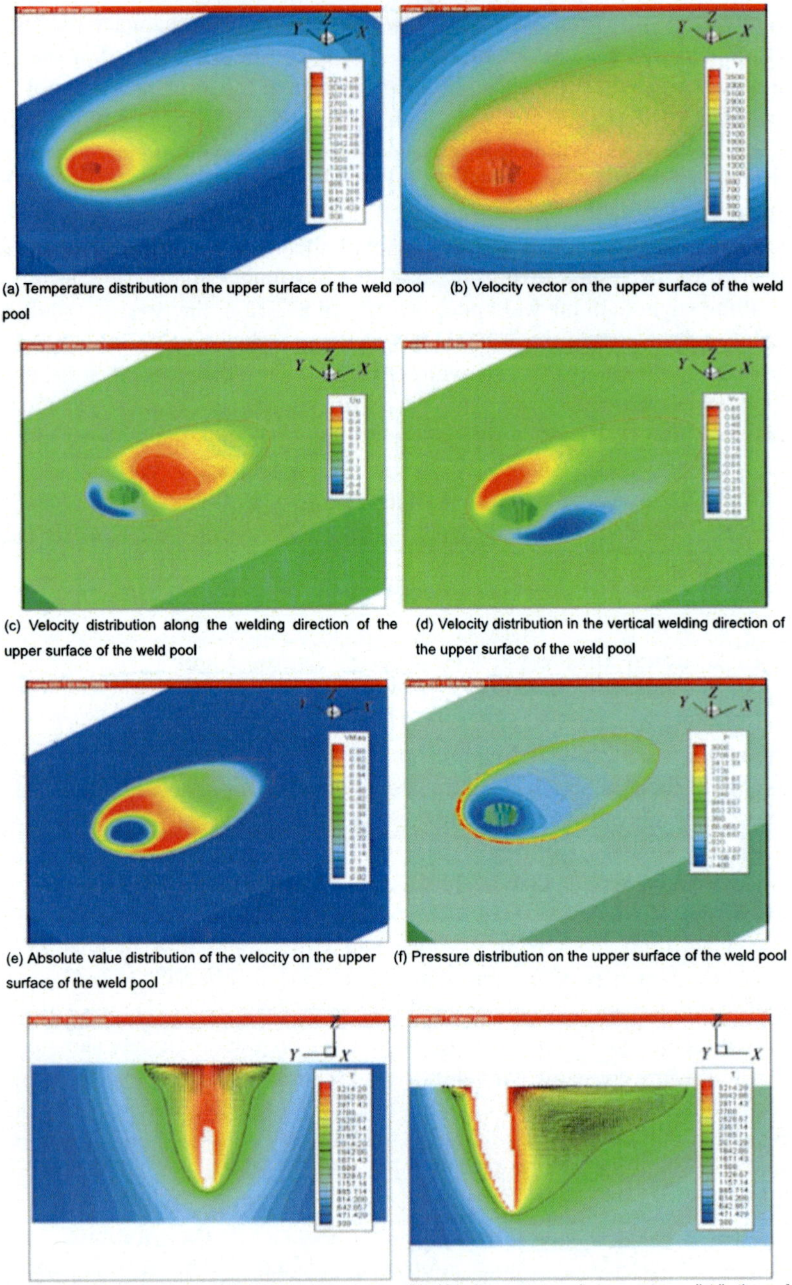

(a) Temperature distribution on the upper surface of the weld pool

(b) Velocity vector on the upper surface of the weld pool

(c) Velocity distribution along the welding direction of the upper surface of the weld pool

(d) Velocity distribution in the vertical welding direction of the upper surface of the weld pool

(e) Absolute value distribution of the velocity on the upper surface of the weld pool

(f) Pressure distribution on the upper surface of the weld pool

(g) Velocity vector of the cross section of the weld pool

(h) Velocity vector and temperature distribution of longitudinal section of the weld pool

Fig. 8.25 Flow and heat transfer characteristics of moving weld pools at the quasi-steady state

Table 8.12 Concentration and behavior characteristics of the alloy element

Wire chemical composition (%)	Parent metal chemical composition (%)	Behavior characteristics of alloy element
Wire as a kind of element 100	0	Such special element does not exist in the parent metal and is diluted by the parent metal
Si 20	0	Wire Si has a higher concentration than parent metal and is diluted by the parent metal
Ti 10	80	Wire Ti has a lower concentration than parent metal, and parent metal is diluted
Al 62	6	Wire Al has a higher concentration than parent metal, and is diluted by the parent metal

Figure 8.26b–f show the transient dilution process of the wire composition to the moving weld pool on the upper surface of the moving weld pool. As it can be seen from the figure that, when the wire composition is in transition to the weld pool at 1.6 ms, the concentration on both sides of the surface moving weld pool is relatively high according to the concentration cloud chart, while the isoconcentration lines in the middle position are almost parallel. By comparing the concentration cloud charts when $t = 1.6$ and $t = 3.2$ ms, it shows that on the weld pool surface, the wire composition moves gradually from the front of the surface weld pool to the rear of the weld pool, and during the process, the weld pool is also gradually diluted by the wire composition. It is observed by comparing the figure at 1.6 ms with the figure at 6.4 ms that, the wire composition on both sides is relatively high, and the isoconcentration lines at the rear of the surface moving weld pool tend to concave towards the front of the weld pool. However, when the quasi-steady state is reached, the isoconcentration lines in the middle of the surface moving weld pool are almost parallel with the cross section direction, and the middle part of the isoconcentration lines at the rear still tends to concave, but not that obvious anymore. It shows that the concentration gradient of the wire composition close to the moving weld pool surface becomes less apparent, but the wire composition in the weld pool micro area is still uneven.

The reason for the above phenomenon is closely correlated to the convection and diffusion factors in the weld pool. At the beginning of the dynamic dilution, due to severe convection and high speed on both sides of the moving weld pool in the upper part, the concentration of the wire composition on both sides of the weld pool in the upper part is high, resulting in concave of the isoconcentration lines in the middle of the upper surface of the weld pool towards the front of the weld pool. Since the speed direction of the weld pool at the rear of the keyhole is from the front to the back as a whole, the wire composition also moves gradually from the front to the rear of the weld pool through convective motion. During the dilution process,

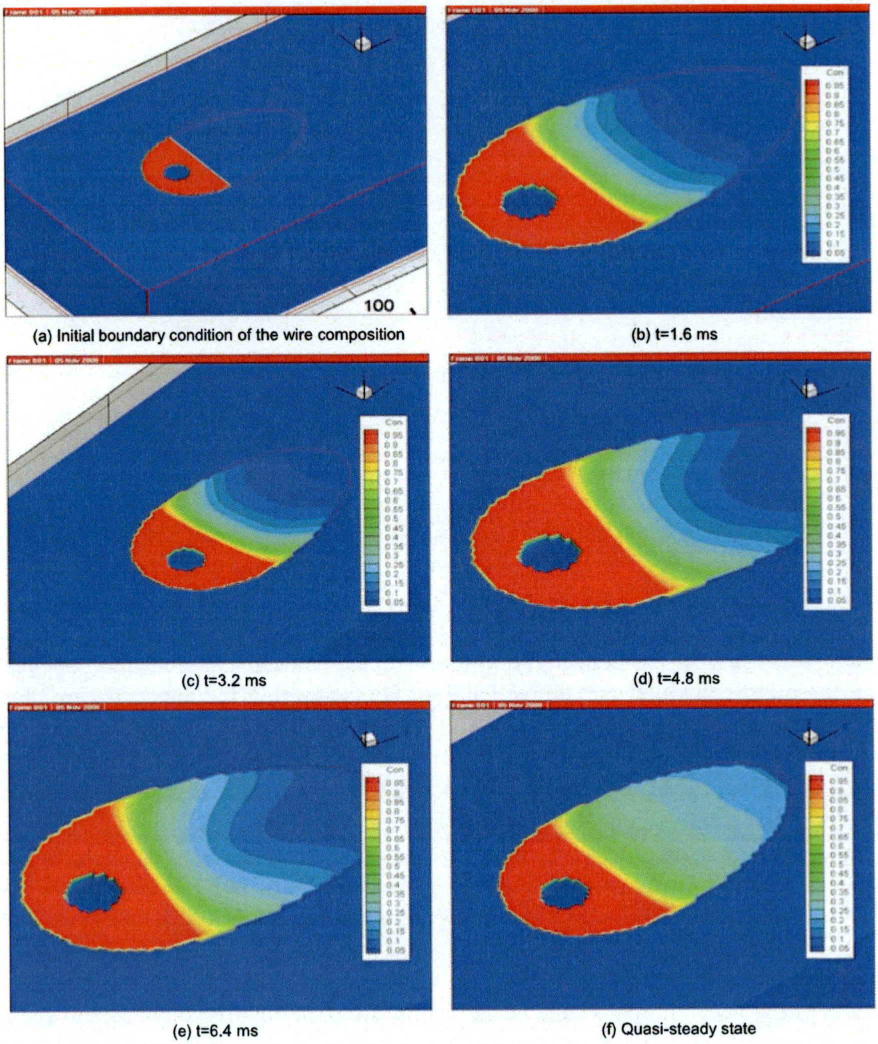

Fig. 8.26 Dilution process of wire composition to the weld pool on the upper surface of the moving weld pool

the wire composition is spreading in the weld pool. Both convection and spreading interact with each other, to finally reach a relatively steady state. Figure 8.27 is the concentration distribution change diagram of wire composition on certain section near the weld joint. The speed characteristics of the moving weld pool on the section are shown in Fig. 8.27d. As shown in the figure that, the wire composition gradually dilutes the parent metal from the edge of the moving weld pool to the depth direction. Due to very low speed, at the vortex of the edge, the concentration value of the wire composition is higher than that in other positions. Generally, in the upper part of

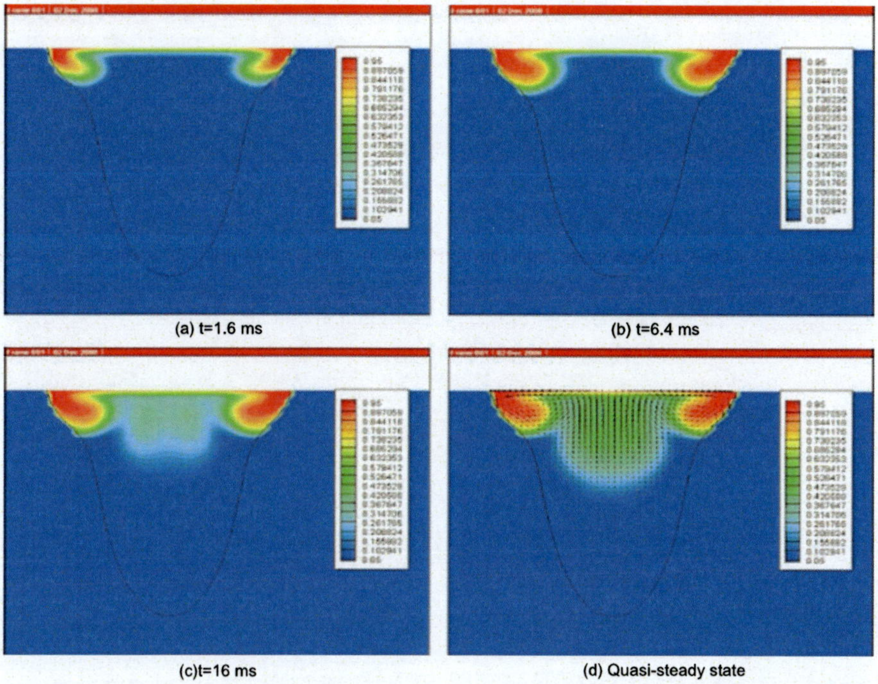

Fig. 8.27 Concentration change process of wire composition on the certain cross section near the weld joint

the moving weld pool, convection is the main factor of the wire composition being diluted by the parent metal.

Figure 8.28 is the concentration distribution diagram of the wire composition in three longitudinal sections of the moving weld pool after the dynamic dilution process is up to the quasi-steady state. Figure 8.28a shows the longitudinal section of the true center of the weld pool, Fig. 8.28c shows the longitudinal section near the edge of the moving weld pool, and Fig. 8.28b shows the longitudinal section between the two interfaces. In the following statements, we might as well refer the three sections shown in Fig. 8.28a–c respectively to Section A, B, and C. It's known from the figure that, the concentration of the wire composition in Section A center is higher than that in Section B center, while the concentration of the wire composition in Section B center is higher than that in Section C center. This proves again that in the upper part of the moving weld pool, wire composition dilutes the weld pool by moving from the edge gradually to the center of the weld pool. Furthermore, in the weld pool behind the keyhole, the concentration of the wire composition at the vortex position is higher than that in the periphery, which further verifies the rule that the concentration of wire composition at the vortex position is relatively high due to low speed.

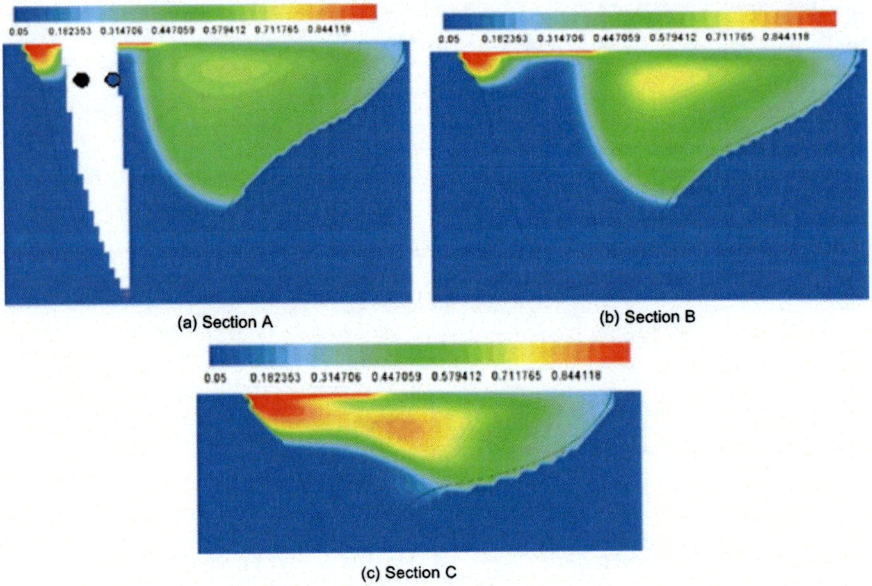

Fig. 8.28 Concentration distribution of the wire composition in different longitudinal sections after the quasi-steady state is reached

To study the dynamic evolution process of the wire composition concentration over time in the upper part of the moving weld pool, two points in the longitudinal section of the true center of the moving weld pool are taken as study objects—Point X (the left dark point in Fig. 8.28a) and Point Y (the right blue point in Fig. 8.28a). As shown in the figure, coordinates of Point X and Point Y only have difference along welding direction; Point X is closer to the keyhole. Figure 8.29 is the evolution

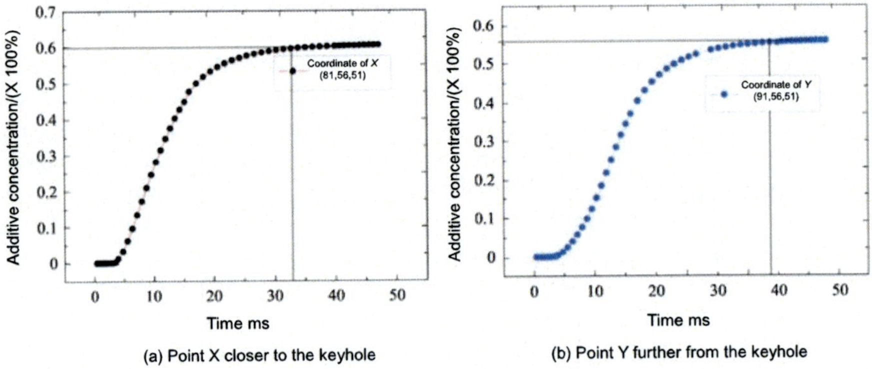

Fig. 8.29 Evolution process of the wire composition concentration at different points in the upper part of the moving weld pool

process diagram of the wire composition concentration of the above two points over time. It can be seen from the figure that, at the two points, the time of Point *X* closer to the keyhole reaching the quasi-steady state is about 32 ms, while Point *Y* further from the keyhole reaching the quasi-steady state is about 38 ms. Therefore, the time of the wire composition reaching the quasi-steady state in the moving weld pool is inconsistent. In addition, when the quasi-steady state is reached, the additive concentration at Point*X* is 60%, while that at the right point is roughly 55%. It shows that the concentration in the moving weld pool is also uneven when the quasi-steady state is reached.

Based on the above discussion, the following conclusions can be made: ① In the upper part of the moving weld pool, the wire composition moves gradually from the front to the rear of the weld pool through convective motion and dilutes the parent metal from the edge of the moving weld pool gradually to depth direction; ② Convection plays a primary role in the dynamic dilution process in the upper part of the moving weld pool; ③ At the places with flowing vortex in the moving weld pool, the wire composition concentration is usually higher than that in the periphery; ④ When the quasi-steady state is reached, the wire composition concentration at different positions in the upper part of the moving weld pool is generally uneven, and the time of reaching the quasi-steady state is also different.

8.5.2.3 Dynamic Dilution of Different Chemical Compositions in the Quasi-Steady Moving Weld Pool

The section will mainly discuss the dynamic dilution behavior of different chemical compositions in the quasi-steady moving weld pool through the numerical simulation. The mentioned chemical compositions in wire alloy are different from those in parent metal alloy. Technical parameters and material parameters used in numerical calculation are consistent with those stated in Sect. 8.5.2.2.

Figure 8.30 is the concentration distribution diagram for Ti chemical composition on two sections of the moving weld pool when the quasi-steady state is reached. Section A is the longitudinal section close to the edge on both sides of the moving weld pool; Section B is the longitudinal section in the center of the moving weld pool. It can be seen from the figure that the concentration distribution trend of Ti chemical composition is just the opposite of the wire concentration trend, i.e., the higher the wire composition is, the lower the Ti chemical composition will be. Therefore, the concentration of Ti chemical composition in the moving weld pool is diluted, and uneven in the moving weld pool, and will reach a local minimum at the vortex.

Figure 8.31 is the concentration distribution diagram of Ti chemical composition in two sections of moving weld pool when the quasi-steady state is reached. Section A is the longitudinal section close to the edge on both sides of the moving weld pool; Section B is the longitudinal section in the center of the moving weld pool. As known in the figure that, Al content in the wire is higher than that in the parent metal, so the Al composition in the wire is diluted by the moving weld pool, and it

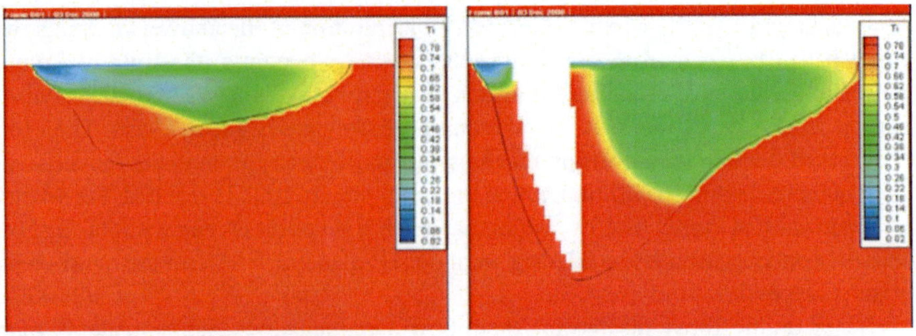

(a) Longitudinal section A **(b) Longitudinal section B**

Fig. 8.30 Concentration distribution of Ti chemical composition in different longitudinal sections when the quasi-steady state is reached

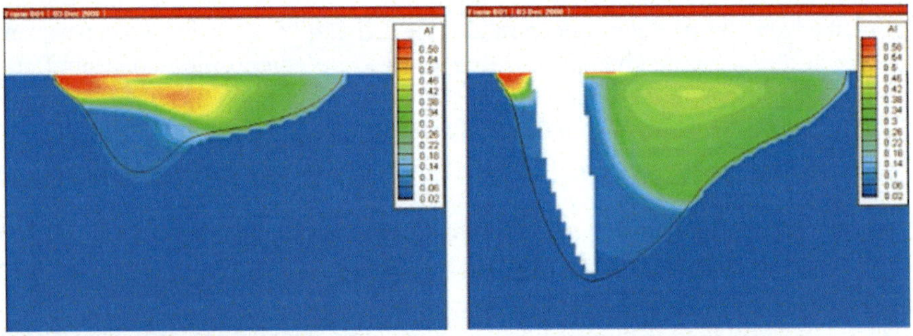

(a) Longitudinal section A (b) Longitudinal section B

Fig. 8.31 Concentration distribution of Al chemical composition in different longitudinal sections when the quasi-steady state is reached

is also uneven in the moving weld pool. At the vortex position in the moving weld pool, Al composition concentration will reach a local maximum due to low speed.

Figure 8.32a and b are respectively the distribution diagrams of Ti and Al chemical compositions in the weld joint section when the dynamic dilution process is just completed, i.e., up to the quasi-steady state. It can be seen from the figure that when the quasi-steady state is reached, due to the convection action of the moving weld pool and vortex flow at two corners in the upper part of the weld joint, both Ti and Al chemical compositions respectively reach concentration local minimum and local maximum.

Figure 8.33 shows the dilution process curve of Ti concentration at two points in the moving weld pool (black point *X* and blue point *Y* in Fig. 8.28a). Here, the concentration gradient is defined as an absolute value of the difference between the concentration of the current chemical composition and the concentration of the wire composition at that position. With the advancing of the dilution process, the concentration gradient of either Point *X* or Point *Y* gradually decreases until the

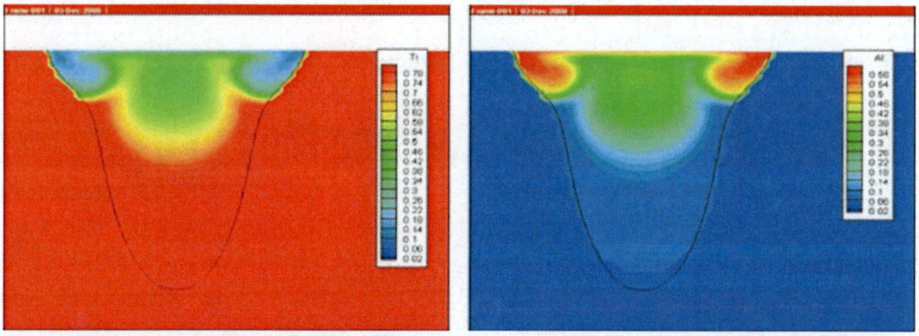

(a) Ti concentration distribution (b) Al concentration distribution

Fig. 8.32 Distribution diagram of the quasi-steady concentration of both Ti and Al elements in the weld joint section of the moving weld pool

region becomes stable, as shown in the figure. This means that the intensity of dilution is constantly weakening with the change of time; the intensity of dilution varies at different positions. Since Point *X* is closer to the keyhole, the speed of movement around is larger. Therefore, Point *X* has a stronger intensity of dilution than Point *Y*. This indicates that the intensity of dilution of wire chemical compositions is generally different at different positions in the weld pool.

Figure 8.34 shows the dilution process curve of Al concentration at two points in the moving weld pool (black point *X* and blue point *Y* in Fig. 8.28a). As shown in the figure, as the time goes on, the Al composition is gradually diluted by the parent metal, and the intensity of dilution also gradually slows down to a steady state; Point *X*, close to the keyhole, takes about 26.5 ms to reach the quasi-steady state, while Point *Y* takes about 30 ms to reach the quasi-steady state. This indicates that the time for Al to reach quasi-steady state is generally different at different positions, and the concentration of Al composition at quasi-steady state is generally non-uniform.

The variation curves of the composition concentration of wire "element", Ti element, Si element and Al element at the position of Point *X* close to the keyhole are analyzed in the same coordinates below, as shown in the Fig. 8.35. According to the figure, it is apparent that the slope of the three tangents of *A* (wire), *B* (Al element) and *C* (Si element) gradually declines. The only possible reason for the difference in slope is the difference in the concentration of elements in the wire and parent metal, namely the concentration gradient, thanks to the basically constant motion velocity distribution of the quasi-steady weld pool, and the little difference in the diffusion rate between various elements and wire "element". This indicates that the larger the concentration difference of an element in the wire and parent metal, the more intense the dynamic dilution process of the element at the initial stage of the dilution process ($t \leq 15$ ms).

In view of the above discussion, it can be concluded preliminarily that: ① If the content of a chemical composition in the wire is higher than that in the parent metal, the chemical composition will be diluted by the parent metal in the quasi-steady

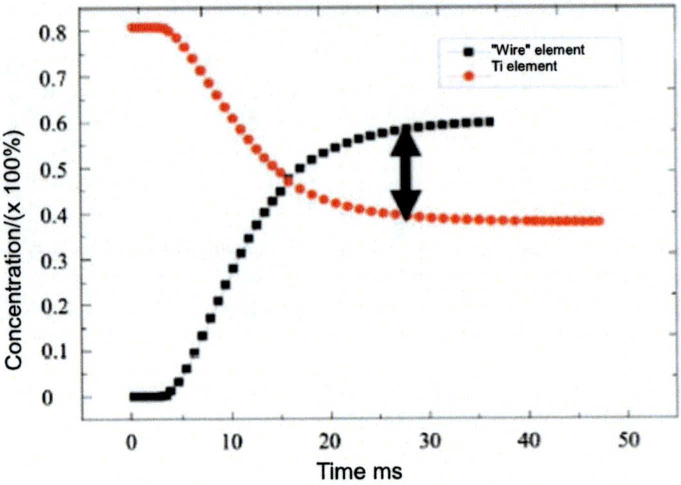

(a) Point *X* closer to the keyhole

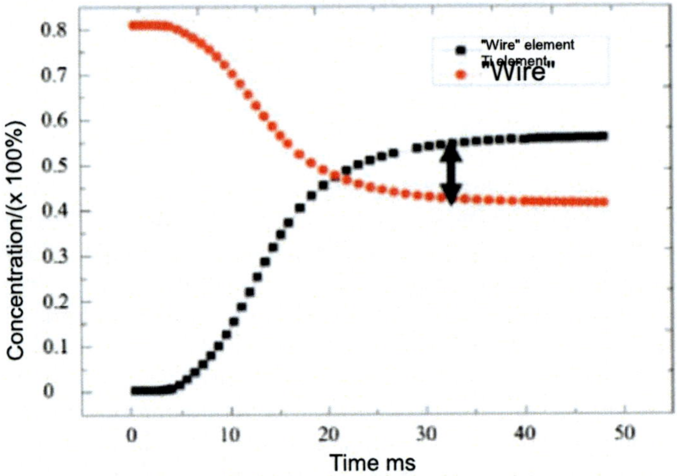

(b) Point *Y* further from the keyhole

Fig. 8.33 Variations in the intensity of dilution of Ti composition at different positions in the upper part of the moving weld pool

moving weld pool; otherwise, the parent metal will be diluted. ② The intensity of dilution of chemical compositions is generally different at different positions, and the time for them to reach quasi-steady state is also inconsistent. ③ The larger the content difference of chemical compositions in the wire and parent metal, the greater the intensity of dilution at the initial stage of dynamic dilution.

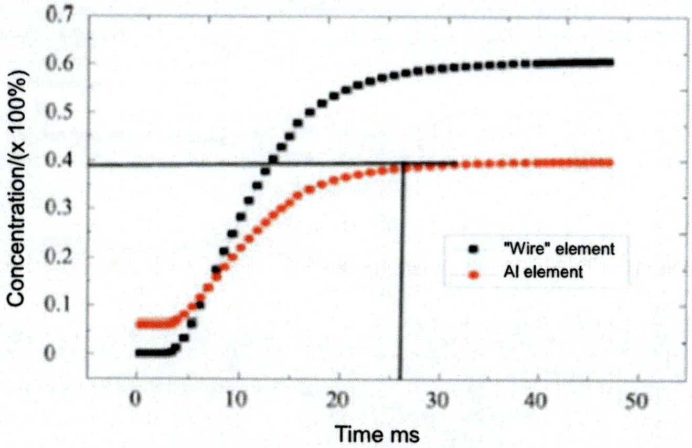

(a) Point *X* closer to the keyhole

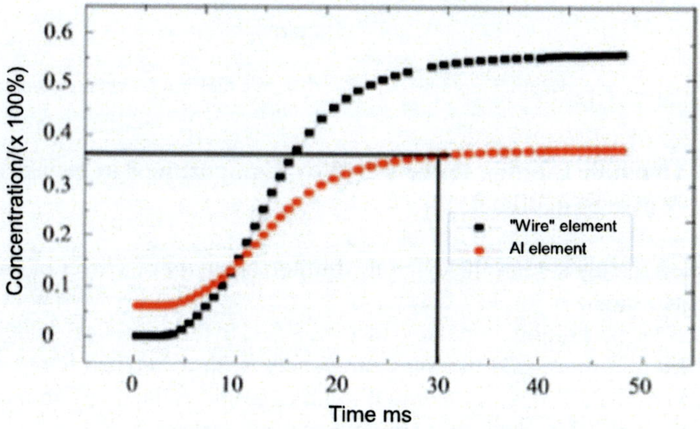

(b) Point *Y* further from the keyhole

Fig. 8.34 Variations in the intensity of dilution of Al composition at different positions in the upper part of the moving weld pool

8.5.3 Dynamic Dilution Behavior in Transient Weld Pools

Both the process of single-beam laser welding and the process of laser welding with filler wires can reach a relatively steady state. But, according to the simulation results in Chap. 4 and this chapter, the keyhole in the two welding processes is in the course of periodic oscillation under most technical conditions. The moving weld pool is always in a high-speed transient change in both laser welding processes, thanks to the close relationship and coupling of the keyhole and moving weld pool behaviors. The dynamic dilution behavior in the quasi-steady moving weld pool is studied systematically in the previous section. This section studies the dynamic dilution behavior in the transient moving weld pool in the course of laser welding.

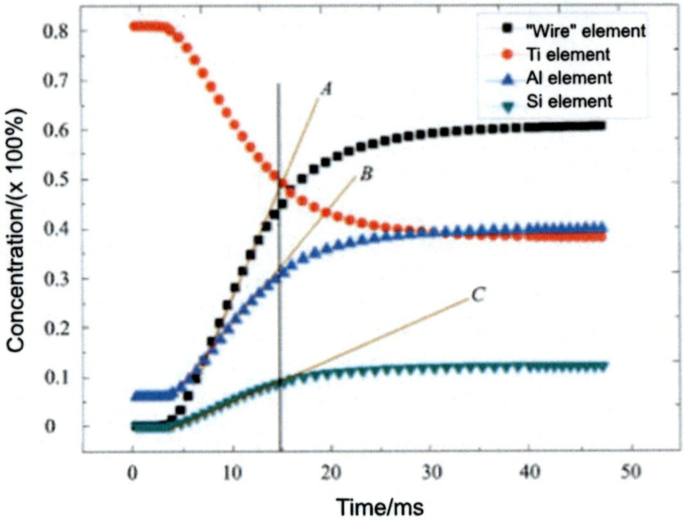

Fig. 8.35 Impact of concentration gradient on the dynamic dilution process of elements

8.5.3.1 Dynamic Dilution Process of Wire Compositions in Transient Moving Weld Pool

This section mainly studies the dynamic dilution behavior of a layer of wire composition coated on the workpiece surface in the transient moving weld pool according to the boundary conditions stipulated in Sect. 8.5.1. In the numerical study, the parent metal is 2.0 mm-thick TC4 titanium alloy sheet, with the coated wire composition layer thickness of 0.1 mm. At the initial welding moment, the wire composition has a concentration of 1.0 on the coating layer, and a concentration of 0 at other positions, as shown in Fig. 8.36. The impact of evaporation and burning loss on concentration is ignored in the numerical calculation process to speed up the calculation speed. The main laser welding parameters, the parent metal thermophysical parameters and the

Fig. 8.36 Schematic diagram of wire composition at the initial welding moment (the coating layer is shown in red)

Table 8.13 Technical parameters of laser welding

Laser power/kW	Laser spot radius/mm	Defocusing amount/mm	Welding speed/ (m/min)
1.0	0.2	0	3

Table 8.14 Numerical simulation parameters

Grid quantity	Spatial step/m	Initial time step/s	Initial temperature of parent metal/K	Ambient temperature/K
100 × 60 × 80	3.0×10^{-5}	3.0×10^{-6}	300	300

numerical calculation parameters used are respectively set out in Tables 8.13, 8.14 and 8.15.

Figure 8.37 shows the simulated evolution process of the concentration of compositions inside the moving weld pool near the keyhole free surface and inside the longitudinal section of the moving weld pool. The three-dimensional free interface shape represents the keyhole morphology, and the black line represents the fusion line. In the color scale, high concentration and low concentration are expressed in red and blue respectively. Figure 8.37a–f respectively show the concentration of compositions near the free interface and inside the cross section of the weld pool after 0.12, 5.04, 9.96, 14.88, 19.8 and 24.72 ms. As can be seen from the figures, the keyhole and weld pool gradually form with the movement of the laser head; in addition, the wire composition coated on the workpiece surface is diluted by convection of the weld pool, and presents a tendency to move to both sides of the weld pool and to the depth direction at the same time. The weld pool dynamic mechanism of the laser welding obtained through study in Chap. 3 indicates that a high-speed downward movement exists near the front wall of the keyhole as a result of recoil pressure. Therefore, a high concentration of wire composition can be found at the weld pool position near the front wall of the keyhole. Besides, according to the conclusions in Chap. 3, there are three typical vortex flows and small speeds at both sides of the upper part, the rear part, and the position of the weld pool near the back bottom of the keyhole. Thus, the concentration of wire composition at these positions is significantly higher than at other positions from 9.96 ms, as shown in Fig. 8.37c–f.

Figure 8.38 shows the evolution process of wire composition at the solidification front and melting front of the transient moving weld pool. According to the figure, with the advancing of the welding process, the wire concentration gradually increases at both sides of the moving weld pool, the bottom of the melting front and the bottom of the solidification front. The foregoing simulation results suggest that after the wire composition enters the moving weld pool from the upper part of the melting front, it bypasses the keyhole to move backward due to the convection on the upper part of the weld pool, and in addition, it moves towards the direction of penetration depth due to the high-speed downward convection caused by the recoil pressure. Figure 8.39 shows the concentration evolution process of the wire composition in the longitudinal section of the weld joint. The figure indicates that, as the welding process proceeds,

Table 8.15 Thermophysical parameters for parent metal

Density /(kg/m^3)	Specific heat capacity/ [J/(kg•K)]	Thermal conductivity /[W/(m•K)]	Liquidus temperature/K	Solidus temperature/K	Latent heat of melting /(J/kg)	Latent heat of evaporation/(J/kg)	Boiling point/K
4000	660	25	1928	1878	3.7×10^5	8.9×10^6	3315

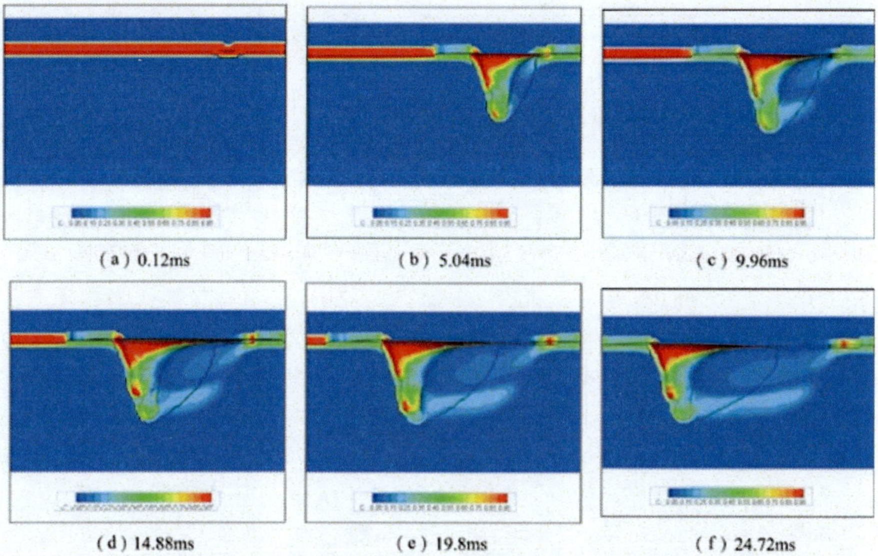

Fig. 8.37 Transient evolution process of wire composition in the weld pool near the keyhole and in the longitudinal section of the weld pool

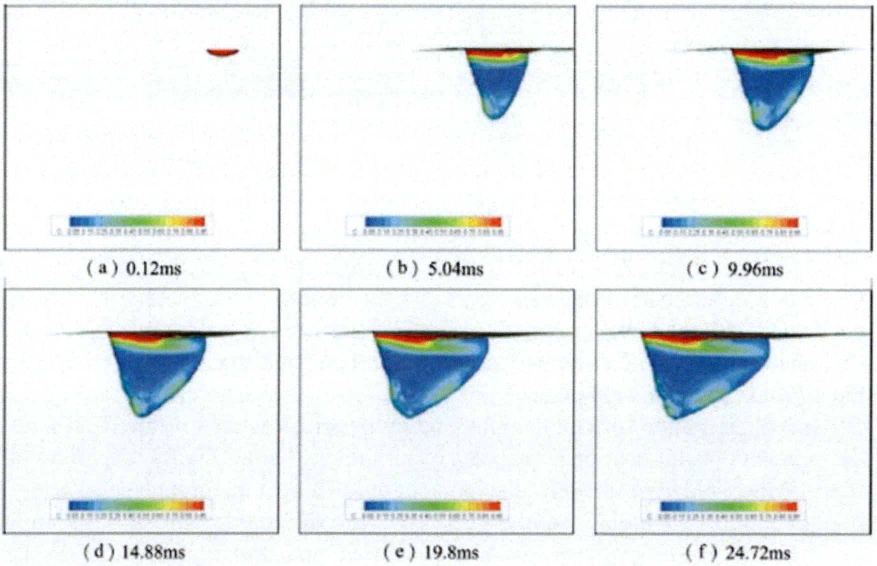

Fig. 8.38 Evolution process of wire composition at the solidification front and melting front of the transient moving weld pool

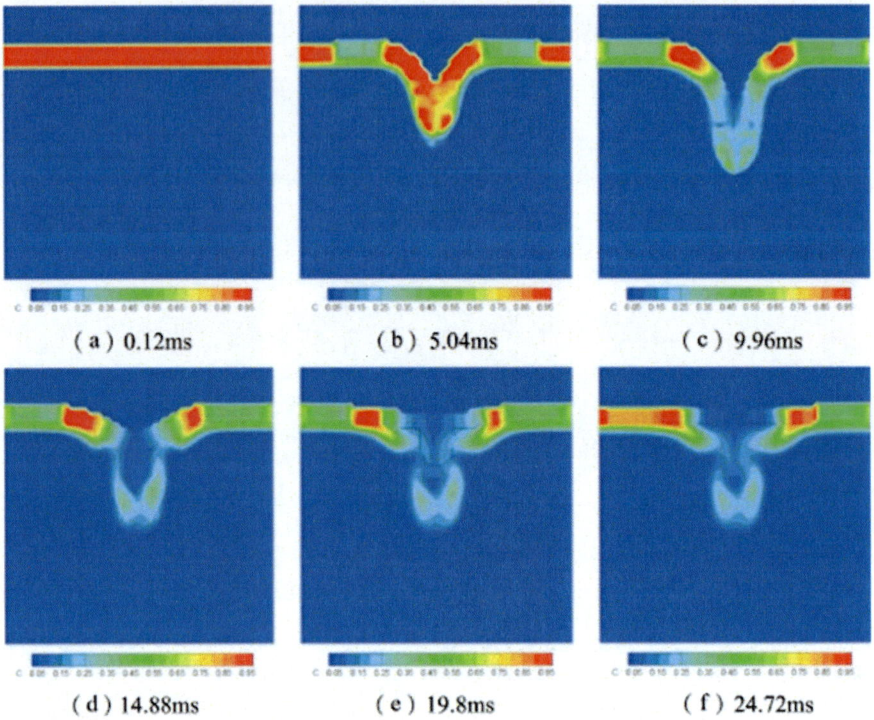

Fig. 8.39 Concentration evolution process of wire composition in the longitudinal section of the weld joint

the wire composition inside the weld joint is gradually diluted by the weld pool. The wire composition tends to move from the edge of the weld pool to the direction of penetration depth. As a result, a relatively high concentration area is gradually formed on either side of the upper part and the bottom of the weld joint, which is contrasted with the low concentration area in the middle of weld joint. The cause of such distribution trend of the wire composition is related to the flow mechanism of the transient moving weld pool.

The concentration of wire composition on the cross section of the weld joint at the moment of solidification is analyzed quantitatively below. Figure 8.39f shows the shape of the weld joint section at the moment of solidification through calculation. In Fig. 8.39f, three positions, namely, A, B and C, are selected for virtual line scanning to analyze the quantitative distribution of wire composition concentration in these positions. The position of A, B and C is 0.12, 0.42 and 0.93 mm away from the surface of the workpiece respectively. Figures 8.40, 8.41 and 8.42 show the distribution curve of wire composition through line scanning at *A, B* and *C* positions inside the weld joint. According to the figures, the wire composition presents a relatively orderly distribution on the three curves, showing a trend of high on both sides and low in the middle.

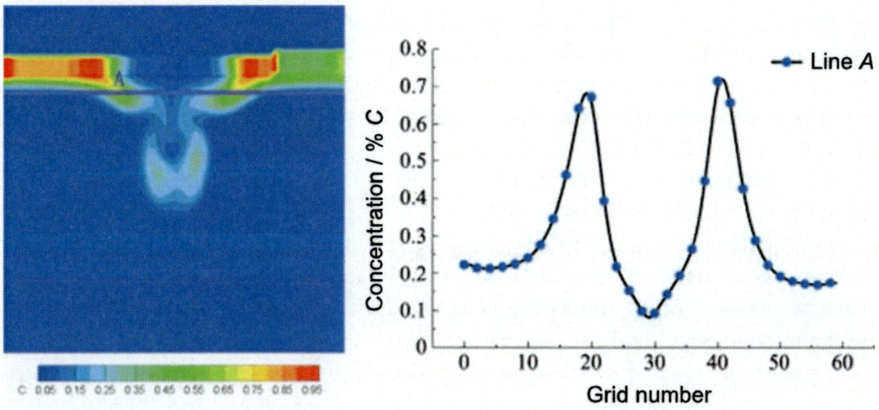

Fig. 8.40 Distribution curve of wire composition through line scanning at the position of *A*

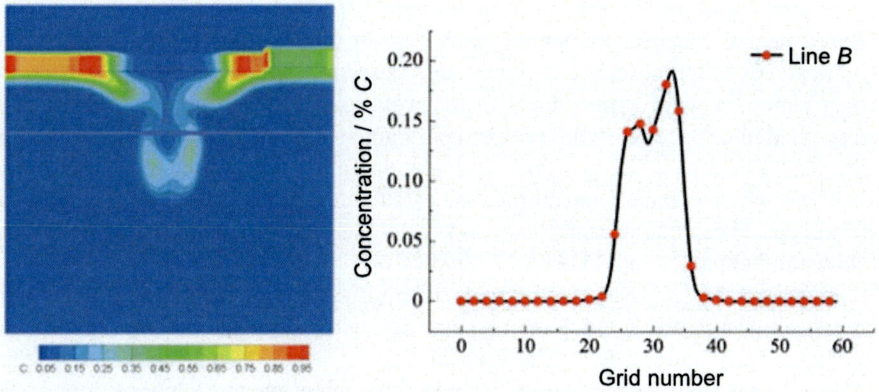

Fig. 8.41 Distribution curve of wire composition through line scanning at the position of *B*

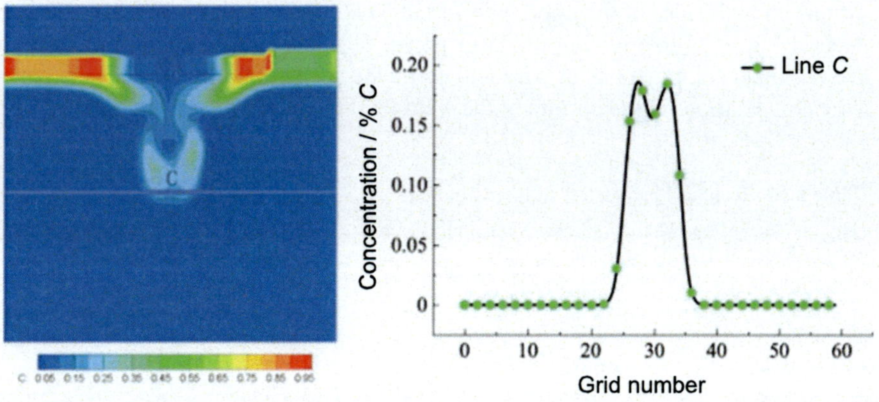

Fig. 8.42 Distribution curve of wire composition through line scanning at the position of *C*

Based on the discussion above, the following conclusions can be drawn: ① The dynamic properties of the downward high-speed fluid caused by the recoil pressure on the wall surface of the keyhole are the primary factors that facilitate the deep permeation of wire compositions into the work piece; ② Similar to the dilution behavior of the quasi-steady moving weld pool, the wire compositions, driven by fluid dynamics that bypasses the keyhole, will concentrate on both sides of the upper part of the weld joint, leading to the formation of local high concentration areas; ③ Under certain conditions, there will be an area with a high concentration of wire compositions on both sides of the middle and lower parts of the weld joint; ④ Convection is still the primary factor in the dynamic dilution of wire compositions in the moving weld pool.

8.5.3.2 Influence of Technical Parameters on Distribution of Wire Compositions in the Weld Joint at the Moment of Solidification

Based on the transient numerical model established in the previous section, the influence of different technical parameters (laser power and welding speed) on the distribution of wire compositions in the weld joint in the solidification instant is discussed in this section. The technical parameters used in the numerical simulation are shown in Table 8.16. It is noticeable that in Table 8.16, the value of laser power is small, which is mainly to improve the calculation speed. The relevant parameters of numerical calculation and the parent metal parameters are consistent with those shown in Table 8.14 and Table 8.15, respectively.

Figures 8.43 and 8.44 are the comparison diagrams of the distribution of wire compositions in the weld joint respectively with different welding speeds and different laser powers. It can be seen in Fig. 8.43 that raising the welding speed will decrease the recoil pressure, making the content of wire compositions in the bottom of the weld joint gradually decline. Meanwhile, according to the principle of mass conservation, raising the welding speed makes the content of wire compositions on both sides of the upper part of the weld joint gradually increase. It can be seen in Fig. 8.44 that raising the laser power will increase the recoil pressure and the chemical compositions of wire will permeate deeper, thus making the content of wire compositions in the bottom of the weld joint increase. Meanwhile, according to

Table 8.16 Technical parameters of laser welding

No	Laser Power/kw	Laser Spot Radius/mm	Defocusing Amount/mm	Welding Speed/ (m/min)
1	1.0	0.2	0	3
2	1.5	0.2	0	3
3	1.0	0.2	0	1.5
4	1.0	0.2	0	6

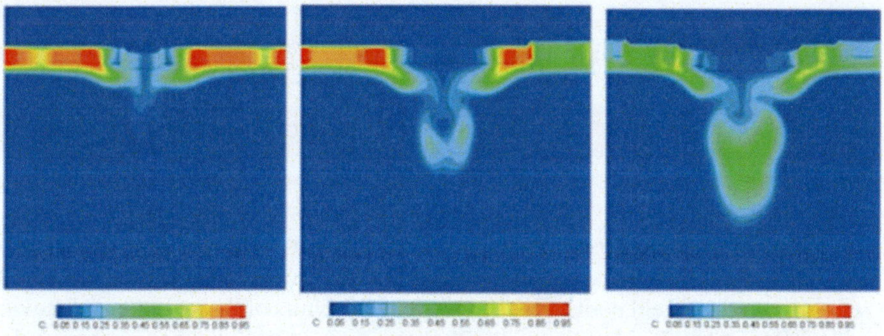

(a) 6m/min (b) 3m/min (c) 1.5m/min

Fig. 8.43 Comparison diagram of the distribution of wire compositions in the weld joint in the solidification instant with different welding speeds (No. 4, No. 1 and No. 3)

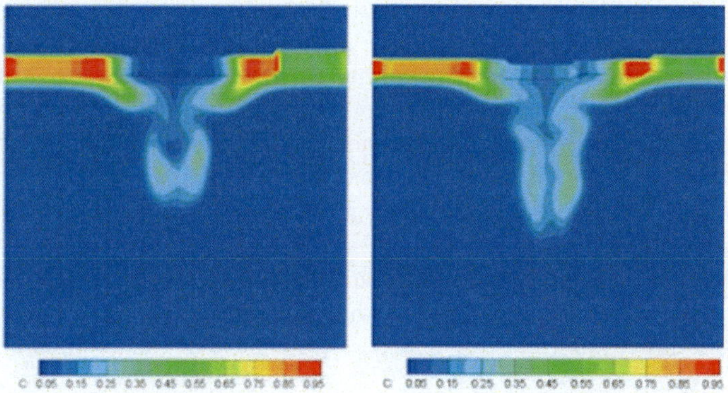

(a) 1.0kW (b) 1.5kW

Fig. 8.44 Comparison diagram of the distribution of wire compositions in the weld joint in the solidification instant with different welding powers (No. 1 and No. 2)

the principle of mass conservation, at this moment the content of wire compositions on the upper part of the weld joint tends to decline.

The influence rule of laser power and welding speed on the wire compositions in the weld joint discussed above can be concluded as the influence of heat input on the wire compositions in the weld joint. Based on the discussion above, the following conclusions can be drawn preliminarily: ① Raising weld heat input will facilitate the permeation of wire compositions to the bottom of the weld joint, and increase the content of wire compositions in the middle and lower parts of the weld joint, thus relatively reducing the content of wire compositions in the middle and upper parts of the weld joint; ② Reducing heat input will reduce the recoil pressure, and can effectively reduce the content of wire compositions in the middle and lower parts of the weld joint, thus relatively increasing the content of wire compositions in the upper part of the weld joint.

8.6 Summary

(1) The 3-D transient mathematical model of laser welding with filler wire is developed, which can describe the coupling effect among the wire, keyhole and moving weld pool. And the transient behaviors of the keyhole and moving weld pool during welding with filler wire respectively under contact transition and free transition are systematically studied. Besides, the influence of the wire feed rate and diameter of wire on the behaviors of the keyhole and the moving weld pool is discussed.

(2) Under certain technical conditions, raising the wire feed rate reduces the maximum velocity fluctuation of the fluid in the weld pool and improves the stability of the weld pool. However, increasing the diameter of wire aggravates the maximum velocity fluctuation of the fluid in the weld pool, thus weakening the stability of the weld pool. For the laser welding with filler wire in case of the wire in front of the keyhole, under general conditions, although the moving weld pool is more stable than that in single-beam laser welding under the same conditions, the keyhole is more unstable than that in the single-beam laser welding.

(3) The mathematical model of the dynamic dilution process of the moving weld pool in non-self-melting laser welding process is developed. With the model, a systematic numerical simulation study on the dynamic dilution behaviors of wire compositions respectively in the quasi-steady moving weld pool and in the transient moving weld pool is conducted, and a laser welding experiment on the powder coated at the surface of parent metal is designed to verify the simulation results. The verification shows that it is feasible to predict the quantitative distribution of the chemical compositions of wire in the moving weld pool in the solidification instant. The establishment of this model provides an idea to control the mechanical properties of a weld joint by precisely controlling the distribution of the chemical compositions of wire in the weld joint.

(4) In the non-self-melting laser welding process, convection is the primary factor for the wire compositions being diluted by the moving weld pool. The high-speed flow towards the depth of penetration caused by the flow that bypasses the keyhole and the recoil pressure is the primary factor for the dilution of the weld pool. The content of the chemical compositions of wire reaches a local maximum or minimum where there is vortex flow in the weld pool.

(5) Raising heat input will facilitate the permeation of wire compositions to the bottom of the weld joint, increase the content of wire compositions in the middle-lower part of the weld joint, and reduce the content in the middle-upper part of the weld joint. Reducing heat input can effectively reduce the content of wire compositions in the middle-lower part of the weld joint, and increase the content of wire compositions in the middle-upper part of the weld joint.

Chapter 9
Dynamical Behaviors of Keyhole and Weld Pool in Vacuum Laser Welding

Abstract This chapter introduces the dynamics behaviors of keyhole and weld pool in vacuum laser welding. Mathematical model for simulating time dependent keyhole and weld pool dynamics in vacuum laser welding is given. The physical behaviors of keyhole dynamics and flow pattern of weld pool under vacuum welding are discussed. The mechanisms of penetration increase under vacuum conditions are explained using numerical simulations.

9.1 Introduction

Vacuum laser welding refers to the welding of workpieces in a vacuum or low vacuum environment by using laser energy. This welding method produces a penetration depth comparable to that in vacuum electron beam welding, and delivers excellent welding quality without radiation pollution. It is perfectly suitable for thick plates. Recent years have seen revolutionary breakthroughs in fiber laser technology. The power of industrial fiber lasers which can work stably for a long term has reached 100 kW. Vacuum laser welding is able to penetrate thick-walled components to a depth of more than 100 mm at a time. It offers a wide range of potential applications in space vehicles, deep-sea equipment, and other fields, and is an important joining technology for the key equipment in the national major strategic engineering projects, such as space engineering and deep ocean engineering.

Similar to laser welding under atmospheric conditions, vacuum laser welding is characterized by rapid gasification and liquefaction of workpieces under a laser beam to form a weld pool, a keyhole, and metallic vapors. However, performed in a vacuum and low vacuum environment, the latter significantly differs from the former in the dynamic characteristics of keyhole and weld pool and the profile and quality of welds. Unlike electron beam which is sensitive to air pollution and requires high degrees of vacuum, laser beam allows high-quality welding under atmospheric conditions by using appropriate welding parameters. In industrial applications, laser welding under vacuum is favored mainly for its ability of achieving a penetration depth several times the depth obtained under atmospheric conditions. Understanding how vacuum degree (ambient pressure) influences penetration depths in laser welding can help

S. Gong et al., *Weld Pool Dynamics in Deep Penetration Laser Welding*,
https://doi.org/10.1007/978-981-16-0788-2_9

design industrial vacuum chambers. This chapter mainly introduces the dynamical behaviors of the keyhole and weld pool in vacuum laser welding, and the effect of ambient pressure on penetration depths in laser welding.

9.2 Transient Coupling Model of Keyhole and Weld Pool in Vacuum

In recent years, experiments have shown that ambient pressure has significant effects on the evaporation process during laser welding. Therefore, ambient pressure must be considered to accurately calculate the force at the evaporation interface. For that reason, the authors have proposed a surface pressure model incorporating the factor of ambient pressure, based on the Soviet Union scientist Anismov's recoil pressure model. The proposed surface pressure model can accurately describe the pressure at the evaporation interface under variable ambient pressures, as shown in Fig. 9.1. It can be expressed as:

$$p_s = \begin{cases} p_{atm} & 0 \le T_{gb} \le T_L \\ \frac{1+\beta_R}{2} p_{atm} \exp\left(\frac{\Delta H_v}{k_B T_v}\left(1 - \frac{T_v}{T_{gb}}\right)\right) & +\infty > T_{gb} \ge T_R \\ p_c(T_{gb}) & T_L \le T_{gb} < T_R \end{cases} \tag{9.1}$$

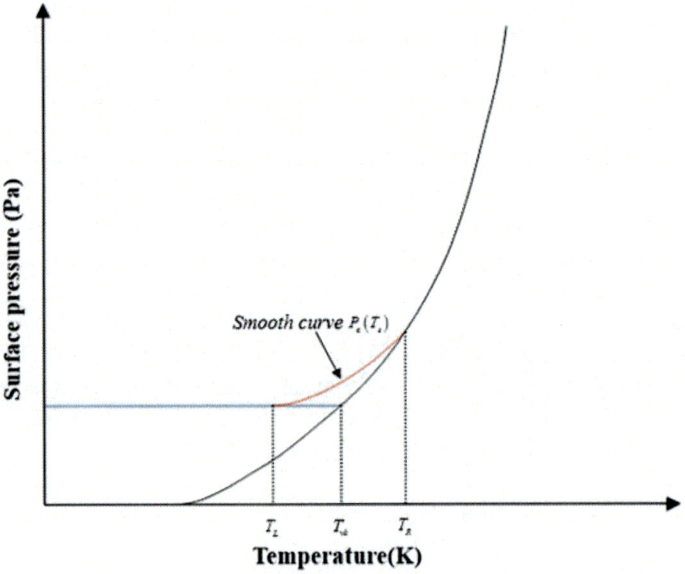

Fig. 9.1 Variation of surface pressure with evaporation interface temperature

The pressure at the free surface during laser welding under vacuum and atmospheric conditions can both be calculated using the surface pressure model. The coupling model of keyhole and weld pool dynamics in vacuum laser welding is similar to that in atmospheric laser welding.

9.2.1 Governing Equations in Vacuum Laser Welding

Similar to laser welding under atmospheric pressure, the mixture model is used to treat the solid–liquid interface. Besides, the fluid flow in weld pool is assumed to be incompressible, and the density of the fluid flow is assumed to change slightly during the solid–liquid phase transition. Thus, the mass conservation of any element on the workpiece (including the weld pool and the unmelted region) can be described as follows:

$$\nabla \cdot \vec{U} = 0 \tag{9.2}$$

where $\vec{U}$ is the three-dimensional velocity vector.

Considering the physical factors including the interface force, viscous force and buoyancy in the melting and solidification regions, the momentum conservation equation of the fluid flow of mixture phases can be expressed as:

$$\rho\left(\frac{\partial \vec{U}}{\partial t} + (\nabla \cdot \vec{U})\vec{U}\right) = \nabla \cdot (\mu_l \nabla \vec{U}) - \nabla p - \frac{\mu_l}{K}\vec{U} - \frac{C\rho}{\sqrt{K}}\left|\vec{U}\right|\vec{U} + \rho\vec{g}\beta(T - T_{ref}) \tag{9.3}$$

where

μ_l dynamic viscosity;
ρ density;
p pressure;
$\vec{g}$ three-dimensional gravitational acceleration vector;
β thermal expansion coefficient;
T_{ref} reference temperature;
K Carman-Kozeny coefficient of the mixture model, also known as permeability coefficient, which is closely related to the liquid fraction f_l in the current grid cell.

Based on the proposed physical model, considering the effect of convection and heat transfer of weld pool, the energy conservation of workpiece during welding can be expressed as:

$$\rho c_p\left(\frac{\partial T}{\partial t} + (\vec{U} \cdot \nabla)T\right) = \nabla \cdot (k\nabla T) \tag{9.4}$$

where

c_p specific heat;
k thermal conductivity;
T temperature.

The Level Set method is used to track the time-varying keyhole free surface. The governing equation of the Level Set method is written as:

$$\frac{\partial \phi}{\partial t} + \vec{U} \cdot \nabla \phi = 0 \tag{9.5}$$

9.2.2 Boundary Conditions in Vacuum Laser Welding

Given the influence of surface pressure, surface tension, and viscous pressure on keyhole free surface, the free surface is subject to a pressure boundary condition consistent with that in atmospheric laser welding. The pressure boundary condition can be expressed as follows:

$$p_f = p_r + \sigma\kappa + 2\mu\vec{n} \cdot \nabla\vec{U}\vec{n} \tag{9.6}$$

where

Subscript f free surface;
p_r recoil pressure;
σ surface tension coefficient;
κ free surface curvature;
μ dynamic viscosity;
$\vec{U}$ three-dimensional velocity vector;
$\vec{n}$ three-dimensional unit normal vector in free surface.

The free surface, is also subject to a viscous stress boundary condition due to the thermal capillary force similar to that in atmospheric laser welding. The viscous stress boundary condition can be expressed as:

$$\begin{aligned}(\mu\nabla\vec{U})_f = {} & \mu(\vec{n}\ \ \vec{t}_1\ \ \vec{t}_2)(\vec{n}\ \ 0\ \ 0)^T(\nabla\vec{U})(\vec{n}\ \ 0\ \ 0)(\vec{n}\ \ \vec{t}_1\ \ \vec{t}_2)^T \\ & + \mu(\vec{n}\ \ \vec{t}_1\ \ \vec{t}_2)(0\ \ \vec{t}_1\ \ \vec{t}_2)^T(\nabla\vec{U}) \\ & - \mu(\vec{n}\ \ \vec{t}_1\ \ \vec{t}_2)(\vec{n}\ \ 0\ \ 0)^T(\nabla\vec{U})^T(\vec{n}\ \ \vec{t}_1\ \ \vec{t}_2)(\vec{n}\ \ \vec{t}_1\ \ \vec{t}_2)^T \\ & + (\vec{n}\ \ \vec{t}_1\ \ \vec{t}_2)\begin{pmatrix} 0 & \nabla_s\sigma \cdot \vec{t}_1 & \nabla_s\sigma \cdot \vec{t}_2 \\ 0 & 0 & 0 \\ 0 & 0 & 0 \end{pmatrix}\end{aligned} \tag{9.7}$$

In addition, considering the Fresnel multiple reflection absorption, heat radiation, and evaporation on the free surface under the vacuum condition, the energy boundary condition can be determined as:

$$k\frac{\partial T}{\partial \vec{n}} = q - \varepsilon_r \sigma_s (T^4 - T_\infty^4) - \rho V_{\exp} T_v \quad (9.8)$$

On other boundaries of the calculated area, given the presence of only thermal radiation, the temperature boundary condition is expressed as:

$$k\frac{\partial T}{\partial \vec{n}} = -\varepsilon_r \sigma_s (T^4 - T_\infty^4) \quad (9.9)$$

In Eqs. (9.7) through (9.9):

Subscript f	keyhole free surface;
$\vec{t}_1, \vec{t}_2$	two unit tangent vectors of the free surface;
$\nabla_s \sigma$	Marangoni force;
k	thermal conductivity;
T	temperature of weld pool;
q	laser energy density absorbed by Fresnel multiple reflection absorption of keyhole wall;
T_∞	ambient temperature;
ε_r	black body radiation coefficient;
σ_s	Stefan-Boltzmann constant;
V_{evp}	keyhole surface recession speed due to the evaporation.

9.3 Behaviors of Keyhole and Weld Pool in Vacuum Laser Welding

9.3.1 Dynamical Keyhole Evolutions

9.3.1.1 Distribution of Keyhole Wall Temperature

This section describes the characteristics of the keyhole and weld pool in vacuum laser welding which is performed with 304 stainless steel at the laser power of 2.0 kW, the welding speed of 3.0 m/min, and the laser spot radius of 0.25 mm. In what follows, the analysis is based on these parameters, unless otherwise stated. Evolutions of the keyhole temperature field during vacuum laser welding are shown in Figs. 9.2 and 9.3. The temperature distribution across the keyhole is non-uniform, primarily between 2300 and 2500 K during welding process. The highest temperature, which is up to 2900 K, often occurs at the hump surfaces on the keyhole wall irradiated directly by the laser beam. In vacuum laser welding, the maximum keyhole temperature is

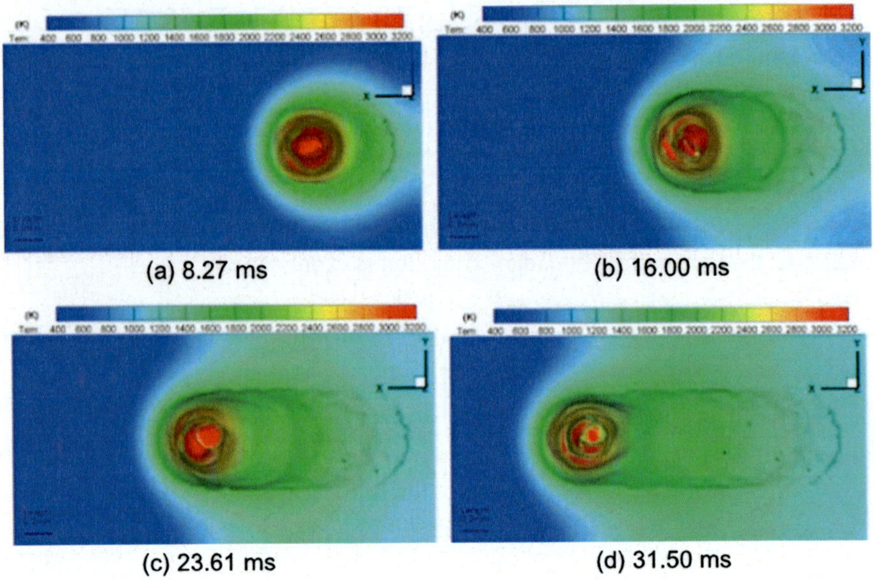

Fig. 9.2 Top view of the evolutions of transient keyhole temperature field during vacuum laser welding

significantly lower than the boiling point of the material at the atmospheric pressure (about 3100 K).

Evolutions of keyhole morphology and temperature field during laser welding under atmospheric pressure are shown in Figs. 9.4 and 9.5. The keyhole wall temperature is also uneven, generally around the boiling point (3100 K) of the material or even higher. The highest temperature, around 3400 K, occurs on the top of the humped keyhole wall, which is directly exposed to laser beam. In short, the average keyhole temperature is around the boiling point of the material or even higher during laser welding under atmospheric pressure. Those predicted results are well consistent with the recent experimental results. This study demonstrates the assumption used for more than 40 years that in laser welding under atmospheric pressure, the keyhole wall temperature is around the boiling point, and at the same time, denies the well-known assumption of Semak and Matsunawa that the keyhole temperature does not need to exceed the boiling point to support the opening of the keyhole during laser welding under atmospheric pressure.

9.3.1.2 Transient Keyhole Instability

Similar to laser welding under atmospheric pressure, laser welding under vacuum can also have its keyhole depth evolution divided into three stages: fast linear increase

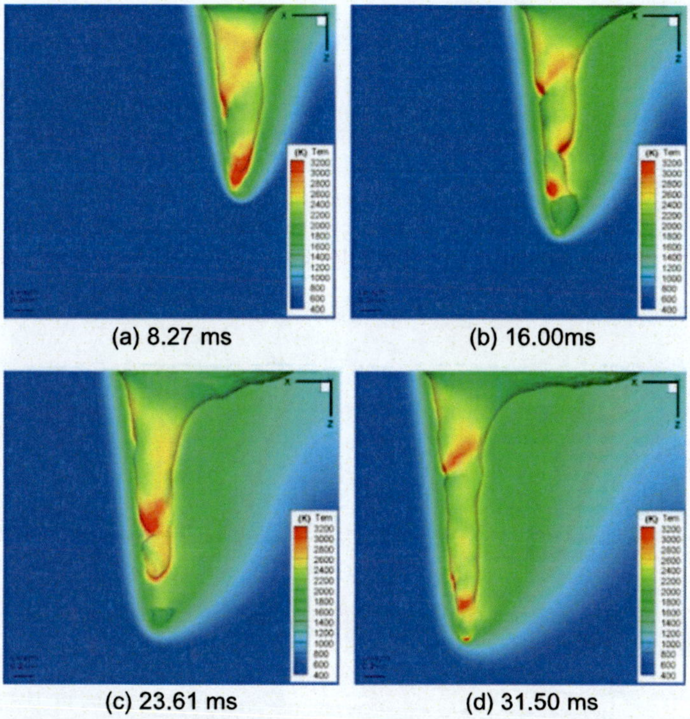

Fig. 9.3 Side view of evolutions of transient keyhole temperature field during vacuum laser welding

stage, slow increase stage with small amplitudes of depth oscillations, and quasi-steady stage with large amplitudes of depth oscillations.

As shown in Fig. 9.6, the keyhole under vacuum is much deeper than that under atmospheric pressure, and can reach 2.5 mm in depth, which is twice the depth (1.2 mm) under atmospheric pressure. Besides,the keyhole oscillation under vacuum is violent, with the oscillation amplitude up to 0.9 mm, compared with 0.5 mm under atmospheric pressure, as shown in Fig. 9.7. Seto et al. found that a deeper keyhole usually exhibits larger amplitude of keyhole depth oscillations using X-ray transmission imaging system. Although laser welding under vacuum has a larger absolute value of the amplitude of keyhole depth oscillation than laser welding under atmospheric pressure, the former has smaller relative amplitude of oscillation, namely, the ratio of keyhole depth oscillation amplitude to mean keyhole depth. As shown in Fig. 9.7b, the ratio for welding under vacuum is $0.9/2.4 = 0.375$, in contrast to $0.5/1.1 = 0.417$ under atmospheric pressure. Therefore, laser welding under vacuum delivers more stable keyhole than laser welding under atmospheric pressure, and also produces fewer spatters. To sum up, despite their difference in keyhole oscillation degree, the two welding methods share the same essential mechanisms for keyhole oscillation.

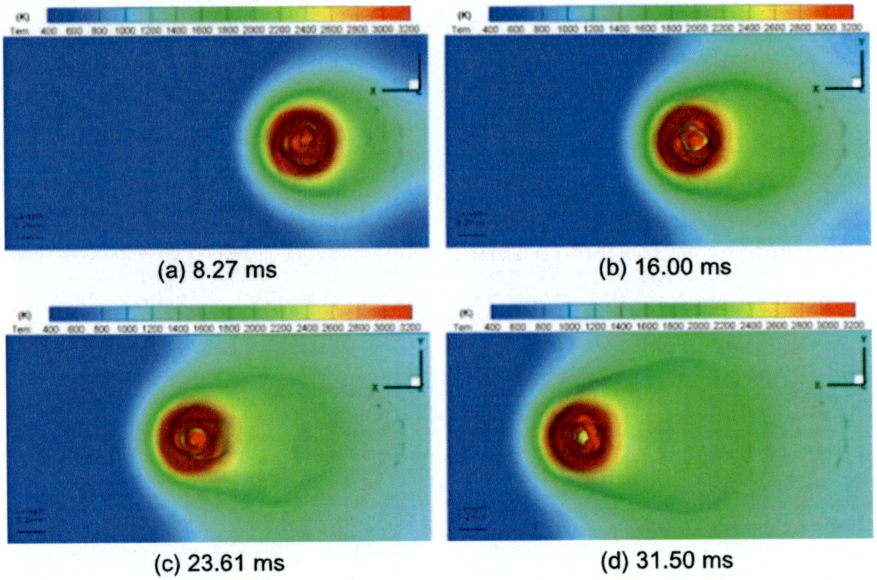

Fig. 9.4 Top view of evolutions of transient keyhole temperature field during laser welding under atmospheric pressure

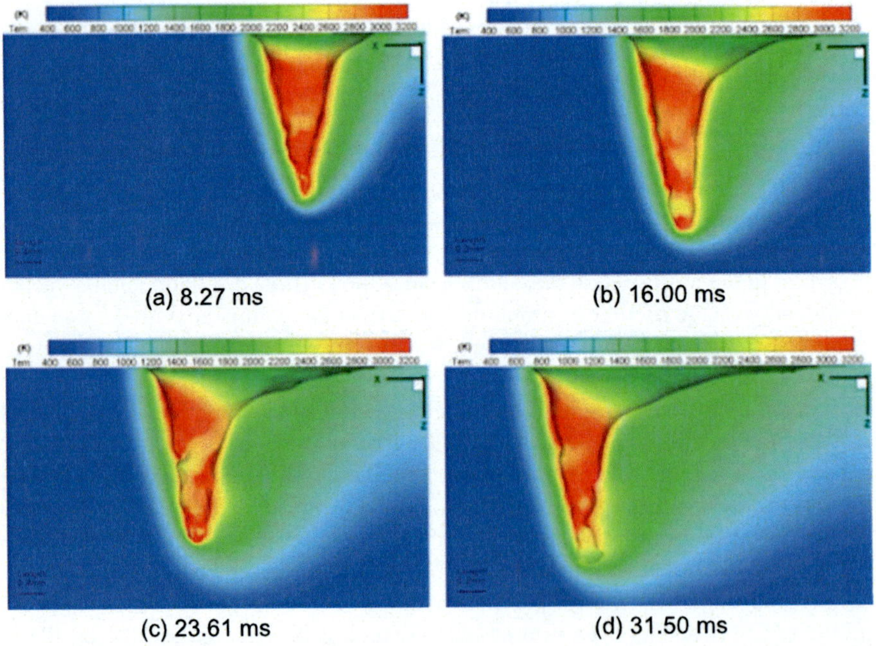

Fig. 9.5 Side view of evolutions of transient keyhole temperature field during laser welding under atmospheric pressure

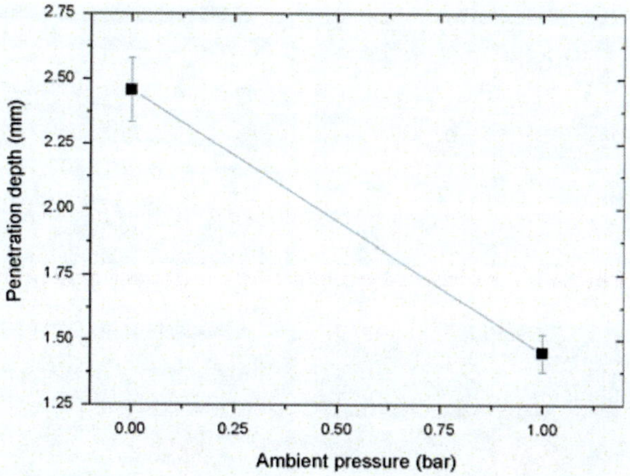

Fig. 9.6 Comparison of keyhole depths in laser welding under vacuum and under atmospheric pressure

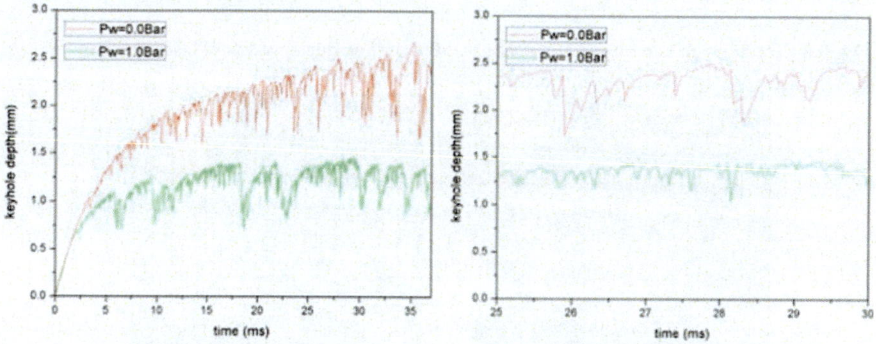

(a) from 0 ms to 35 ms

(b) from 25 ms to 30 ms

Fig. 9.7 Evolution curves of keyhole depths in laser welding under vacuum and under atmospheric pressure

9.3.2 Moving Weld Pool Flow Patterns

Evolutions of weld pool flow behaviors during vacuum laser welding are shown in Figs. 9.8 and 9.9. The flow behaviors, which are quite similar to those during laser welding under atmospheric pressure (see Figs. 9.10 and 9.11), are mainly as follows. In areas near the keyhole opening on the weld pool surface, driven by recoil pressure and Marangoni force, the fluid flows from the keyhole opening to the peripheral region of weld pool. These flows can be always observed during the welding process. The speed of these flows under vacuum is about 0.4 m/s, significantly smaller than the speed (about 1.5 m/s) under atmospheric pressure. Besides, high-speed downward flows driven by recoil pressure can be found near the keyhole wall. The magnitude

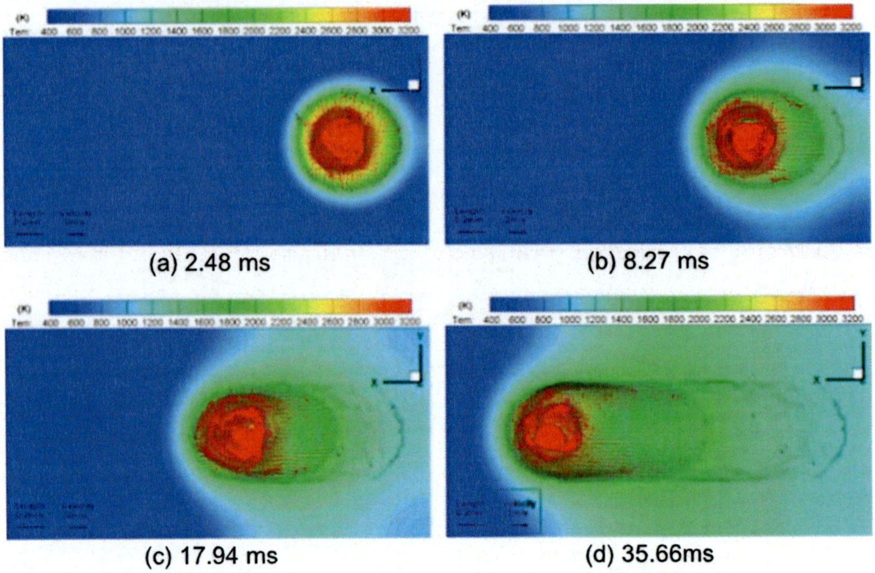

Fig. 9.8 Top view of evolutions of surface weld pool flow during vacuum laser welding

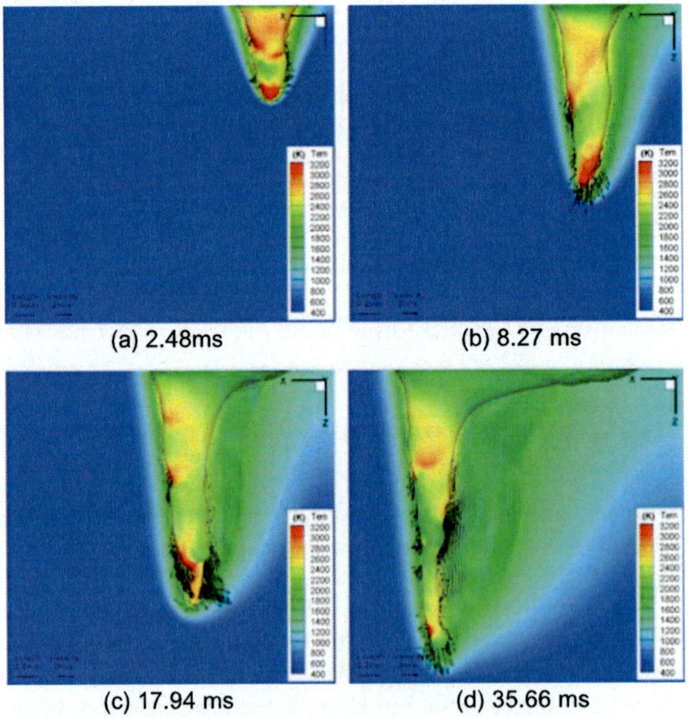

Fig. 9.9 Side view of inner weld pool flow during vacuum laser welding

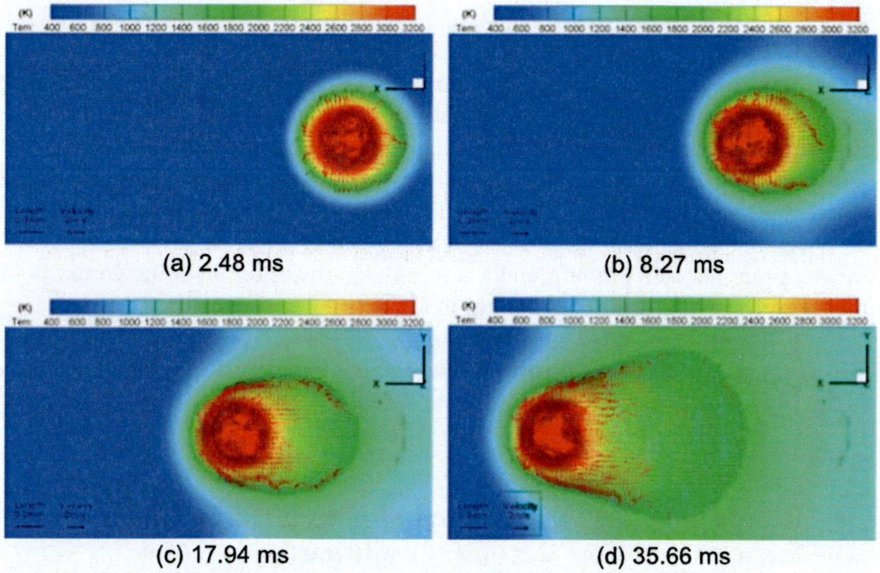

Fig. 9.10 Top view of evolutions of surface weld pool flow during laser welding under atmospheric pressure

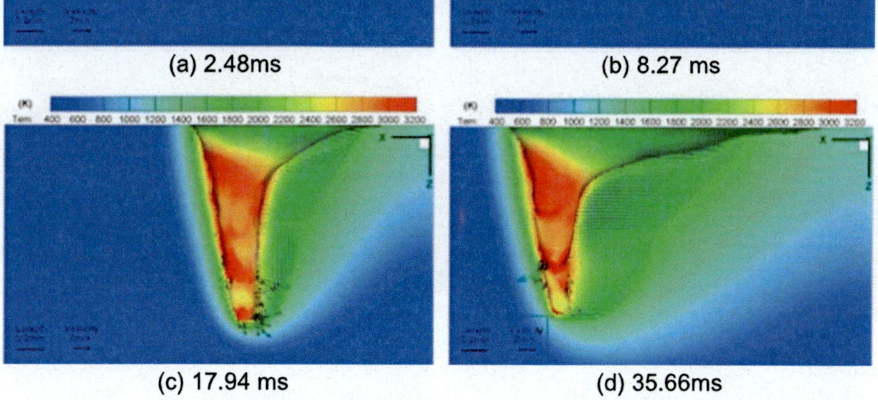

Fig. 9.11 Side view of inner weld pool flow during laser welding under atmospheric pressure

of these flows is more than 10 m/s, which is equivalent to that under atmospheric pressure (about 15 m/s). From the above discussions, it can be seen that the intensity of the flows in weld pool during vacuum laser welding, especially in surface weld pool, is more gentle and stable than that during laser welding under atmospheric pressure. This is in line with the experiments in recent years. Additionally, this also demonstrates the mechanism that vacuum laser welding could efficiently reduce or suppress the spatter defects. Moreover, the weld pool flow behaviors during vacuum laser welding differ from those during laser welding under atmospheric pressure. For example, in laser welding under atmospheric pressure, vigorous vortex flows are observed on the weld pool behind the bottom of the keyhole; in vacuum laser welding, however, such flows are not found in this area. The disappearance of this vortex flow is good for the floating of the bottom bubbles, which may reduce the possibility of the formation of the porosity defects.

9.4 Penetration Depth Increase in Laser Welding Under Vacuum and Low Vacuum Conditions

9.4.1 Mechanism for Penetration Depth Increase with Ambient Pressure Decrease

Both simulations and experiments (Fig. 9.12) found that the penetration depth in vacuum laser welding increased compared with that in atmospheric laser welding. Over the years, as to the mechanism for penetration depth increase in vacuum laser welding, there are mainly the following views: under vacuum condition, (1) the inverse bremsstrahlung (IB) absorption effect of plasma is reduced; (2) the scattering

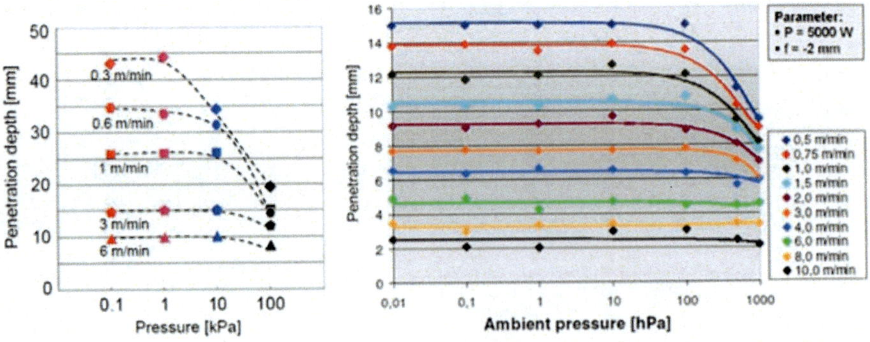

(a) Experimental results of Abe et al. (b) Experimental results of Börner et al.

Fig. 9.12 Experimental results of penetration depth curves in laser welding under different ambient pressures and refraction effects of metallic vapor plume are reduced; (3) the decrease of the average keyhole wall temperature leads to more laser energy acting on the keyhole wall in the penetration direction.

9.4.1.1 IB Absorption Effect of Plasma

During long wavelength laser welding (e.g. CO_2 laser with a wavelength of 10.6 μm) process, highly ionized laser induced plasma is produced. The induced plasma can absorb the long wavelength laser beam passed through. This will largely reduce the laser energy acting on the keyhole wall, causing an obvious decrease of penetration depth compared with that under atmospheric pressure. 10 kW CO_2 laser welding is used as an example (laser spot diameter: 0.5 mm). He is used to measure IB absorption coefficient using the spectral analysis method. The ideal gas equation and Saha equation can be described by the formulas (9.10) and (9.11), as follows:

$$P_{atm} = (N_0 + N_i + N_e)kT_e \approx N_0 kT_e \tag{9.10}$$

$$\frac{N_e N_i}{N_0} = C_1 T_e^{3/2} \exp\left(\frac{-E_i}{kT_e}\right) \tag{9.11}$$

Combine the above two formulas and substitute the parameters into the combined formula:

$$N_e = C_2 \sqrt{P_{atm}} T_e^{1/4} \exp\left(\frac{-(45.10)^3}{T_e}\right) \tag{9.12}$$

The electric density of He is shown in Fig. 9.13.

Thus, the IB absorption coefficient is obtained:

$$\alpha = C_3 N_e^2 \frac{\lambda^2}{T_e^{1/2}} \approx 0.4\,\text{cm}^{-1} \tag{9.13}$$

The IB absorption coefficient is about 20%, and even larger in the inner keyhole. Therefore, during the long wavelength laser welding (e.g. CO_2 laser welding) process, the IB absorption effect under atmospheric condition is very strong. This effect has been largely reduced during vacuum laser welding, allowing more laser energy to be used for the increase of penetration depth. As a result, the reduced IB absorption effect of plasma during vacuum laser welding is an important reason for the increase of penetration depth of the long wavelength (e.g. CO_2 laser) laser welding.

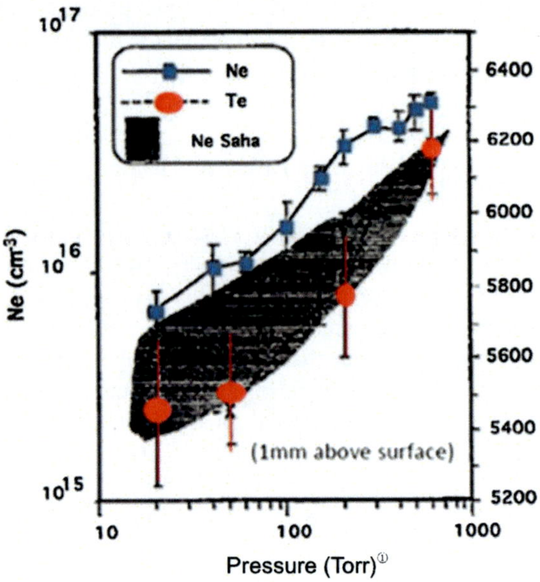

Fig. 9.13 Variation of He electric density with temperature and pressure (①1 Torr = 133.3 Pa.)

9.4.1.2 Scattering and Refraction Effects of Metallic Vapor Plume

According to the relationship between the IB absorption coefficient and the laser wavelength (Eq. (9.13)), a shorter laser wavelength indicates smaller absorption. Thus, the ionization of the metallic vapor in short wavelength laser (e.g. the commonly used fiber laser with a wavelength of 1.07 μm and YAG laser with a wavelength of 1.06 μm) welding is very small, even if the laser power reaches 10 kW during laser welding. The IB absorption coefficient is also correspondingly not obvious. However, with superfine particles from coagulated vapor or unmelted metals, metallic vapor plume exerts considerable scattering and refraction effects on laser beam, as shown in Fig. 9.14.

The scattering effect of metallic vapor plume mainly includes Rayleigh scattering and Mie scattering. The degree of the Rayleigh scattering is inversely proportional to the fourth power of the incident laser wavelength. Thus, the Rayleigh scattering mainly occurs in the short wavelength laser welding process. When the wavelength of the incident laser is equivalent to the size of the superfine particles inside the metallic vapor plume, the Mie scattering is quite obvious. Because the size of the superfine particles inside the metallic vapor plume in the keyhole cannot be accurately measured, the absorption and diffusion coefficients of the scattering are also difficult to be accurately obtained.

According to previous literature data, the attenuation of the high-power fiber laser energy is about 10–20% due to the scattering effect of metallic vapor plume. Under vacuum condition, the density of the metallic vapor plume decreases largely. The Rayleigh and Mie scattering are both not obvious. Thus, the penetration depth can

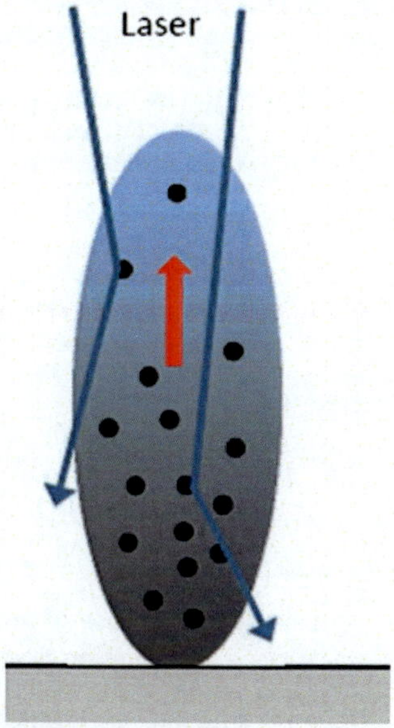

Fig. 9.14 Scattering and refraction effects of metallic vapor plume on short wavelength laser beam

increase by 10–20% compared with that under atmospheric condition. However, the penetration depth can double or even increase more during vacuum laser welding compared with that under atmospheric condition, according to the existing experimental and simulation data. Therefore, the scattering effect from metallic vapor plume is not the main reason for the penetration depth increase in short wavelength laser welding under vacuum.

9.4.1.3 Boiling Temperature Reduction Effect

Under vacuum conditions, the materials irradiated by laser beam are easier to be evaporated, and their boiling temperature decreases obviously. The Calusius-Clapeyron equation is used to calculate the pressure of metallic vapor when the gas–liquid phase is in equilibrium state:

$$P_{cc}(T_s) = P_0 \exp^{(B(1-T_v/T_s))} \tag{9.14}$$

where P_0 (=1 bar) is the metallic vapor pressure corresponding to evaporation temperature T_v under the standard atmospheric pressure. The evaporation temperature of iron (B = 13.57) is calculated using Eq. (9.14). A decrease of about 800 K occurs when the welding condition changes from the atmospheric pressure (1 bar) to the vacuum pressure (1 kPa). Fabbro et al. observed the free surface temperature of Ti alloys irradiated by laser under different ambient pressures, using an optical pyrometer. They found that the surface temperature is more than 3400 K (boiling point 3310 K) under the atmospheric condition (100 kPa). The surface temperature

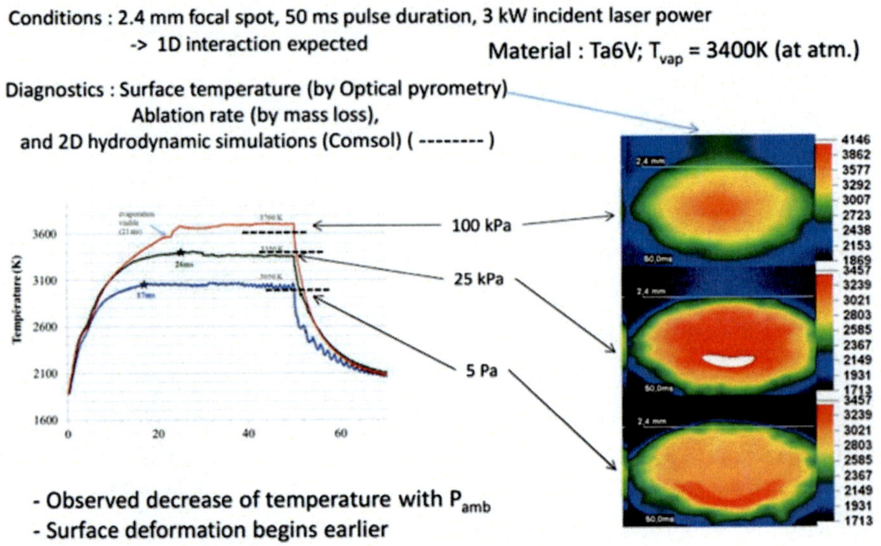

Fig. 9.15 Experimental results of the surface temperature distributions under different ambient pressures

decreases to about 3200 K as the ambient pressure is about 25 kPa. When the ambient pressure decreases to 5 kPa, the surface temperature is only about 2800 K, which is 600 K lower than that under atmospheric condition. This is consistent with the calculated results (Fig. 9.15).

Based on the 2D model $q = \frac{P}{L_{KH}} \approx \frac{6.3K(T_v - T_0)}{Ln(\frac{2.25\kappa}{V \cdot r})}$ put forward by Fabbro et al., when the welding speed V is 2 m/min and the radius r is 0.25 mm, the penetration depth in vacuum laser welding of iron increases by 40% compared with that in laser welding under atmospheric pressure. Additionally, by using the recent ambient pressure dependent surface pressure model (Eq. (9.1)) and the laser welding model of the computational fluid dynamics, simulations of the laser welding of 304 stainless steel are also made. When the laser power is 1.5 kW, the welding speed is 3 m/min, and the laser spot radius is 0.25 mm, the penetration depth in vacuum laser welding has nearly doubled compared with that in laser welding under atmospheric pressure, as shown in Fig. 9.16.

To sum up, the boiling temperature of the material decreases under vacuum, allowing more laser energy to act on the weld pool penetration direction. This demonstrates that the decrease of the boiling temperature during vacuum laser welding (especially with short wavelength) is the main reason for the increase of the penetration depth.

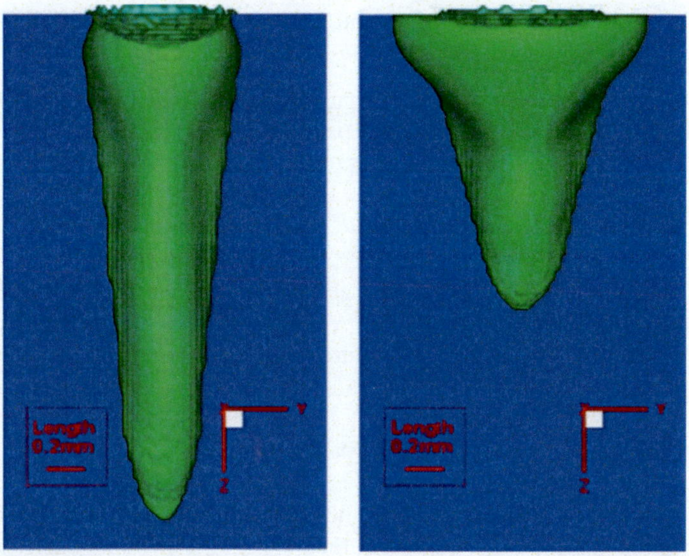

Fig. 9.16 Comparisons of weld pool cross section in fiber laser welding of 304 stainless steel under vacuum and atmospheric conditions

9.4.2 Finite Increase of Penetration Depth

In the experiments, if the ambient pressure falls below some critical ambient pressure (about 1 kPa as shown in Fig. 9.12) and continues to fall, the penetration depth almost stops increase. This shows that when the vacuum laser welding technology is used to obtain large penetration depth in industry, a lower vacuum degree of the vacuum chamber (or local vacuum equipment located at the processing area) does not mean a better result, but instead, the vacuum degree should reach an optimal theoretical value,. To obtain the optimal theoretical value during welding process, the physical phenomenon of the limited increase of penetration depth due to the reduction of ambient pressure is theoretically studied. To keep the keyhole open during laser welding, the total pressure P_0 that must exist inside the keyhole can be written as:

$$P_0 = P_{amb} + P_c + P_h + P(V) \tag{9.15}$$

where

P_{amb}	the ambient pressure;
$P_c = \sigma/r$	the closing pressure of the keyhole defined by the surface tension σ of the liquid and its radius r (which is of the order of magnitude of the focal spot radius);
$P_h = \rho_l g L_{KH}$	the maximum hydrostatic pressure resulting from a liquid thickness corresponding to the keyhole depth (ρ_l is the liquid density, g is the gravity, L_{KH} is the keyhole depth);

$P(V)$ a function that takes into account the displacement of the keyhole with a welding speed V and increases with it.

At low welding speed (1 m/min as shown in 9–12), the effect of $P(V)$ can be neglected. For usual values of different parameters involved in Eq. (9.15), one can also neglect $P_h(P_h \approx 1.6\,\text{kPa}$ for $L_{KH} \approx 20\,\text{mm})$. So, Eq. (9.15) becomes:

$$P_0 \approx P_{amb} + P_c \tag{9.16}$$

From the above equation, it is clear that the minimum pressure inside the keyhole that keeps it open is its closing pressure $P_c(=\sigma/r)$, which is obtained under complete vacuum. Hence, the lower limit of evaporation temperature inside the keyhole will be defined by this pressure P_c. So one can define a critical ambient pressure:

$$P_{CR} \approx P_c/10 \tag{9.17}$$

Below this critical ambient pressure, the effect of ambient pressure can be considered negligible in Eq. (9.16). Using this theory, the $P_{CR1}(\sigma \approx 2\,\text{N/m}, r \approx 0.2\,\text{mm})$ in No. 120 reference is about 1.0 kPa. It can be seen in Fig. 9.12a that the experimental saturated ambient pressure is about 1.0–2.0 kPa, which well agrees with the predicted result. Using the same method, the $P_{CR1}(\sigma \approx 2\,\text{N/m}, r \approx 0.2\,\text{mm})$ in No. 121 reference is about 1.2 kPa, while the experimental saturated ambient pressure is about 1.0 kPa as shown in Fig. 9.12b. To sum up, in laser welding under vacuum conditions, it is unnecessary to lower the degree of vacuum to a minimum possible value. A suitable vacuum degree can be calculated using the critical ambient pressure formula (Eq. (9.17)) to obtain a big penetration depth.

When the welding speed increases, the effect of $P(V)$ in Eqs. (9.16) and (9.17) must be considered. Thus, it is easy to know that the critical ambient pressure P_{CR} increases with the welding speed, though this behavior is not obvious. This phenomenon is shown in Fig. 9.12. Additionally, as shown in Fig. 9.17, the keyhole wall temperature increases as the welding speed increases. When the increments are the same, the variations of keyhole wall temperature under lower ambient pressure is larger. This demonstrates that in high-speed welding, the penetration depth is more sensitive to the welding speed under lower ambient pressure, which is consistent with the experimental results of Abe et al. and Börner et al. (Fig. 9.12) and the simulation results of Pang et al. (Fig. 9.18).

It is interesting to notice that a keyhole can be considered as a nozzle that ejects a vapor inside an ambient atmosphere. Based on the aerodynamic theory, the threshold condition for vapor to eject from the keyhole to the environment with ambient pressure of P_{amb} at a supersonic speed is given by the relation:

$$\frac{P_0}{P_{amb}} = K = \left(\frac{\gamma+1}{2}\right)^{(\gamma/\gamma-1)} \tag{9.18}$$

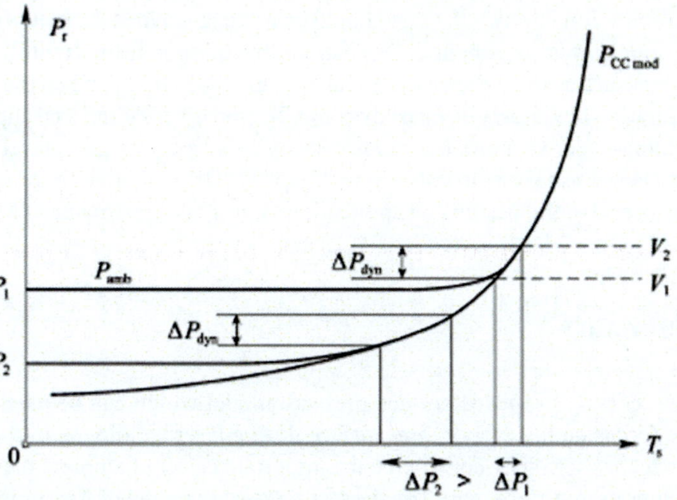

Fig. 9.17 Variation of keyhole wall temperature with welding speed under different ambient pressures

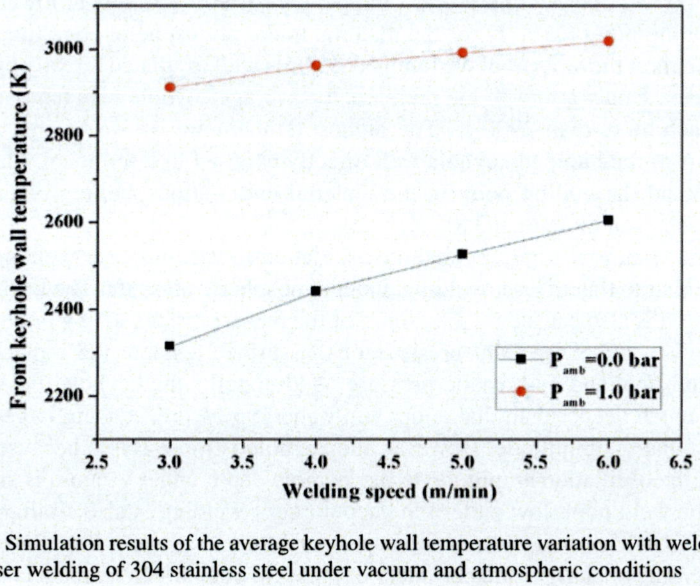

Fig. 9.18 Simulation results of the average keyhole wall temperature variation with welding speed in fiber laser welding of 304 stainless steel under vacuum and atmospheric conditions

where the specific heat ratio γ is 5/3, K is 2.05. The supersonic ejection condition can be determined from Eq. (9.16):

$$P_{amb} \leq P_c \tag{9.19}$$

If this condition is verified, there should be a shock wave at the exit of the keyhole and the vapor flow is then choked. One can see that this condition is fulfilled when the ambient pressure is lower than the critical one (Eq. (9.17)), i.e., when the penetration depth saturation previously discussed occurs. To sum up, for industrial applications of laser welding under vacuum, it is not necessary to have a very good level of vacuum, as in the case of usual electron-beam welding. The critical ambient pressure theory should be useful in the design and application of proper industrial vacuum chambers.

9.5 Summary

(1) The recently proposed surface pressure model which can be used to describe the forces on the evaporation surface during laser welding is introduced. This model has seriously considered the important effect of ambient pressure on the evaporation surface, based on the recoil pressure model of Anismov. Therefore, the surface pressure model can be used to calculate the surface pressure on the evaporation surface in laser welding under various ambient pressures.

(2) The keyhole wall temperature in vacuum laser welding generally ranges from 2300 to 2500 K. The highest temperature (2900 K), 200 K lower than the boiling point (3100 K) of the material under atmospheric conditions, usually occurs at the top part of the humped keyhole wall irradiated directly by the laser beam. Under atmospheric pressure, the average keyhole wall temperature can reach more than 2900 K. The highest temperature is also located at the top part of the humped keyhole wall directly exposed to laser beam. The value is around the boiling point of the material under atmospheric pressure, or can reach even more than 3400 K.

(3) The feature of transient keyhole oscillation in vacuum laser welding is much similar to that in laser welding under atmospheric pressure. The keyhole oscillation is also caused by the failure of the recoil pressure acting on the keyhole wall to offset the joint pressure of the surface tension, the impact force of fluid flow and hydrostatic pressure. Additionally, the keyhole under vacuum is much deeper than that under atmospheric pressure, causing larger absolute oscillation amplitude. However, the keyhole is more stable because the ratio of the oscillation amplitude to the keyhole depth under vacuum is smaller.

(4) The weld pool flow pattern in vacuum laser welding is also similar to that in laser welding under atmospheric pressure. Specifically, the fluid flows from the keyhole opening to the periphery of the weld pool driven by the recoil pressure and Marangoni force, and flows rapidly downward near the keyhole wall driven by the recoil pressure. However, the fluid flow in weld pool in vacuum laser welding is more moderate than that under atmospheric pressure. The speed of flow in surface weld pool under vacuum is only about 0.5 m/s, compared to 1.5 m/s under atmospheric pressure. This could be the important reason that the spatter defect is reduced in laser welding under vacuum. Additionally, compared with the violent vortex at the keyhole bottom in the rear of weld pool

under atmospheric pressure, there is no obvious vortex under vacuum, which is good for the reduction of porosity defect.

(5) The penetration depth increases obviously under vacuum compared with that under atmospheric pressure. The mechanisms proposed recently for penetration depth increase are systematically introduced and clarified: the boiling temperature of materials under vacuum decreases obviously, which allows more laser energy to be used in the penetration direction; this is the main reason for the increase of the penetration depth; the inverse bremsstrahlung absorption effect mainly takes an effect on the long wavelength laser welding (e.g. CO_2 laser welding); the scattering and fraction of metal vapor plume is an important factor in penetration depth increase in the short wavelength laser welding under vacuum, especially in the high-power laser welding; but this important factor only accounts for only 10–20% of penetration depth increase and is thus not the main reason for the increase.

(6) The penetration depth will not continue to increase as the ambient pressure decreases to a certain value. The physical phenomena and mechanism of the finite increase of penetration depth under reduced ambient pressure are introduced and clarified. The critical ambient pressure theory and formula for the vacuum laser welding process are also introduced. These findings provide guidance for the design of industrial vacuum chambers, and promote the development and industrial application of vacuum laser welding technology.

Bibliography

1. W. Wang, L. Zhu, X. Shang, A discussion on the testing technology of welding temperature field. Phys. Exam. Test. **6**, 25–28 (1998)
2. D.T. Swift-Hook, A.E.F. Gick, Penetration welding with laser. Weld. J. **52**, 492–499 (1973)
3. T. Chande, J. Mazumder, Estimating effects of processing conditions and variable properties upon pool shape, cooling rates and absorption coefficient in laser welding. Appl. Phys. **56**(7), 1981–1986 (1984)
4. S. Kou, Weld pool convection and its effect. Weld. J. **65**(3), 63–70 (1986)
5. S. Kou, Computer simulation of convection in moving arc weld pools. Metall. Trans. **17A**(12), 2271–2277 (1986)
6. A. Duan, *A Study on the Characteristics and Related Mechanism of Unstable Perforation Process of CO_2Deep Penetration Laser Welding* (Huazhong University of Science and Technology (HUST), Wuhan, 2006)
7. J. Dowden, The flow of heat and the motion of weld pool in penetration welding with a laser. Appl. Phys. **57**(9), 4474–4479 (1985)
8. J. Dowden, M. Davis, P. Kapadia, Molten-region temperature distribution in laser welding. Phys. D: Appl. Phys. **18**, 1987–1993 (1985)
9. J.A. Goldak, M.J. Bibby, J.E. Moore et al., Computer modeling of heat flow in welds. Met. Trans. B **17**(B), 587–600 (1986)
10. O. lchiko, N. Hamada, H. Soga, Development of the simulation model for 15 kW CO_2 laser materials processing, in *Laser Advanced Materials Processing-Science and Application (LAMP'87)*, Osaka, Japan (1987), pp. 31–36
11. P.S. Wei, M.D. Shian, Three dimensional analytical temperature field around the welding cavity produced by a moving distributed high-intensity beam. ASME J. Heat Transf. **115**, 848–856 (1993)
12. P.S. Wei, M.D. Shian, Three dimensional analytical temperature field and its application to solidification characteristics in high-of-low-power-density-beam welding. Heat Mass Transf. **40**(10), 2283–2292 (1997)
13. L. Liu, M. Chi, R. Huang et al., Infrared measurement and numerical simulation of welding temperature field of magnesium alloy AZ31. Sci. China (Ser. E) **36**(1), 29–38 (2006)
14. R. Huang, L. Liu, M. Chi, Infrared measurement and numerical simulation of TIG welding temperature field of magnesium alloy laser. Trans. China Weld. Inst. **27**(10), 89–93 (2006)
15. Y. Lei, B. Zhu, J. Wang et al., Three-dimensional dynamic simulation and infrared temperature measurement of temperature field of tungsten argon arc welding. J. Jiangsu Norm. Univ. (Nat. Sci. Ed.) **29**(4), 308–311 (2008)
16. A. Duan, L. Chen, Characteristics of keyhole and molten pool during laser welding of TC4 Ti-alloy. Laser Eng. **22**(5–6), 361–369 (2011)
17. S.V. Patankar, *Numerical Heat Transfer and Fluid Flow* (trans. Z. Zheng) (Science Press, Beijing, 1989), pp. 12–16

S. Gong et al., *Weld Pool Dynamics in Deep Penetration Laser Welding*,
https://doi.org/10.1007/978-981-16-0788-2

18. W. Tao, *Numerical Heat Transfer*, 2nd edn. (Xi'an Jiaotong University Press, Xi'an, 2001)
19. K. Pericleous, C. Bailey, Study of marangoni phenomena in laser melted pools, in *Modeling of Casting, Welding and Advanced Solidification Processes*, vol. VII (1995), pp. 91–100
20. G.A. Taylor, H. Michael, S. Nadia, P. Koulis, Finite volume methods applied to computational modeling of welding phenomena, in *Second International Conference on CFD in the Minerals and Process Industries* (CSIRO, Melbourne, 1999), pp. 6–8
21. S. Wu, H. Zhao, Y. Wang et al., New heat source model in numerical simulation of high energy beam welding, **25**(1), 91–94 (2004)
22. J. Goldak, A. Chakravarti, M. Bibby, A new finite model for welding heat source. Metall. Trans. B **15B**, 299–305 (1984)
23. H. Wang, *A Study on Fluid Dynamics of Deep Penetration Laser Welding Process* (Beijing University of Technology, Beijing, 2007)
24. J. Goldak, M. Bibby, J. Moore et al., Computer modeling of heat flow in welds. Metall. Trans. B **17B**, 587–600 (1986)
25. R. Courant, E. Issacson, M. Rees, On the solution of non-linear hyperbolic differential equations by finite differences. Pure Appl. Math. **5**, 243–269 (1952)
26. R.A. Gentry, R.E. Martin, B.D. Daly, An Eulerian differencing method for unsteady compressible flow problems. J. Comput. Phys. **1**, 87–91 (1966)
27. H.Z. Barakat, J.A. Clark, Analytical and experimental study of transient laminar nature convection flows in partially filled containers, in *Proceedings of Third International Heat Transfer Conference*, vol. 2 (1966), pp. 87–91
28. A.K. Runchal, M. Wolfshtein, Numerical integration procedure for the steady state Navier-Stokes equations. J. Mech. Eng. Sci. **11**, 445–453 (1969)
29. P.J. Roache, *Computational Fluid Dynamics* (trans. X. Zhong, X. Liu) (Science Press, Beijing, 1983)
30. C. Chen, *Computational Fluid Mechanics* (Chongqing Publishing House, Chongqing, 1984)
31. K. Kuwahara, A novel finite difference method for flow simulation and visualization, in *Proceedings of the Symposium of Energy Engineering in the 21th Century*, vol. 1, ed. by P. Chen (2000), pp. 149–160
32. B.P. Leonard, A survey of finite differences with upwinding for numerical modeling of the incompressible convective diffusion equations. Comput. Tech. Transient Turbul. Flows **2** (1981)
33. B.P. Leonard, A stable and accurate convective modeling procedure based on quadratic upstream interpolation. Comput. Mech. Appl. Mech. Eng. **29**, 59–98 (1979)
34. W. Tao, *Progress of Computational Heat Transfer* (Science Press, Beijing, 2000)
35. C.J. Freitas, Prospective: selected benchmarks from commercial CFD codes. J. Fluids Eng. **117**, 208–218 (1995)
36. F.H. Harlow, J.E. Welch, Numerical calculation of time-dependent viscous incompressible flow of fluid with free surface. Phys. Fluids **8**(12), 2182–2189 (1965)
37. L.S. Caretto, R.M. Curr, D.B. Spalding, Two numerical methods for three-dimensional boundary layers. Comput. Methods Appl. Mech. Eng. **1**, 39 (1972)
38. L.S. Caretto, A.D. Gosman, S.V. Patankar et al., Two calculation procedures for steady, three-dimensional flow with recirculation, in *Proceedings of the Third International Conference on Numerical Methods for Fluid Dynamics*, Paris, vol. 11 (1972), p. 60
39. S.V. Patankar, D.B. Spalding, A calculation procedures for heat mass and momentum transfer in three-dimensional parabolic flows. Int. J. Heat Mass Transf. **116**, 369–380 (1994)
40. L. Zhang, *Study on the Forming Mechanism of Metallic Material Deep Penetration Laser Welding* (Xi'an Jiaotong University, Xi'an, 2008)
41. S.V. Patankar, D.B. Spalding, A computer model for three-dimensional flow in furnaces, in *Proceedings of 14th Symposium on Combustion* (1973), p. 605
42. D. John, J.R. Anderson, *Computational Fluid Application: The Basis with Application* (Tsinghua University Press, Beijing, 2002)
43. S.V. Patankar, *Numerical Heat Transfer and Fluid Flow* (Hemisphere, New York, 1980), p. 202

44. K.C. Mills, Recommended values of thermophysical properties for selected commercial alloys, in *ASME International* (2002), p. 189
45. C.J. Smithells, E.A. Brandes, *Metals Reference Book*, 7th edn. (Reed Educational and Professional Publishing Ltd., 1992)
46. J.M. Jouvard, K. Girard, O. Perret, Keyhole formation and power deposition in Nd: YAG laser spot welding. J. Phys. D: Appl. Phys. **34**, 2894–2901
47. Z. Wenhai, Modeling the formation and collapse of a keyhole during laser welding process. *Ph.D. thesis* (University of Missouri-Rolla, 2002), pp. 125–127
48. M. Mizutani, S. Katayama, Keyhole behavior and pressure distribution during laser irradiation on molten metal, in *ICALEO*, Section A (2003), pp. 149–167
49. S. Katayama, N. Seto, J. Kim et al., Formation mechanism and suppression procedure of porosity in high power laser welding of aluminum alloys, in *ICALEO*, Section C (1998), pp. 24–33
50. N. Seto, S. Katayama, A. Matsunawa, A high-speed simultaneous observation of plasma and keyhole behavior during high power CO_2 laser welding, in *ICALEO*, Section E (1999), pp. 19–27
51. A. Matsunawa, S. Katayama, J. Kim et al., Dynamics of keyhole and molten pool in high power laser welding. J. Laser Appl. **10**(6), 247–254 (1998)
52. N. Seto, A. Matsunawa, S. Katayama et al., High-speed simultaneous observation of plasma and keyhole behavior during high power CO_2 laser welding: effect of shielding gas on porosity formation. J. Laser Appl. **12**(6), 245–250 (1998)
53. N. Seto, S. Katayama, M. Mizutani et al., Formation mechanism of porosity in high power YAG laser welding, in *ICALEO*, Section A (2000), pp. 16–25
54. C. Tix, G. Simon, A transport theoretical model of the keyhole plasma in penetration laser welding. J. Phys. D: Appl. Phys. **26**, 2066–2074 (1993)
55. J. Wang, X. Wang, J. Shen, Calculative pattern of laser welding stability for titanium alloys. Trans. China Weld. Inst. **21**(4), 13–16 (2000)
56. J. Greses, P.A. Hilton, G.Y. Barlow et al., Plume attenuation under high power Nd: yttrium-aluminum-gamet laser welding. J. Laser Appl. **16**(1), 9–15 (2004)
57. D. Lacroix, G. Jeandel, C. Boudot, Solution of the radiative transfer equation in an absorbing and scattering Nd: YAG laser-induced plume. J. Appl. Phys. **84**, 2443 (1998)
58. J. Zhou, H.L. Tsai, T.F. Lehnhoff, Investigation of transport phenomena and defect formation in pulsed laser keyhole welding of zinc-coated steels. J. Phys. D: Appl. Phys. **39**(24), 5338–5355 (2006)
59. S. Pang, *A Study on the Transient Keyhole and Moving Weld Pool Behaviors and Mechanisms of Deep Penetration Laser Welding* (Huazhong University of Science and Technology (HUST), Wuhan, 2011)
60. H. Ki, P.S. Mohanty, J. Mazumder, Modeling of laser keyhole welding: Part I. Mathematical modeling, numerical methodology, role of recoil pressure, multiple reflections, and free surface evolution. Metall. Mater. Trans. A **33**, 1817–1830 (2001)
61. H. Ki, P.S. Mohanty, J. Mazumder, Modeling of laser keyhole welding: Part II. Simulation of keyhole evolution, velocity, temperature profile, and experimental verification. Metall. Mater. Trans. A **33**, 1831–1842 (2001)
62. J.H. Cho, S.J. Na, Three-dimensional analysis of molten pool in GM A-laser hybrid welding. Weld. J. **88**(2), 35–43 (2009)
63. J.-H. Cho, S.-J. Na, Implementation of real-time multiple reflection and fresnel absorption of laser beam in keyhole. J. Phys. D: Appl. Phys. **39**(24), 5372–5378 (2006)
64. S. Osher, R. Fedkiw, *Level Set Methods and Dynamic Implicit Surfaces* (Springer, Berlin, 2002)
65. A. Matsunawa, V.Y. Semak, The simulation of front keyhole wall dynamics during laser welding. J. Phys. D: Appl. Phys. **30**(5), 798–809 (1997)
66. V. Semak, A. Matsunawa, The role of recoil pressure in energy balance during laser materials processing. J. Phys. D: Appl. Phys. **30**(18), 2541–2552 (1997)

67. H. Zhao, A fast sweeping method for Eikonal equations. Math. Comput. **74**(250), 603–627 (2005)
68. M. Sussman, E.G. Pucket, A coupled level set and volume-of-fluid method for computing 3D axisymmetric incompressible two-phase flows. J. Comput. Phys. **162**(2), 301–337 (2000)
69. D. Enright, R. Fedkiw, J. Ferziger et al., A hybrid particle level set method for improved interface capturing. J. Comput. Phys. **183**(1), 83–116 (2002)
70. B.D. Nichols, C.W. Hirt, R.S. Hotchkiss, *SOLA-VOF: A Solution Algorithm for Transient Fluid Flow with Multiple Free Boundaries* (Los Alamos Scientific Lab, 1980)
71. A.J. Chorin, Numerical solution of the Navier-Stokes equations. Math. Comput. **22**(104), 745–762 (1968)
72. D.S. Kershaw, The incomplete Cholesky—conjugate gradient method for the iterative solution of systems of linear equations. J. Comput. Phys. **26**(1), 43–65 (1978)
73. S. Fujinaga, H. Takenaka, T. Narikiyo et al., Direct observation of keyhole behaviour during pulse modulated high-power Nd: YAG laser irradiation. J. Phys. D: Appl. Phys. **33**(5), 492–497 (2000)
74. A. Matsunawa, J.D. Kim, N. Seto et al., Dynamics of keyhole and molten pool in laser welding. J. Laser Appl. **10**(6), 247–254 (1998)
75. M. Mizutani, S. Katayama, Keyhole behavior and pressure distribution during laser irradiation on molten metal, in *ICALEO 2003*, Jacksonville, FL, United States (2003)
76. J. Kroos, U. Gratzke, M. Vicanek et al., Dynamic behaviour of the keyhole in laser welding. J. Phys. D: Appl. Phys. **26**(3), 481–486 (1993)
77. T. Klein, M. Vicanek, J. Kroos et al., Oscillations of the keyhole in penetration laser beam welding. J. Phys. D: Appl. Phys. **27**(10), 2023–2030 (1994)
78. T. Klein, M. Vicanek, G. Simon, Forced oscillations of the keyhole in penetration laser beam welding. J. Phys. D: Appl. Phys. **29**(2), 322–332 (1996)
79. J.Y. Lee, H.K. Sung, D.F. Farson et al., Mechanism of keyhole formation and stability in stationary laser welding. J. Phys. D: Appl. Phys. **35**(13), 1570–1576 (2002)
80. S. Katayama, M. Mizutani, A. Matsunawa, Development of porosity prevention procedures during laser welding, in *Proceedings of SPIE*, Osaka, Japan (2002), pp. 281–288
81. S. Katayama, Y. Kawahito, Elucidation of phenomena in high power fiber laser welding, development of prevention procedures of welding defects, in *Proceedings of SPIE*, vol. 7195 (2009), p. 71951R
82. V.V. Semak, W.D. Bragg, B. Damkroger et al., Transient model for the keyhole during laser welding. J. Phys. D: Appl. Phys. **32**(15), L61–L64 (1999)
83. H. Ki, P.S. Mohanty, J. Mazumder, Modeling of laser keyhole welding: Part I. Mathematical modeling, numerical methodology, role of recoil pressure, multiple reflections, and free surface evolution. Metall. Mater. Trans. A **33**, 1817–1830 (2001)
84. H. Ki, P.S. Mohanty, J. Mazumder, Modeling of laser keyhole welding: Part II. Simulation of keyhole evolution, velocity, temperature profile, and experimental verification. Metall. Mater. Trans. A **33**, 1831–1842 (2001)
85. X. Jin, P. Berger, T. Graf, Multiple reflections and Fresnel absorption in an actual 3D keyhole during deep penetration laser welding. J. Phys. D: Appl. Phys. **39**(21), 4703–4712 (2006)
86. R. Fabbro, S. Slimani, F. Coste et al., Study of keyhole behaviour for full penetration Nd-Yag CW laser welding. J. Phys. D: Appl. Phys. **38**(12), 1881–1887 (2005)
87. S. Pang, A three-dimensional sharp interface model for self-consistent keyhole and weld pool dynamics in deep penetration laser welding. J. Phys. D: Appl. Phys. **44**(2), (2011)
88. H. Ki, J. Mazumder, P.S. Mohanty, Modeling of laser keyhole welding: Part II. Simulation of keyhole evolution, velocity, temperature profile, and experimental verification. Metall. Mater. Trans. A **33**(6), 1831–1842 (2002)
89. J. Zhou, H.L. Tsai, T.F. Lehnhoff, Investigation of transport phenomena and defect formation in pulsed laser keyhole welding of zinc-coated steels. J. Phys. D: Appl. Phys. **39**(24), 5338 (2006)
90. N.S. Hosseini Motlagh, P. Parvin, M. Jandaghi et al., The influence of different volume ratios of He and AT in shielding gas mixture on the power waste parameters for Nd: YAG and CO_2 laser welding. Opt. Laser Technol. **54**, 191–198 (2013)

91. M. Kang, R.P. Fedkiw, X.D. Liu, A boundary condition capturing method for multiphase incompressible flow. J. Sci. Comput. **15**(3), 323–360 (2000)
92. J.Y. Lee, S.H. Ko, D.F. Farson et al., Mechanism of keyhole formation and stability in stationary laser welding. J. Phys. D: Appl. Phys. **35**, 1570–1576 (2002)
93. S. Pang, K. Hirano, R. Fabbro et al., Explanation of penetration depth variation during laser welding under variable ambient pressure. J. Laser Appl. **27**(2), (2015)
94. H. Ki, J. Mazumder, P.S. Mohanty, Modeling of laser keyhole welding: Part I. Mathematical modeling, numerical methodology, role of recoil pressure, multiple reflections, and free surface evolution. Metall. Mater. Trans. A **33**(6), 1817–1830 (2002)
95. W. Tan, N.S. Bailey, Y.C. Shin, Investigation of keyhole plume and molten pool based on a three-dimensional dynamic model with sharp interface formulation. J. Phys. D: Appl. Phys. **46**(5), (2013)
96. E.H. Amara, A. Bendib, Modelling of vapour flow in deep penetration laser welding. J. Phys. D: Appl. Phys. **35**(3), 272 (2002)
97. A.K. Dasgupta, J.L. Mazumder, Physics of zinc vaporization and plasma absorption during CO_2 laser welding. J. Appl. Phys. **102**(5), (2007)
98. M. Zhang, *Study on the Behavior of Metallic Vapor Plume and Defects Control During Deep Penetration Laser Welding of Thick Plate Using 10-kW Level High Power Fiber Laser* (Hunan University, Changsha, 2013)
99. A. Kaplan, A model of deep penetration laser welding based on calculation of the keyhole profile. J. Phys. D: Appl. Phys. **27**(9), 1805–1814 (1994)
100. Y. Su, *A Study on the Thermal Efficiency in Laser Welding* (Beihang University, Beijing, 2000)
101. W.W. Duley, *Laser Welding* (Willy, New York, 1999)
102. M.H. Glowacki, The effects of the use of different shielding gas mixtures in laser welding of metals. J. Phys. D: Appl. Phys. **28**(10), 2051–2059 (1995)
103. F. Remy, S. Sonia, D. Ivan et al., Experimental study of the dynamical coupling between the induced vapour plume and the melt pool for Nd-Yag CW laser welding. J. Phys. D: Appl. Phys. **39**(2), 394–400 (2006)
104. K. Kenichi, I. Takashi, Y. Kouzou et al., Prevention of welding defect by side gas flow and its monitoring method in continuous wave Nd: YAG laser welding. J. Laser Appl. **14**(3), 136–145 (2002)
105. L. Zhang, J. Zhang, R. Wang et al., Effects of side assist gas on the CO_2 laser welding process of thin stainless plate. Appl. Laser **25**(4), 217–221 (2005)
106. L. Zhang, J. Zhang, R. Wang et al., Effects of side assist gas on weld formation in laser welding of thin stainless plate. Rare Metal Mater. Eng. **35**(Suppl. 2), 39–44 (2006)
107. H. Ki, J. Mazumder, P.S. Mohanty, Modeling of laser keyhole welding: Part I. Mathematical modeling, numerical methodology, role of recoil pressure, multiple reflections, and free surface evolution. Metall. Mater. Trans. A **33**, 1817–1830 (2002)
108. J. Zhou, H.L. Tsai, T.F. Lehnhoff, Investigation of transport phenomena and defect formation in pulsed laser keyhole welding of zinc-coated steels. J. Phys. D: Appl. Phys. **39**(24), 5338–5355 (2006)
109. S. Osher, R. Fedkiw, *Level Set Methods and Dynamic Implicit Surfaces* (Springer, Berlin, 2002)
110. A. Matsunawa, Y.V. Semak, The simulation of front keyhole wall dynamics during laser welding. J. Phys. D: Appl. Phys. **30**(5), 798–809 (1997)
111. V. Semak, A. Matsunawa, The role of recoil pressure in energy balance during laser materials processing. J. Phys. D: Appl. Phys. **30**(18), 2541–2552 (1997)
112. H. Ki, P.S. Mohanty, J. Mazumder, Modeling of laser keyhole welding: Part I. Mathematical modeling, numerical methodology, role of recoil pressure, multiple reflections, and free surface evolution. Metall. Mater. Trans. A **33**, 1817–1830 (2001)
113. H. Ki, P.S. Mohanty, J. Mazumder, Modeling of laser keyhole welding: Part II. Simulation of keyhole evolution, velocity, temperature profile, and experimental verification. Metall. Mater. Trans. A **33**, 1831–1842 (2001)
114. J. Xie, Dual beam laser welding. Weld. J. **81**, 223–230 (2002)

115. J. Xie, Weld morphology and thermal modeling in dual-beam laser welding. Weld. J. **81**, 283–290 (2002)
116. T. Iwase, H. Sakamoto, K. Shibata et al., Dual-focustechniqne for high-power Nd: YAG laser welding of aluminum alloys, in *Proceedings of the International Conference on High-Power Lasers in Manufacturing*, vol. 3888 (SPIE, Osaka, 2000), pp. 348–358
117. S. Mishara, S. Chakraborty, T. DebRoy, Probing liquation cracking and solidification through modeling of momentum, heat, solute transport during welding of aluminum alloy. J Appl. Phys. **97**, (2005)
118. Y. Yu, *Study on the Technology and Filler Wire Melting Dynamics during the Laser Welding with Filler Wire [D]* (Huazhong University of Science and Technology (HUST), Wuhan, 2010)
119. S. Pang, L. Chen, Y. Yin et al., Simulation of behavior of transient keyhole and moving weld pool in deep penetration laser welding. Trans. China Weld. Inst. **31**(2), 71–74 (2010)
120. Y. Abe, M. Mizutani, Y. Kawahito et al., Deep penetration welding with high power laser under vacuum, in *Proceedings of the ICALEO 2010* (LIA, Anaheim, CA, 2010), pp. 648–653
121. C. Bomer, K. Dilger, V. Rominger et al., Influence of ambient pressure on spattering and weld seam quality in laser beam welding with the solid-state laser, in *Proceedings of the ICALEO 2011* (LIA, Orlando, FL, 2011), pp. 621–629
122. R. Fabbro, M. Hamadou, F. Coste, Metallic vapor ejection effect on melt pool dynamics in deep penetration laser welding. Laser Appl. **16**, 16–19 (2004)
123. S. Katayama, Y. Kobayashi, M. Mizutani et al., Effect of vacuum on penetration and defects in laser welding. Laser Appl. **13**, 187–192 (2001)
124. A. Verwaerde, R. Fabbro, G. Deshors, Experimental study of continuous CO_2 laser welding at subatmospheric pressures. Appl. Phys. **78**, 2981–2984 (1995)
125. Y. Kawahito, M. Mizutani, S. Katayama, Elucidation of high-power fiber laser welding phenomena of stainless steel and effect of factors on weld geometry. J. Phys. D: Appl. Phys. **40**, 5854–5859 (2007)
126. P. Shcheglov, A. Gumenyuk, I. Gomushkin et al., Experimental investigation of the laser-plume interaction during high power fiber laser welding, in *30th International Congress on Applications of Lasers and Electro-Optics, ICALEO 2011*, 23–27 October 2011 (Laser Institute of America, Orlando, FL), pp. 637–645
127. S.I. Anisimov, Vaporization of metal absorbing laser radiation. Sov. Phys. JETP **27**, 182–183 (1968)
128. R. Fabbro, K. Chouf, Keyhole modeling during laser welding. Appl. Phys. **87**, 4075–4083 (2000)